BEYOND THE CORAL SEA

Michael Moran was born and educated in Australia and has led a varied and colourful life. He spent his twenties wandering the islands of Polynesia and Melanesia, finally settling among the descendants of the Bounty mutiny on Norfolk Island, where – as broadcasting officer – he helped set up a radio station. Later he trained as an English teacher in London and studied the piano and harpsichord professionally.

An abiding interest in the German Pacific Empire precipitated his latest return to the South-West Pacific. A Fellow of the Royal Geographical Society and an incessant traveller, today he lives and works in Warsaw, Sydney and London.

Flamingo
An Imprint of HarperCollins*Publishers*
77–85 Fulham Palace Road,
Hammersmith, London W6 8JB

www.harpercollins.co.uk

Published by Flamingo 2004
9 8 7 6 5 4 3 2 1

First published in Great Britain by
HarperCollins*Publishers* 2003

www.michael-moran.net

A catalogue record for this book
is available from the British Library

ISBN 0-00-655235-8

Printed and bound in Great Britain by Clays Ltd, St Ives plc

BEYOND THE CORAL SEA

*Travels in the old Empires of the
South-West Pacific*

MICHAEL MORAN

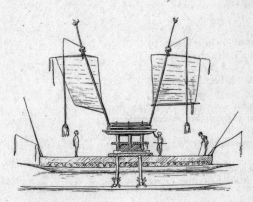

Flamingo
An Imprint of HarperCollinsPublishers

For my mother
who saw this voyage begin but not end
and
the children of Papua New Guinea
so full of energy and eternal delight

Contents

List of Illustrations

Colour photographs (by the author)
Replica of *Endeavour* at Trial Bay, New South Wales.
Samarai Island in China Strait.
Mother and son at Bilbil, near Madang.
Children from Ohu or 'Butterfly Village'.
Demas Kavavu, Paramount Chief and former Premier of New
  Ireland.
Sorcerer from New Hanover.
Slaughtered pigs, New Ireland Province.
Two children from New Ireland Province.
Dancing at the *malagan* ceremony, New Ireland Province.
The active volcano Tavurvur.
Male cult figures, East New Britain Province.
'Daisy' from Buka Island, North Solomons Province.
A *waga* drawn up at Sim Sim.
Lone fisherman at dawn, Kaibola Beach.
Paramount Chief Pulayasi's House, Trobriand Islands.
'Millton', guide to the thermal springs of Dei Dei.
The author and his favourite shirt, Tovorua Plantation.

Black-and-white illustrations (from the author's collection)
Count Luigi Maria D'Albertis (1841–1901).
The steam launch *Neva*.
Charles W. Abel of Kwato.

Memorial to Charles Abel.

Carving by master carver Mutuaga.

Tedworth House, Wiltshire, boyhood home of C.T. Studd.

The 'set of Studds' (three brothers).

The 'Cambridge Seven' missionaries.

Rev. James Chalmers (1841–1901).

Charles 'Cannibal' Miller.

Men performing penis-swinging dance on Manus Island.

Charles Bonaventure du Breil, Marquis de Rays (1832–93).

'Queen' Emma (1850–1913).

Gunantambu, Queen Emma's favourite mansion.

Melanesians of German New Guinea.

Wreckage of a Japanese 'Betty' bomber, Rabaul.

Men from the island of New Hanover.

Shark callers from New Ireland.

Bronisław Malinowski (1884–1942).

Malinowski with a Tukwaukwa girl.

Malinowski's photograph of skull cave, Trobriand Islands.

The site of Malinowski's tent.

'Baron' Nikolai Miklouho-Maclay (1846–88).

Errol Flynn (1909–59).

German graves at Alexishafen, near Madang.

Black-and-white line drawings

iii Two-masted Balangut Vang canoe, Ragetta, Astrolabe Bay, Madang Province; 1, 369, 375 Finial from a Trobriand Island canoe, Milne Bay Province; 220 The Castle of Quimerc'h, Brittany, the original home of the Marquis de Rays; 5 H. M. Bark *Endeavour* under sail; 168, 189, 207, 232 *Tatanua* mask from New Ireland, New Ireland Province; 301 Decorated canoe paddle from Buka and Bougainville, North Solomons Province; 148, 271 Device of the German *New Guinea Compagnie*; 130 Telum ancestral figure from Astrolabe Bay, Madang Province; 15 Typical 'crab-claw' sail from Port Moresby, National Capital District; 30 Raising the *Reichsflagge* at Mioko in the Bismarck Archipelago, 4 November 1884 (detail); 245 Loading a Junkers G31 of *Guinea Airways* in the 1930s, Boroko, Madang Province; 47, 66, 96, 112, 328, 337, 356 *Tabuya* (prowbroad) and *lagim* (splashboard) from a Trobriand Island *masawa* (kula trading canoe); 252, 286 *Kavat* mask of the Baining people of East New Britain Province.

Acknowledgements

The journey traced in this book took place in two episodes, a result of the difficulties inherent in travelling around the Melanesian islands at present. They remain some of the most remote places on earth. In three months of roaming I met one other European 'traveller'.

The book is the culmination of many years of a misspent youth wandering the dream world of Oceania. The selection of obscure places for the mill of the travel writer has become increasingly problematical as the planet shrinks and manufactured litter appears, even in Patagonia. The island provinces of Papua New Guinea remain pristine through their extreme isolation. They are a rich mine of colonial history and their distinct cultures are still evolving in surprising ways. The colourful and incomprehensibly neglected history of the German Empire in the Pacific drew me to the Bismarck Archipelago. The reading of works by the anthropologist Bronisław Malinowski on frosty nights in Zakopane, deep in the Polish Tatra mountains, drew me to the tropical warmth of the Trobriand 'Islands of Love'.

I could not have begun this ambitious journey without the boost given to my self-confidence by the travel writer, critic and broadcaster Robert Carver. My commissioning editor at HarperCollins, Lucinda McNeile, possessed the perfect degree of civilised empathy as difficulties mounted and she remained a supportive focal point when persistence failed in the face of increasing adversity and frustration. Sorcery from the 'Land of the Unexpected' even managed to touch her from afar as the book was nearing its end. Richard Johnson took up the reins and offered the most intelligent advice with the tact, humour and perception that only great experience can bring. The President of the Oceanic Art Society of Australia, Harry Beran, inspired me with his passionate exegesis of the art of Mutuaga, the great Massim carver. His own remarkable collections and specialised knowledge fired my enthusiasm to visit their place of origin. The urbane and philanthropic David Baker was equally influential in his passionate defence of the 'feverish nightmares' of the New Ireland art of the Malagan and the genius of the master carver Edward Sale. Dr David Warrell, Professor of Tropical Medicine at Oxford, gave generously of his advice on malaria and other horrors. Prue Watters calmed my exaggerated fears of medical emergencies.

Sir Kina Bona, the Papua New Guinea High Commissioner to London, encouraged me in the early uncertain days and selflessly shared with me his memories of growing up on Samarai, Kwato and the Abel mission station. Ilaiah Bigilale, the Principal Curator of Natural History at the PNG National Museum, assisted my planning for the islands of Milne Bay Province in Port Moresby. Michael Young, Visiting Fellow in the Department of Anthropology, Research School of Pacific Studies at the Australian National University, gave me invaluable advice on judging the relevance and appropriateness of the sometimes sensational historical account of the Trobriands and cannibals of Dobu. He is presently engaged on a biography of Bronisław Malinowski and generously checked the material relating to him on the Trobriand Islands. The fine Trobriand anthropologist Linus Digim'Rina from the University of Papua New Guinea corrected my interpretations of the complex metaphysical world of the Massim. My linguistic limitations were assisted by his son, Tokovaria, who provided the contemporary 'Port Moresby' Tok Pisin (pidgin) translations. Dr John Ombiga and his wife Dianne provided even more variations on this vibrant and entertaining language. Ray Aitken facilitated the vital initial stages of my journey and persuaded me without difficulty that not all logging companies are exploiters and ravagers of the environment. Musje at Melanesian Tourist Services made all manner of rare travel opportunities available to me with the greatest friendliness and attention to detail.

The strong Polish and German dimension to the literature could not have been explored without the help of Barbara Adam, Senior Lecturer in the Language Studies Centre at the Warsaw School of Economics. Her unstinting emotional support defeated the attacks of malicious witchcraft everywhere I travelled. The librarians at the Royal Geographical Society in London, maintaining a remarkable resource on Papua New Guinea, were always enormously patient with my compulsive fossicking. Robert Mathews helped me greatly in researching rare volumes on the gentlemen of the Cambridge Seven in the Cambridge University Library. His delicious wit made arduous hours pass effortlessly.

On a more personal note, the death of my mother in Australia in the middle of this enterprise created difficulties both practical and emotional. The gifted photographer and authority on the deeper metaphysics of the Trobriand islands, Jutta Malnic, offered the calming and spiritual hand of a Trobriand avatar.

I cannot thank enough all the people mentioned by name in this book and regret if in some fashion I have misconstrued their words or excessively filtered their experience and opinions through my own personality. Above all I thank the intense and beautiful children of Papua New Guinea, so full of love and laughter, who gently took my hand. They helped me to relearn that joyful delight in life they possess in abundance which I thought had forsaken me forever. It is to them that this book is respectfully dedicated.

Michael Moran
Woolloomooloo Hill
Sydney 2002

Author's Note

The difficulty of transposing the diverse oral languages of Papua New Guinea into written English with a standardised spelling cannot be overestimated. Local variations abound, particularly in the spelling of the names of rivers, roads, villages and people now that indigenous titles have supplanted 'colonial' ones in so many cases. Pidgin (Tok Pisin), too, is subject to local variation. I apologise in advance for the mistakes that are bound to have occurred, but I did my best.

I have always thought the situation of a Traveller singularly hard. If he tells nothing that is uncommon he must be a stupid fellow to have gone so far, and brought home so little; and if he does, why – it is hum – aya – a tap of the Chin; – and – 'He's a Traveller.'

WILLIAM WALES
Astronomer and Meteorologist
Captain Cook's Second Voyage in the *Resolution*
Journal 13 May, 1774

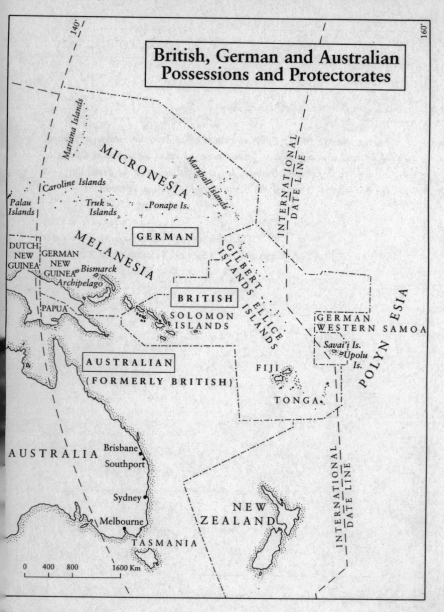

British, German and Australian
Possessions and Protectorates

The old empires of the South-West Pacific in July 1914

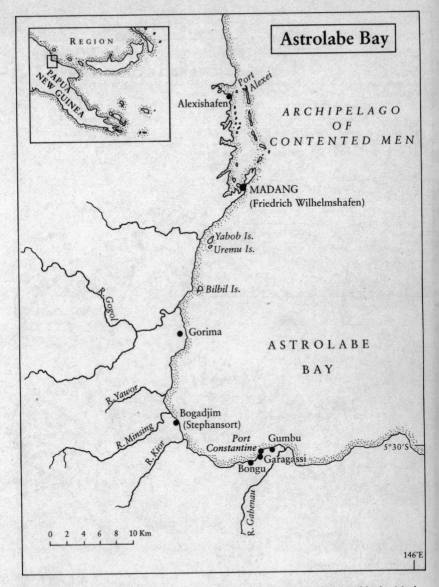

Astrolabe Bay and the surrounding areas frequented by 'Baron' Nikolai Miklouho-Maclay

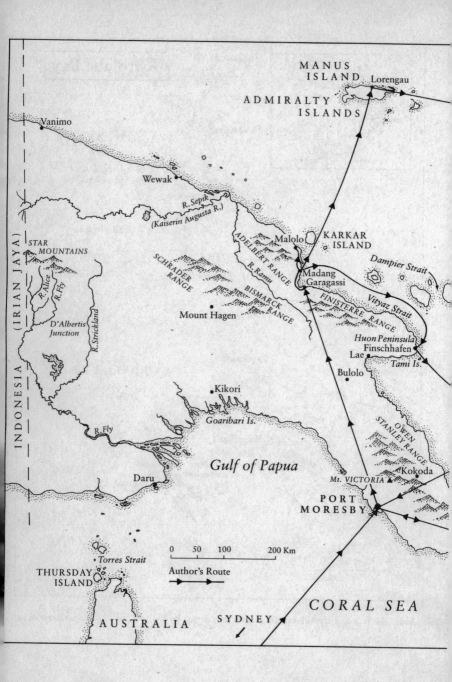

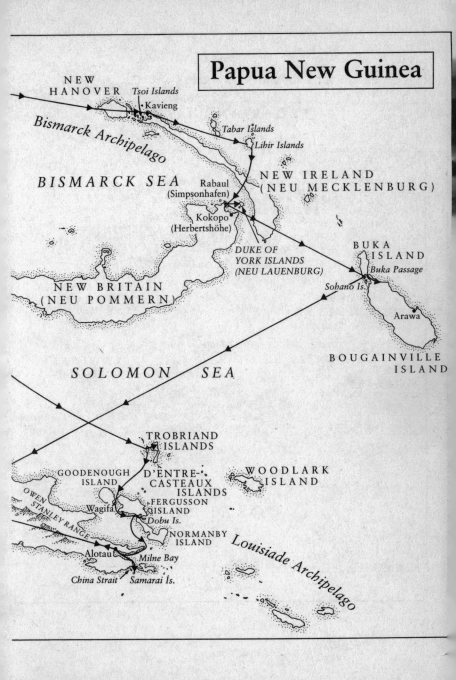

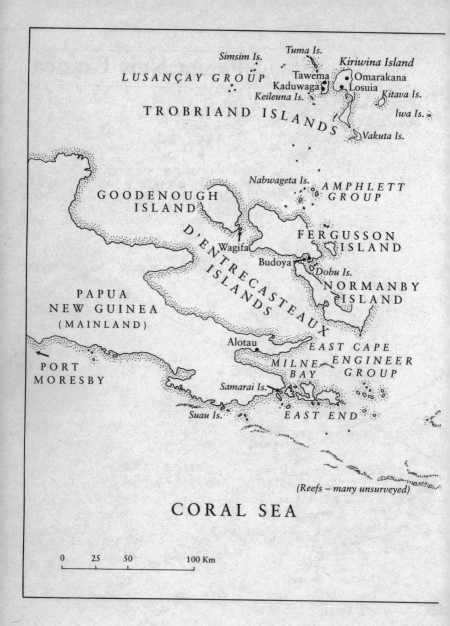

Simsim Is.

Tuma Is.

LUSANÇAY GROUP

Kiriwina Island

Tawema
Omarakana
Kaduwaga
Losuia

Keileuna Is.

Kitava Is.

TROBRIAND ISLANDS

Iwa Is.

Vakuta Is.

Nabwageta Is.

AMPHLETT GROUP

GOODENOUGH
ISLAND

FERGUSSON
ISLAND

D'ENTRECASTEAUX

Wagifa

Budoya

Dobu Is.

NORMANBY
ISLAND

ISLANDS

PAPUA
NEW GUINEA
(MAINLAND)

Alotau

EAST CAPE

ENGINEER
GROUP

PORT
MORESBY

MILNE
BAY

Samarai Is.

Suau Is.

EAST END

(Reefs – many unsurveyed)

CORAL SEA

0   25   50   100 Km

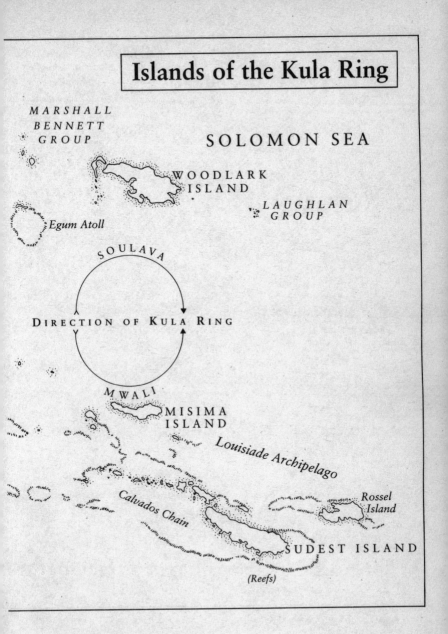

Islands of the Kula Ring

MARSHALL
BENNETT
GROUP

SOLOMON SEA

WOODLARK
ISLAND

LAUGHLAN
GROUP

Egum Atoll

SOULAVA

DIRECTION OF KULA RING

MWALI

MISIMA
ISLAND

Louisiade Archipelago

Calvados Chain

Rossel
Island

SUDEST ISLAND

(Reefs)

Prologue

'If you dress well, they won't eat you!' Wallace said.

He shuffled the cards with the stump of his right arm, beginning another interminable game of patience. The light was failing, the atmosphere oppressively hot and humid as the cards flapped on the bare table. Local boys glanced in darkly as they passed the flyblown screens covering the louvred windows. They were interested in the visitor and craned for a better view. A wretched poster of Bill Clinton greeting King Harald V of Norway hung at a crazy angle from the flaking wall.

'We thought you were Gods.'

His rippling, grey hair caught the sun and he smiled, teeth showing the past ravages of chewing betel nut. Wallace Andrew was a distinguished personage with a heart of gold. This virtue had brought him many misfortunes in life. He began to hum the hymn 'All Things Bright and Beautiful'.

'Such a lovely tune, don't you think? Young people today have abandoned proper hymns.'

The ceiling fan was motionless, the air thick and still. A pretty village woman with an ancient profile began to hurriedly set the table for dinner, laying out cutlery, bananas, pineapple and some lurid green cordial in a glass jug. She covered it with mesh. Malarial mosquitoes had already begun to ride the last shafts of sunlight in the dusk. 'Napoleon will be here at seven. They will come directly from the chamber and then go out again,' she said in excellent English, clearly for my benefit. They generally spoke the Suau language in the islands around Milne Bay in Eastern Papua New Guinea.

'Fine men. Like my grandfather, a fine man,' Wallace noted sadly, another fast game of patience in progress in the gloom. He adopted a consistently high moral tone in all his conversations and talked often of selfless Christians.

'Charles Abel, one of the first English missionaries, always wore a bow tie, white shoes, starched shirt and trousers. He was never *kai kai*'d[1] because they respected him. His wife came from England too. She delivered a village baby after they landed and her white dress was soon covered in blood. They didn't eat her. She helped them.'

Wallace was, after all, the grandson of a cannibal and an expert on matters of cannibal etiquette.

Two men carrying folders dragged open the grill on the front door and entered the main room. They glanced quickly and expectantly at the deserted bar but it had been some time since any festivities of an alcoholic or social kind had taken place there. They greeted Wallace. He stood up full of respect and pleasure that government ministers had chosen to be guests at his establishment.

'We go up, then come down to eat, then go out.'

The brevity of their speech was almost aggressive as they noticed the white stranger in their midst. The assertive masculinity of Melanesian culture. Their dark features could scarcely be seen as they climbed the central flight of a once-grand staircase that branched into two wings of remarkable austerity and dilapidation. Their bare feet made only the slightest sound like large cats padding about. Floorboards creaked overhead and doors slammed. Silence apart from the worn cards softly slapping one over the other. Wallace scarcely glanced at the deck as he deftly adjusted his amputated arm, leaning slightly to one side, gathering them in.

---

[1] Pidgin for 'food' is *kai kai*. Here it is used as a passive verb – to be *kai kai*'d is to be eaten.

'You can walk around the whole island in the moonlight. It's beautiful. Even if you are drunk nothing will happen to you here – not like the hell of Alotau!'

Wallace was full of trust in his fellow man yet he had suffered many betrayals. Tropical foliage spun by the moon appealed to my sense of romance, but this particular night was pitch black.

Fluorescent lights cruelly illuminated the dining room. The Kinanale Guesthouse was in desperate need of refurbishment. During colonial days it had been the single accommodation for white employees of the Steamships Trading Company.[1] Paintings of sailing ships and bush huts with strange watchtowers covered the larger cracks. A small lounge opened off the main room like a builder's afterthought. Geckos darted in erratic motion across the stained walls. Dinah removed the mesh from the table. She laid out fish and taro on platters together with a jug of crystal-clear iced water. A solitary bell sounded the hour over the football pitch, former cricket ground, former malarial swamp that lay before this once select building in the centre of the island. An air of abandonment and futility gave rise to a curious sense of threat and lethargy.

The government officials had changed into crisp shirts for the evening session and padded over to the table. Wallace, perhaps sensing their shyness, decided to introduce me.

'This is Mr Michael from England. He is a famous man and wrote me a letter,' searching the while in a battered briefcase. He produced the creased relic and began to read out loud, to my acute embarrassment. 'Dear Mr Andrew, your name was given to me by Sir Kina Bona, the High Commissioner in London and I . . .'

Their fierce expressions changed at once to broad smiles of

---

[1] Steamships Group is one of the largest public companies in Papua New Guinea with many diverse business interests apart from shipping. The original Australian company was established in adverse circumstances in Port Moresby in 1924 by a retired sea captain, Algernon Sydney Fitch, the first branch opening on Samarai in 1926.

extreme friendliness. But the visitors must always make the first move.

'Wallace has been telling me all about your important government work. What are you doing on the island?' I was tactfully pouring a glass of the luminous cordial so as to avoid appearing overly inquisitive. Wallace beamed from his proprietor's perch.

'I'm Napoleon, Assembly Clerk for the Milne Bay Province and this is the Principal Adviser to the Provincial Government. He's from Morobe Province. We are running a seminar for local councillors. Welcome to our difficult and beautiful country.' The introductions seemed overly formal, even odd, in this place that had clearly seen better days.

I was on Samarai, a tiny island in China Strait that lies off the southeastern tip of Papua New Guinea, described before the Great War as 'the jewel of the Pacific'. It was the original port of entry to British New Guinea and had been the provincial headquarters before Port Moresby. This gem lay on the sea route between China and Australia. The tropical enchantment cast by Samarai was loved by all who visited it. Destroyed by the Australian administration in anticipation of a Japanese invasion that never happened, it was now more like the discarded shell of the pink pearls still harvested nearby.

Having dinner with the descendant of a cannibal, a man who spoke reverentially and compulsively of the shedding of the blood of Christ whilst humming 'All things bright and beautiful, all creatures great and small', was just the beginning of a cultural adventure through the largely unknown islands of Eastern Papua New Guinea.

1. Forsaking Pudding Island

It was raining heavily as I clambered out of the taxi in the Mall and ran up the grand flight of steps past the Duke of York column into Waterloo Place. The statues of the explorers Sir John Franklin and Captain Scott looked stern and Olympian. I was heading for the High Commission of Papua New Guinea through a forest of history and high culture, umbrella up, head down. The high classicism of Nash's Via Triumphalis, former site of the Regent's wanton and ruinous Carlton House, could not have contrasted more strongly with the musky odour in the corridor of pagan carvings that led to the High Commissioner's office. Grimy windows overlooked Waterloo Place. The national flag wrapped around its pole badly needed cleaning. Papua New Guinea time and GMT were indicated by rough signs on mismatching clocks. This was clearly the lair of a culture unconcerned with cosmetic niceties. His Excellency Sir Kina Bona, the High Commissioner, was chewing gum and watching the Rugby World Cup as I wandered in. He had an instantly likeable face and seemed unaffected by his diplomatic status.

'How do you do, sir?' I held out my hand respectfully.

'Much better if I could get out of here mate! Do you like rugby? What can I do you for?'

The gum thunked into the bin. Rugby was the furthest thing

from my mind, but this was a promising beginning. He had a refined, educated air and wore fine-rimmed glasses. Underneath the banter I felt a moral outlook at odds with the modern political world.

We sat down and began to talk. An islander from Kwato in Milne Bay Province, he had attended the mission school as a child, secondary school in Sydney, studied Law at the University of Papua New Guinea and was Crown Prosecutor at the Public Prosecution Office until 1994 when he was subsequently appointed to the post of High Commissioner. Married to a 'Lancashire lass', he should have left London two years ago.

'Britain has little interest in PNG but all the Commonwealth High Comms cooperate very well.'

Grotesque Sepik river masks grinned down like a nightmare from another world.

'Where are you off to?' he enquired vaguely, settling uncomfortably into a leather chair.

'I'm planning a trip around the islands next year. I don't intend to go to the Highlands at all. Far too violent. Just the islands.'

'Yes – the violence. Moresby is pretty bad. The police are so under-funded that corruption is rife . . . the jungle hardly lends itself to strict policing. Not like Surrey!'

He laughed with a hint of derision at the ease of civilised life.

'I used to live on Norfolk Island off the coast of Australia. Home to the descendants of the *Bounty* mutiny.'

I was fighting to establish a rapport with this fellow islander, some common ground. The masks seemed threatening in Waterloo Place. The contrast was suffocating.

'Really? Islands are special places. I miss the sweet waves of Samarai and Kwato on moonlit nights. Cities, well . . .'

He drifted off into an unexpected romantic reverie. I explained myself.

'I got bored with catching the seventy-three bus down Oxford Street to Victoria every morning. Threw it up in the end. The job I mean.'

'What were you doing?'

'Teaching languages.'

'I had fine English teachers at the Mission. The very best.' He paused. 'Now we only bash the missionaries during election time!' He grinned broadly.

After some desultory chat about the independence movement in East Timor and the excitement of family life in Hampstead Garden Suburb I rose to leave.

'I'll send you some family contacts and useful people to look up. They'll look after you, or eat you!' More good-natured laughter.

I signed the visitor's book and left the office. I was heading for Berry Bros in St James's to collect a good bottle of red Graves. A final farewell to civilisation. A feeling of exhilaration passed over me as I glanced back through rain-lashed Waterloo Place at the windows harbouring that alien world. For a moment I watched the beads of water running off the polished bonnet of his midnight-blue diplomatic Jaguar.

I was about to escape from Pudding Island.

The original idea of sailing a copra schooner called *Barracuda* around the islands of Eastern Papua New Guinea in the spirit of Robert Louis Stevenson had faded, as do so many boyhood dreams. I had always wanted to sail the old vessels of the past on those remarkable voyages of discovery. My friends at the Royal Papua Yacht Club laughed at the idea and told me that all the old ketches and schooners had rotted in the mud. The price of copra had collapsed and the corpse of the industry was barely twitching. No one would dream of wasting money building or even repairing an old copra schooner. There were no more sailing ships plying the islands. Traditional sailing canoes like the majestic *lakatoi* of Port Moresby with their towering crab-claw sails and multiple hulls had by now almost completely disappeared. Chartering a vessel as an individual was prohibitively expensive. Even if I had sailed my own yacht I could easily become a victim of unfavourable trade winds or worse, piracy. Unless I was prepared to wait for unreliable boats from Thursday Island in the far north of Australia, it was impossible legally to enter Papua New Guinea except by air through Port Moresby or Mount Hagen in the Highlands. I was disappointed but determined to sail at least part of the Australian coastline in the old style, completely dependent on the vagaries of wind and weather.

A rare opportunity arose to 'take passage' on the replica of Captain Cook's ship *Endeavour* as a supernumerary member of the crew. That it was sailing in precisely the opposite direction to my intended destination did not disturb me. I would experience sailing a tall ship along the New South Wales coast for a week from Southport (near Brisbane) to Sydney. A suitably nautical frame of mind would then enable me to jet off to Port Moresby with equanimity.

Endeavour is a handsome vessel and a magnificent replica of the original ship. It was constructed as Australia's flagship from 1988–94 in Fremantle in Western Australia. She is built of jarrah and has upper sides of varnished pine, finished in the Royal Navy colours of blue, red and yellow. I took Sir Joseph Banks's cabin on the after fall deck – a small space that I later discovered was occupied not by Sir Joseph himself but by his dogs – a bitch spaniel called Lady used as gun dog, and a greyhound taken on board to run down game.

Joseph Banks was only twenty-five when word reached him on 15 August 1768 that *Endeavour* was ready to take him aboard on a great adventure to the South Seas. He was at the opera in London with Miss Harriet Blosset, a French ward to whom he was engaged and in love, but with whom he lamentably lacked the French language to communicate. Confessing himself to be of 'too volatile a temperament to marry', and unable to explain the meaning of his imminent departure, he drank heavily in a romantic funk the night before he left London for Plymouth. Poor weather delayed the sailing until 25 August.

Banks's father was an MP and the family were wealthy and well-connected, living at Revesby Abbey in Lincolnshire. He had

8

been to Eton (which trained him no doubt for the rigours of the voyage but not for the travails of love), spent seven years at Oxford studying botany, and worked at the British Museum in London. In February 1768, the Royal Society decided that observing 'the passage of the Planet Venus over the Disc of the Sun . . . is a Phaenomenon that must . . . be accurately observed in proper places'. The Admiralty decided on Tahiti as a place of observation, and James Cook was appointed chief observer of the transit. He selected a Whitby 'cat'[1] called the *Earl of Pembroke* as the most suitable vessel for such a voyage, refitted and renamed her the *Endeavour*.

As a Fellow of the Royal Society, Banks contributed ten thousand pounds to purchase a vast quantity of both practical and elegant equipment for the voyage, and transported a comprehensive library of some one hundred and fifty volumes. A party of nine made up his gentleman's entourage, all trained in the techniques of collecting and preparing specimens. He became almost more famous than Cook himself, but remained dogged by the unfortunate repercussions of the 'caddishly abandoned' Miss Blosset ('Miss Bl: swooned &c', his journal coolly observes).

Cook had a complement of some ninety-four souls together with chickens, pigs, a cat and a milch goat that had already circumnavigated the globe. In a letter to Banks in February 1772, Dr Johnson included a Latin elegy for the celebrated animal, part of which runs:

> *In fame scarce second to the nurse of Jove,*
> *This Goat, who twice the world had traversed round,*
> *Deserving both her master's care and love,*
> *Ease and perpetual pasture now has found.*

---

[1] The English north-country vessel known as a 'cat' was a Whitby collier. This was the type of working vessel on which James Cook learnt his calling. In 1771 he wrote in admiration of her handling, 'No sea can hurt her laying Too under a Main Sail or Mizon ballanc'd.'

His adventurous friend James Boswell records that Johnson was sceptical of what a traveller might learn by taking long voyages, despite on one occasion when dining with the Reverend Alexander Grant at Inverness, divertingly 'standing up to mimic the shape and motions of a kangaroo' and making 'two or three vigorous bounds across the room'.

By 16 August 1770, Cook had reached the Great Barrier Reef, courageously searching for the elusive passage between New Guinea and Australia. The passage had originally been discovered in 1606 by the great Spanish navigator Luis Vaes de Torres, but strategic secrecy was paramount for Spain and on many charts the two land masses appeared joined. Capricious winds drove Cook and his crew ineluctably towards disaster. '. . . a speedy death was all we had to hope for . . . ,' reiterated Banks. But by 21 August, the *Endeavour* had threaded its way around Cape York, the northern extremity of Australia, and passed through the Endeavour Straits or as it is now named, Torres Strait.

On 27 August they set sail for New Guinea. They voyaged past the south coast of the island, almost a year now after the visit to Tahiti where the officers had observed the transit of Venus and Mr Banks the slow and painful tattooing of a girl's bottom. Two days later they came to a landfall fringed by dense vegetation and mangrove swamps. Banks wrote: 'Distant as the land was a very Fragrant smell came off from it realy in the morn with the little breeze which blew right off shore . . .' The water was warm, muddy and shallow, keeping them away from the coast until 3 September when they waded in to land. Banks collected a few specimens but remained curiously unimpressed. They found human footprints which caused them as much consternation as that felt by Robinson Crusoe. They proceeded with caution until they came to a hut in a grove of coconut palms. Three warriors suddenly rushed them from the jungle, throwing spears and incendiary devices, shouting hideously. A hundred naked Papuans appeared around a promontory. It was time to leave.

The following excerpts are taken from my voyage diary:

Endeavour, 7.30 p.m.
4 September 2000

Have come off Afternoon Watch and had dinner. A nerve-wracking and terrifying day. Rose early 6.00 a.m. Troubled sleep – excitement, nerves and information overload. Claustrophobia with the cabin door shut. Prolix talk of 'bunts', 'clews', 'belaying lines', 'bracing the yards' and generally hauling on any of the innumerable ropes in sight. Mind-snapping terms of the sea is assumed knowledge – understood absolutely nothing.

Time to 'go aloft'. Terrified. My group designated Foremast Watch. Forced by bravado to climb the 'ratlines'. Felt decidedly like a rat. The lines are angled up to a platform called the 'tops'. Remainder of the thirty-three metre mast towers above. Palms sweating. Shuffling along the yard (to which sails are furled) on rope not much thicker than a garden hose. 'Stepping on!' is the brisk instruction. 'Falling off!' screamed as you crash to the deck. Managed that, then. Is this my future for the next seven days? Much preparation casting off.

Very calm day, brilliant sunshine with light NE wind.

'Stand by for cannon!' shouts the ship's carpenter, a handsome, blond Cornishman, responsible for construction of the replica and loved by all the girls.

'Fire in the hole!' He lights the powder.

Boom! Replica four-pounder carriage gun recoils, acrid smoke rolls across the deck. A terrific report, too close to some nautical types sipping Pimms on the deck of their chromium cruiser. They fell backwards off their chairs as shredded paper and smoke engulfed them.

'Haul on the halyards! Ease on the bunts and clews!'

Felt like easing myself but not permitted until further out. Hauled on lines until palms sore. Not seasick but a visit to the heads (mariners' term for onboard toilet) could bring it on. Open grey valve, pump up water, do your business, keep your balance as you have a good look while you pump out, repeat three times, close grey valve under pain of castration. Voyage will be no picnic. Comforting smell of tar.

Came off Afternoon Watch at 4.00 p.m. and resume First Watch at 8.00 p.m. Ship glides slowly and is deeply restful. In perfect harmony with the sea. Progress about 3 knots – a stately speed which would have given Banks and his party ample time to draw, read, discuss and describe their collections. Sun setting through the stern sash windows of the Great Cabbin. Storm lanterns lit, secretive plashing of water at the stern and creaking of the ship. Absolutely magical and poetic.

Sailing at night on the *Endeavour* is like taking part in a Wagnerian opera, the Flying Dutchman, perhaps. On watch, time to gaze up at the moon through the swaying rigging, silhouetted against the myriad stars of southern latitudes. A shadowy helmsman guides us across the deep. Silence on deck. Ship groans quietly as it folds through the sea. Watching the phosphorescence at the bow I was suddenly transfixed by the appearance of silver tunnels and comet trails cut by porpoises as they dodged and played before the ship. Captain ordered us to 'wear ship' – rudely-broken reverie. Had to set the sails and belay lines (fasten the coils of rope around wooden pins) in the dark. After stress and furious activity, lying on my back in front of the helm watching the masts arch like giant pointers across the constellations. Dreamed of the discovery of New Guinea on a ship such as this.

Endeavour
6 September 2000

Morning Watch began at 4.00 a.m. Ungodly hour to be on deck. A still night with feathery winds and countless stars, the moon intensely bright. Silhouettes of the crew on watch float like wraiths. Dawn a glowing rind of orange before sun breaks the horizon. Red Ensign flies from the stern mast and stern lantern glints in the dawn.

Later in the morning a hump-backed whale breached – spectacular arch of patent-leather black and white. Barometer falling. 'It's coming all right,' crackled Captain Blake ominously on the weather deck. During night watches he often comes on deck barechested in a maroon sarong. Seems to sense any unnatural movement of the ship through his sleep. Catapults from the companion to bark orders in eighteenth-century style.

Mainmast Watch took in sail at Trial Bay off coast of New South Wales. Landing from surf boats. Moving reconciliation ceremony with Aboriginal community. Exhausted from climbing and hauling – aching and stressed by vast quantity of strange sailing nomenclature.

Endeavour
7 September 2000

Oppressive lowering sky and ominous calm. Hardly slept for last three days. On watch and took the 'brains' side of the wheel.[1] Wind strength increased towards evening, gusting to 40 knots. Bow ploughed into the 1.5m swell but the ship felt strong. Sea a magnificent expanse of breaking waves, wind tearing the lashing foam. Shrieks rent the rigging. Wheel duty in this weather madly exhilarating. Maintaining course fraught with problems, arms aching, slow response to helm. Bow rises to frightening heights before ploughing back down into the troughs.

Lines lashed. Many seasick. Going aloft 25 metres in these conditions to take in t'gallant sails not for the fainthearted. Respect for the old mariners boundless – their achievement unimaginable until you sail a tall ship. Vessel utterly at the mercy of wind. So tired cannot sleep. Eating little.

---

[1] Two seamen are normally at the wheel – the 'muscles' on the port side who only helps turn it, and the 'brains' on the starboard side who turns and maintains the course, calling out the setting and watching the instruments.

> Great Cabbin, *Endeavour*
> 8 September 2000

Physically impossible to write. Force 8 gales. Taking in all sail. On verge of throwing up. Gorge rising. Ship lurched and shuddered through night. Roped myself into the fixed cot.

> *Endeavour*
> 9 September 2000

Wind sufficiently abated to write a journal entry. Warm sun as we sail along the coast of New South Wales and begin to set sails again after the storm. Activity everywhere.

'Hauling on the halyards! Easing on the bunts and clews! Bracing the yards!'

'Two, six . . . heave!' we hauled on the lines.

'Two, six . . . heave!'

'Belay all lines.' Signs of relief.

Leaned against the capstan and idly looked at a jetliner high above, slicing across the sky leaving a glittering trail of ice crystals; the eighteenth century contemplating the twenty-first century. The original exploration of the Black Islands of New Guinea was on ships such as this. My own journey to that fabled land would be on an aircraft such as that.

2. The Eye of the Eagle

Final approach to Port Moresby in the dry season is over arid, brown hills bereft of vegetation and a polished turquoise sea. The colonial terminal at Jacksons airport has faded lettering on the fibro huts, the modern terminal a bland feel, new paint already peeling in the heat. An Air Force Dakota without engines lies abandoned on one side of the runway, a reminder of the first commercial flights. The blast of desiccated air as you disembark is like a physical punch, gusts of the south-east trades dry the mouth. Certainly you are no longer referred to as *masta* as your bags are unloaded. Only one officer is on duty at passport control to process the entire jetload of passengers. Welcome to modern Papua New Guinea, 'land of the unexpected'.

Captain John Moresby may have been the first white man the native people had ever seen when he sailed into the harbour aboard the HMS *Basilisk* in 1873. He spent some time trading with the villagers of the local Motu tribe. He wrote that civilisation seemed to have little to offer this culture. The London Missionary Society were settling in a year later and by 1883 there were five resident Europeans in Port Moresby including the Reverend James Chalmers, a gregarious character who was eventually murdered and eaten by cannibals on Goaribari Island in Western New Guinea. Despite its isolation and absence of road connections, Moresby has remained the capital.

The taxi driver informed me that the bullet hole in the corner of the cracked windscreen was from *raskols* – a misleadingly benign Pidgin word meaning 'violent criminal'. They had attempted to hold him up on the way to 'Town', the centre of the city. He was a Highlander with an ambiguous smile somewhere between a welcome and a nasty threat. I began to glance anxiously at passing cars.

'*Wanpela sutim mi nogut tru lon hia. Olgeta bakarap.*'[1]

'Were you hurt? Did they take your money?'

'Took everything but I drive away quick. Back at work next day. *Mosby em gutpela ples.*'[2]

This made no logical sense at all to me so I fell silent until we reached the hotel. It was a dusty drive with colourful children and resentful adults crowding the roadsides. There had been a drought for the last seven months. The usual Western corporate signage had been bleached by the savage sun. A car had collided with a truck bearing the company name 'Active Demolition'. I glimpsed the original Motuan stilt village of Hanuabada, fibro huts replacing the traditional bush materials. A few cargo ships lay becalmed in the port.

A midday stroll among the sterile office blocks, slavering guard dogs and confectioner's nightmares thrown up by financial institutions did not appeal, so I headed south for Ela Beach, an inviting stretch of sand facing Walter Bay. Trucks cranked past with men crammed like sardines in the back and small PMV[3] buses smoked happily by like toys. Seaweed, cans and other detritus marred the shore, but kiosks gaily painted in Jamaican style lifted the spirits. A rugby side were training on the sand, running forward through a line of plastic traffic cones and then suddenly reversing through them. Many who were overweight fell over during the difficult backward manoeuvre but there was

---

[1] 'Somebody shot at me. Everything around here's pretty bad. It's completely buggered up!'

[2] 'Port Moresby's a good place.'

[3] Public Motor Vehicle – these minibuses are considered to be dangerous for visitors, but in my experience they were a source of all my best conversations and friendships with local people.

no laughter, just embarrassment. Papua New Guinea is the only country in the world that has rugby as its national sport and every aspect of it is taken seriously. Training was interrupted by the capture of a turtle on the breakwater. A long time was spent inspecting and discussing the prize. Some of the boys scribbled graffiti on its shell in luminous paint and then released it back into the bay, fins flapping, neck craning. Training resumed.

Palm trees with slender trunks curved over the bay in front of international high-rise apartments. I walked past a group of suspicious-looking youths sitting under some trees outside a café and strolled out onto the disintegrating breakwater. A family were competing with each other, skimming pebbles across the surface of the water. The five children, father and mother were screaming with delight at this simple game that seemed to bond them so intimately.

Visitors are warned by expatriates not to approach, in fact to walk away from groups of youths but I decided to wander over to the cluster beneath the casuarina trees. They were chewing betel nut and spitting the blood-red juice in carefully-aimed jets. They were shocked when I greeted them, but smiled almost immediately. The smile on a Melanesian face is like the unexpected appearance of a new actor on the stage.

'*Monin tru, ol mangi. Yupela iorait?*'[1]

'*Orait tasol, bikman.*[2] Where do you come from?' They stood up, even respectfully I thought.

'England. I live in London. My name's Michael.' I held out my hand which was shaken softly. They shuffled about looking at the ground, showing signs of amazement by spitting fast red gobbets in the dust.

'And you've come here! Do you like our country?'

'Everyone seems pretty friendly to me. What's your name?' I asked a boy with the most intense black skin I had ever seen –

---

[1] 'Good morning, boys. How're you?'

[2] 'Fine thanks, Sir.'

it was almost blue. He had dreadlocks, perfect white teeth and eyes like an eagle. He appeared highly intelligent, but melancholic shadows fleetingly crossed his features.

'Gideon. I'm from Buka.' His open face smiled engagingly.

'Really! I hope to go there. I'm visiting the islands.'

'It's beautiful on Buka, but no work. The Bougainville war destroyed everything. I came to Moresby but can't get a job. I've got my electrician's certificate.' The shadows were well established.

'*Mipela ino inap lon bikpela skul*,'[1] said a fierce lad with broken, heavily-stained teeth. It looked as though a bomb had gone off in his mouth. He was angry.

'I come from the Sepik. I have no parents and no money.' He looked savagely at the ground and started violently peeling a new nut.

'Are you all unemployed?' I asked, already knowing the answer. They nodded dreamily.

'Don't you miss your family?'

Silence.

'I've heard that some boys break into houses and steal. Is that true?' I was living dangerously, considering it was my first afternoon.

'Yes, but they're not bad boys, sir. We're not *raskols*! We need the money to eat. We want to work but we can't get a job.'

'That's not really a good reason to steal. You could go to prison. Ruin your life.'

'Corrupt politicians have ruined our country. You don't see them going to prison.' Gideon offered this as a challenge for me to refute.

'No one gives us a chance. We're on the outside looking in.'

'Where are you from?'

'Chimbu.' The older man seemed to stand apart from the rest and was more deeply resentful.

'Where do you live in Moresby?'

'Are you a priest?' It was an aggressive answer. 'Ragamuga.

---

[1] 'We can't afford university.'

Six-Mile Dump. You'll visit?' He smirked and spat into the dust.
I had only read about this desperate migrant settlement situated
behind a large rubbish dump. No, I did not intend to go there.

'But what about your future?' I was moving into a dead-end.

'So? No one gives a shit about us. Politicians just want money.'
Isaiah was from East New Britain and had parallel tattoos on
his cheeks.

'We're bored and no future. That's the problem.'

I realised with alarm that my group had grown into a small
crowd that surrounded me. They pushed forward not to attack
but desperate to explain, to justify themselves, expecting me to
provide an explanation, an instant solution.

'Everyone hates us. No tourists come because the newspapers
report so much violence.'

I said nothing but the headline in the newspaper in my hotel
screamed of the rape of three nurses at Mt Hagen Hospital and
the theft of an ambulance. The thieves were demanding compensation for the return of the vehicle or they would torch it.

'Violence is terrible in the Highlands. That's why I chose the
islands.' I was out of my depth.

'You're lucky. You have money to travel,' said a boy from
Oro Province wearing a bedraggled feather in his hair.

Beavis and Butthead cartoons flickered on the screen behind the
heavily-barred windows of the 'Jamaica Bar'. Papuan reggae music
was playing somewhere. I was a distraction but not a solution.
Some drifted away and sat under the trees again. Large spots of
rain from an afternoon storm kicked up the reddened dust.

'Well, I'd better be going. Nice to have met you. Gideon,
Isaiah . . .' I shook their hands.

'*Tenk yu tru long toktok wantaim yu, bikman.*[1] All the way
from London! Enjoy our country.'

They remained standing and smiling as I headed back. This
encounter with wasted potential, cynicism, and the crushed

---

[1] 'Thanks very much for talking to us, Sir.'

optimism of youth left me feeling depressed and impotent. The cultural diversity of the country meant that there was tension between youths from many regions thrown together by unemployment. In traditional villages in the past, fear of neighbouring peoples and respect for the authority of the elders would have limited the freedom of the young. The notion of respect had almost disappeared, but not only in Papua New Guinea. London and Sydney were similar, but this country was poor and the politicians corrupt.

Many observers blame the present law and order troubles on the premature commitment Australia made to the granting of independence in 1975. At the outset, an inappropriate Westminster-style two-party democracy was imposed on the country with legislative power vested in a national parliament. The national government devolved power through nineteen provincial governments. The country joined the Commonwealth with the Queen as Head of State, and a governor-general appointed as her representative. Papua New Guinea covers a vast area (it is the second largest island in the world) and possesses such extreme cultural diversity that the growth of a properly integrated strategy for development has remained a perennial challenge. The first Prime Minister, Sir Michael Somare, was an able and popular politician, but the many emergent parties have become increasingly unable to establish clear ideological principles. Political candidates pursue personal or regional goals at the expense of party policies.[1]

---

[1] Sir Michael Somare was born in 1936 in Rabaul, East New Britain. He led the Pangu Pati (Party), the largest and most influential political party in the move towards independence in 1975. He became the first Prime Minister of independent Papua New Guinea from 1975–80 and again from 1982–5. His membership of the Pangu Pati ended in 1997 and he formed the National Alliance Party which won a comfortable majority in the violent 2002 elections. After seventeen years, Sir Michael Somare, 'the father of the nation', was elected Prime Minister for a remarkable third term.

Most Papua New Guineans still live a subsistence lifestyle in villages quite separate from the influence of the cash economy. The villagers have become convinced that, at both the provincial and national levels, politicians are self-serving and uninterested in their welfare. Traditional social arrangements had already begun to disintegrate under colonial rule. The adoption of an inappropriate Western legal system has only exacerbated the agony of cultural fragmentation. Tribal fighting has resumed in the Highland provinces, but under more murderous rules than in the past. As the traditional society in the village disintegrates, many young people flee to the urban ghettoes of Port Moresby and Lae to face almost certain unemployment followed by a descent into crime. The challenge remains to evolve a system that combines the strengths of traditional leadership with the ideals of modern government, giving due legal weight to the fraught claims of land ownership by the numerous clans.

On my return to the 'executive floor' of my hotel, I passed psychedelic kiosks and wrecked cars in the oppressive heat and scalding rain. Luxury expatriate enclaves seemed to be going up everywhere on higher ground. Segregation has made this a city divided against itself. The cool, spacious lobby transported me to a different planet to that inhabited by my 'new friends' on Ela Beach, and the grim reality of their settlement homes. This was the arena where exploitation and 'aid' were strategically planned by company generals. The sunset from the elevation of the executive floor was sublime; copper and tarnished brass shot through with blue. This luxurious scene was decidedly different from the wild 1920s when Tom McCrann's hostelry in Moresby displayed a notice in the saloon:

Men are requested not to sleep on the billiard table with their spurs on.

At dinner there was an astounding mixture of guests. A tattooed Scot was having dinner with an Asian engineer.

'Glad you're on the fuckin' project, Wang. You've got a degree.'

A German trio who had run out of time were attempting to negotiate a price for the ethnic decorations on the hotel walls. A heavily-tattooed Pacific islander in a black sleeveless singlet, chiselled black beard and jeans patched with grandmother's chintz was eating soup and tugging at his pearl earring. A Belgian photographer with a ponytail was talking to a glamorous Parisian collector of artefacts from the Maprik region who had a gallery in Aix-en-Provence.

'Every week I 'ave ze fever on ze exact same day!' she exclaimed in desperation.

A newly-rich Highlander was eating a roast chicken, juggling greasy drumsticks in both hands and attempting to talk on a mobile phone. Pallid Englishmen and tanned Australians were earnestly discussing football and drink. They had the weak eyes and the furtive mouths of social casualties, bolstering their own false optimism or drowning betrayals in liquor.

'The free drinks are from five thirty to six thirty. Don't come after or we'll have to pay.'

'Right, mate!'

'They're tough men the South African rugby team!'

'Blood oath! Fuckin' tough!'

'Hides like a rhinoceros!'

'More like a fuckin' elephant, mate!'

'Fuckin' tough.'

'Yeah. Fuckin' tough, real men.'

'Fuckin' tough!'

'Yeah, fuckin' . . .' and so on, endlessly, whilst downing bottle after bottle of South Pacific lager.

A huge butterfly enamelled in iridescent blue battened against the glass door leading out to the swimming pool. A Chopin nocturne floated across the lounge from the Papua New Guinean pianist playing a grand piano. I wandered over at this unexpected appearance of European culture and spoke to him.

'You're playing Chopin,' I rather pointlessly observed.

'Yes. I studied classical music for many years. Do you have a request?'

'Not classical. Jazz. Can you play "Misty"?'

'Sure. If you like jazz you might like my novel. It's on the music stand.'

A small pile of paperbacks entitled *The Blue Logic: Something from the Dark Side of Port Moresby* by Wiri Yakaipoko was stacked on one side above his fluent fingers.

'What's your novel about?'

'It's a crime novel about Moresby. Plenty of it around here to write about.' I could hardly disagree.

Chopin, jazz and crime are an odd mixture. Unexpected conjunctions and unpredictable outcomes were to become a feature of all my travels in Papua New Guinea. I went to bed suffering from a blinding headache which seemed to come from the combination of the anti-malarial drug Lariam[1] and alcohol.

The next day the usual horrors were introduced quietly under the door of my room via the dailies.

A youth was chopped to death and two houses burnt down in the Kaugere suburb of Port Moresby over the weekend.

Under the banner headline 'Patients Hungry' we learn that patients' food was stolen from Modilon Hospital in Madang.

A thirteen-year-old sex worker said, 'My aunt kicked me out because she said I slept with her husband. Prostitution is fun and I get a lot of money.' Tribal fighting now takes place with homemade guns, grenade launchers and Kalashnikovs rather than spears.

The city looks more attractive on my birthday. Red and mauve

---

[1] Mefloquine or Lariam (the trade name) is the most powerful of the anti-malarial prophylactics. Unlike other drugs, it protects against the fatal strain of cerebral malaria. It can have disturbing psychological side-effects.

bougainvillea are flowering, Ela Beach looks inviting and the frangipani spiral down in pink and white. I decide to go for a walk. Outside the US Embassy I am almost arrested for writing down the sign *NOKEN PARK LONG HIA* meaning 'No Parking' in Tok Pisin (Pidgin).

The evolution of Melanesian Pidgin (or bêche-de-mer English, as it was popularly known in colonial days) was complex. There are many regional varieties of this colourful and witty language which originated on the Pacific plantations of Queensland, Samoa and New Caledonia in the early eighteenth century. It had emerged fully formed by about 1885 and is still evolving in rich referential complexity. Around eight hundred or one seventh of the world's languages are spoken in Papua New Guinea. Some two hundred are Austronesian spoken in the coastal and island regions, and the remainder are Papuan spoken in the Highland areas. There are three *lingua francas* – English, Motu (spoken in Port Moresby) and Tok Pisin.

Outside the Westpac Bank a huge Alsatian and armed guard in a baseball cap stand in the centre of three signs warning 'Beware of the Dog'. The brooding atmosphere of male unemployment hangs about like a miasma, and I have not seen a white face in three hours. Huge holes in the pavement and deep storm-water channels offer possibilities of serious injury. The light burst of a glowering Melanesian face suddenly smiling. At the Port Moresby Grammar School, children in pale uniforms are caged up in a security tunnel hung with plants waiting to go home. Fishing trawlers of unbelievable decrepitude are moored at the wharves. Thick, black smoke pours from their funnels, the idle crews lounging in the shade of tarpaulins or carrying huge tuna by the gills. One boy drops a plastic bag full of silver sprats that cascade over the wharf like pirate's treasure. Six mothers breast-feeding babies inexplicably sit in the broiling sun on a concrete platform raised above a potholed road. I slip into the shade of the Port Moresby Public Library. An eerie silence reigns, but people greet my unexpected presence with smiles of surprise.

Useful titles such as *Australian Imperialism in the Pacific* and *Tuscan Cuisine* grace the shelves.

A friend, John Kasaipwalova, had invited me for a birthday dinner. He is a prominent and controversial Papua New Guinean poet and writer and was a student rebel during the drive for Independence in the 1970s. He is also chief of the Kwenama clan on Kiriwina Island in the Trobriands, one of my destinations. I was collected in a Mitsubishi Pajero with gigantic bull-bars, a fantastically cracked windscreen and peeling sun filters. John has a round friendly face framed by a halo of tightly curled hair, his sensibility a rich repository of poetic image and symbolic knowledge. But entrepreneurial activities tend to preoccupy him these days, as he attempts to balance the claims of individual business and his responsibility to his own clan community. He was accompanied by Mary, his attractive Malaysian wife, and 'Uncle Sam', who drives the Pajero with fearsome spirit, thundering over unsealed roads past striped drums marking dark detours. While avoiding a cavernous pothole, he asked me to guess his nationality. His mother turned out to be from Sri Lanka and his father an unusual mixture of Dutch, Portuguese and Australian Aboriginal. 'Dad's family moved about quite a bit.' Under an Australian bush hat he had the long grey beard of a swami and spoke with a slight Indian accent. It was a striking face, a colonial cocktail.

The shopping precinct that housed the Chinese restaurant was protected by a high security fence with bars two inches thick, armed guards, slavering dogs and a searchlight.

'It's a gourmet place!' explained Sam as we parked among a crowd of children.

We were shown into a private room with intense fluorescent lighting. Geckos erupted into life on the walls like surrealistic wallpaper. Sam's gold rings glinted on his slender fingers and the cutlery was reflected in his melancholic eyes.

Delicious coconut prawns, chilli crab and coral trout with tender asparagus appeared like magic. The conversation ranged lethargically over many topics, as if we were in an island village. They were shocked to learn of my walking alone in Moresby and even more surprised when I mentioned the young boys.

'I'm hoping to go to the Trobriands quite soon, John.' I briefly outlined my island itinerary.

'You've made the best decision in choosing the islands. How come the Trobriands?'

'Well, it's a short story that's taken a long time to complete. I bought a *tabuya*[1] or wave-splitter from Kiriwina in an artefact shop many years ago. It's been in my music room in London for ages, and I've always wanted to visit where it was made.'

'I can tell you that the *tabuya* has been watching you. The design symbolises *bulibwali* or the eye of the sea eagle [osprey]. You had to come. His eye never sleeps, you know. In an instant he decided on you as his particular fish. That's why you came. It's very simple.'

'Do you really believe this?'

'Of course. You're a person who possesses concentration. You plan and attend to detail. Am I right?'

'Actually, yes. I drive people mad with it.'

'There you are!' John reached for more coconut prawns in an ebullient mood. He continued his arcane explanations with some seaweed poised between chopsticks in midair. I wanted to hear an account of the famous *kula* trading ring from the chief of a clan. I was anxious to know if the classical descriptions were still accurate.

'Tell me something about *kula*, John.'

'Well, first you must understand the mystery of *Monikiniki* or the Five Disciplines of Excellence.'

'Sounds a bit complicated.'

'Never! It's simple! The disciplines are symbolised in the five compartments of a Trobriand mollusc shell. Each compartment represents one of the senses and is represented by a bird, plant

---

[1] A *tabuya* is the prowboard of a Trobriand canoe.

or even a grasshopper. The eye is represented by the *bulibwali* or the sea eagle.'

We had moved into the realm of myth and magic for which these islands are famous, rather daunting for a European unused to the sharing of mystical experience.

'But what is *kula* exactly?' I was impatient as usual.

'That's not easy to answer, but basically it's an activity of giving and receiving between people that results in them growing spiritually.'

'But doesn't it involve trading valuable *soulava* or necklaces in a clockwise direction around certain islands and *mwali* or arm shells in a counter-clockwise direction?'

'Of course, but they're only the outward manifestation of the activity, in fact the consummation of it. The objects accumulate power as they pass from hand to hand over time. Some might even kill you. But it's the quality of this experience that's important.'

I began to be drawn irresistibly into the rich mythological world of the Trobriand Islands, so unlike the sterility of my own empirical society where success seemed the sole criterion. I began to look forward to my trip with keen anticipation. A couple of lines of a poetic song concerning the *kula* came to mind.

> *Scented petals and coconut oil anoint our bodies*
> *We're ready to sail with the south-east wind*

John fell silent and took some more chilli crab. The mood had become serious yet our state of mind was happy and free.

'I've never been to the yam festival in the Trobes. Never managed to get there. God knows why.' Sam trailed off and adjusted his hat to a more comfortable position. He reached for some more coral trout.

'God's saving you, Sam, from a long period of self-abuse,' John observed. Everyone laughed heartily. The yam festival is famous for its ecstatic expression of sexual freedom in celebration of the harvest and the end of ten months hard gardening.

Myth and magic give life meaning in the islands. We discussed the weighty word *kastom*. It is an essential Pidgin concept that derives from the English word 'custom' but with a more complex Melanesian meaning and multifarious connotations. It is normally used in reference to traditional culture that has come under threat from aggressive European development. But *kastom* cannot be simply translated. There are many contradictions within this multilayered concept. The idea has led to a strong cultural revival as regional identities become increasingly diluted. People are always talking about the loss of it. Closeness to nature and the traditional sense of belonging to a community are being replaced by the desire for individual consumption. European technology dominates modern life in the cities, yet a profound need remains for the unseen worlds of magic and religion. A further complication is the extreme cultural diversity of the country. Many distinct cultures have been wilfully cobbled together into the artificial political entity known as Papua New Guinea. Cultural differences are ignored, or worse, attempts are made to diffuse them.

'More chilli crab?' Sam spun the lazy susan.

'Do you know there is a ruined temple on the top of Egum Atoll?' John said, secretively.

'Yes, and flat stones with magical properties on Woodlark Island,' his wife whispered.

It was getting late. We emerged from the restaurant into the glare of security searchlights. The massive gates swung open and we drove out of the compound. Uncle Sam began to sing the praises of Port Moresby as we drove back into town. Mansions surrounded by high fences topped with glistening razor wire, signs painted with cartoon-like dogs and guards posturing with guns, spun through the headlights. Dark hills sprinkled with twinkling lights reared on either side of the highway.

'Nothing is as beautiful as this in the world!' Sam suddenly exclaimed with great feeling.

I spent a restless night poring over maps, anxious to leave the place. Papua New Guinea can be broadly divided into the mountainous interior, the coastal regions, great rivers and the island provinces. My decision to explore the islands had come from their extreme isolation, their reputation for beauty, tranquillity and the preservation of their ancient cultures. Near Moresby, the start of the Kokoda trail had been closed by tribal fighting. There were reports of a white, female bushwalker who had been raped even though she was with a local guide. This constant threat of violence in the capital had begun to depress me. I was tired of being holed up for safety in a luxury hotel with paranoid expatriate businessmen planning the disintegration of a culture for profit. My jumping-off point for the islands would be Alotau, the capital of Milne Bay Province at the eastern extremity of the mainland. From there I could leap aboard a banana boat[1] to Samarai, the traditional gate to the old empires.

---

[1] The term 'banana boat' has nothing to do with bananas or their transport. It refers to the shape of the innumerable fibreglass dinghies fitted with forty-horsepower outboard motors that ply the islands and coast of PNG like noisy water insects. They have taken the place of the elegant sailing canoes of the past, which have almost completely disappeared. They sometimes carry suicidal numbers of passengers, often travel enormous distances across open ocean, and never take a single life jacket. Many simply disappear, the occupants lost to drowning or sharks.

3. 'No More 'Um Kaiser, God Save 'Um King'

Australian Military Proclamation 1914

East of Java and West of Tahiti a bird of dazzling plumage stalks the Pacific over the Cape York Peninsula of Australia, her head almost touching the equator, tail looping above. In her wake she spills clusters of emeralds on the surface of the sea. These are the unknown paradise islands of the Coral, Solomon and Bismarck Seas, the islands lying off the east coast of Papua New Guinea.

As a child I had been captivated by the monolithic Moai statues of Easter Island. Painstakingly, I built a balsa replica of the *Kon-Tiki* raft on which Thor Heyerdahl tested his theories of the migration of the Incas and their sun-kings across the Pacific to Polynesia two thousand years ago. As I carved, lashed and rigged my diminutive vessel, I dreamed the boyhood dreams of distant voyages to the South Seas with only a green parrot for company. My seafaring uncle, Major Theodore Svensen,[1] a former naval draughtsman born in Heyerdahl's own Norwegian coastal town of Larvik, was a veteran of the Boer War and the Gallipoli

---

[1] My great-uncle, then Lieutenant N. T. Svensen, was an officer in the 15th Battalion (Queensland) 4th Infantry Brigade, Australian Imperial Force, that landed at Anzac Cove on the Gallipoli Peninsula on the evening of 25 April 1915. He made an entry at 6.47 p.m. in his meticulously-detailed

Campaign in 1915. He stoked my imagination with tales of the sea and foreign campaigns, his budgie chirruping on his shoulder, a large tropical butterfly tugging against the thread that tethered it to a palm trunk in his garden.

'Useless to read books m'boy! Head for the front line! Go to the islands – that's the last virgin land. Sail before it's too late!' he would thunder as he waxed his magnificent moustache, jabbing with a finger at yellowing maps. Many years were to pass before I could attempt such a voyage to Melanesia, and in many ways it turned out to be sadly too late.

The geographical term 'Melanesia' originates from the Greek *melas* meaning 'black' and *nesos* meaning 'island'. The region was known up to the late nineteenth century as the 'Black Islands', a reference to the strikingly dark skin colour of the indigenous population and their former formidable reputation for cannibalism and savagery. Melanesia is situated in the South-West Pacific, south of Micronesia and west of Polynesia, occupying an area about the size of Europe and containing mainland Papua New Guinea, the Bismarck Archipelago, the Solomon Islands, Vanuatu, New Caledonia, Fiji and the innumerable intervening islands. The extreme cultural diversity of the region evades neat categorisation and facile generalisations remain suspect. It can be observed, however, that Melanesian society is more egalitarian and the qualities of leadership more achievement-oriented than in Polynesia and Micronesia, where power is largely based on inheritance.

Melanesian marsupials have been more deeply studied than the origins of 'Melanesian Man'. The Australian Aborigines and the Negrito populations of South-East Asia are distant relatives from the Pleistocene era some 50,000 years ago. There were two main migratory waves, the ancient Papuan (from the Malay

---

diary, written by moonlight on a torpedo boat heading for the beach. 'We are under shrapnel fire and two or three men have been hit already, one bullet within 18 inches of my foot.' He distinguished himself in the campaign, was wounded and repatriated to Cairo.

papuwah meaning 'frizzy-haired') extending over many thousands of years, and the more recent Austronesian.[1] The intervening millennia have witnessed enormous cultural intermixing. These movements have given rise to the two main cultural traditions in evidence in Melanesia today, the Papuans being the most numerous.

Geographically, New Guinea provided some of the greatest natural obstacles to exploration encountered in any country, with little prospect of gold or cargoes of spices as reward for the sacrifices of the voyage. Nature runs riot in the hot, humid and wet climate. Superlatives abound – over 700 species of birds, 800 distinct languages, the largest butterflies and beetles in the world, five times the species of fish in the Caribbean. Thomas Carlyle idly observed, 'History, distillation of rumour.' He could scarcely have known how appropriate his comment would be regarding expeditions to this fabled land.

The earliest surviving sketches of Pacific peoples were four rather crude drawings of warriors observed off the southern shores of New Guinea made in 1606 by the Spaniard Diego Prado de Touar. My destination, the coast and islands of what was to become German New Guinea, were mapped almost lethargically by a procession of European voyages of discovery. The Spanish and Portuguese were followed by the Dutch, who were succeeded by the English and the French. In 1700, that colourful buccaneer-explorer William Dampier aboard HMS *Roebuck* (a true exotic who mentions in his journal consuming 'a dish of flamingoes tongues fit for a prince's table') found a strait between New Britain and New Guinea. He navigated the coasts of New Ireland and

---

[1] The term 'Austronesian' refers to one of the two major language groups in Melanesia. The other, older and more complex group is known as Non-Austronesian or Papuan. Austronesian may have originated in South-East Asia and all of its languages have a family resemblance, unlike the enormous diversity of the Papuan languages. Austronesian languages are spoken in Indonesia, Polynesia, Micronesia, the Philippines, Malaysia, Vietnam and Taiwan.

named the larger island Nova Britannia. He was the first European to be recorded as discovering and anchoring in the Bismarck Archipelago, formerly regarded as an integral part of New Guinea.

The French, too, have a distinguished history of New Guinea exploration. In 1768 the French Comte de Bougainville charted New Ireland, the Admiralty Islands, Buka and Bougainville. Louis XVI was an enthusiast for exploration and helped to plan and support the ill-fated expedition of Jean-François Galaup, Comte de Lapérouse. Although an aristocrat, the Comte had remained a darling of the revolution as he had married beneath him for love. For the time, this enlightened navigator held radical views on exploration. He observed in his journal:

> What right have Europeans to lands their inhabitants have worked with the sweat of their brows and which for centuries have been the burial place of their ancestors? The real task of explorers was to complete the survey of the globe, not add to the possessions of their own rulers.

He disappeared in the Pacific after leaving Botany Bay in 1788. Louis despatched a search party under the command of Antoine Joseph Bruny d'Entrecasteaux. Part of this voyage of the *Recherche* and the *Espérance* in 1792 contributed to the accurate mapping of the Solomon Sea and the Trobriand Islands. This expedition remained the last significant exploration of the Bismarck Archipelago.

Captain John Moresby in HMS *Basilisk* discovered Port Moresby harbour in April 1873 naming it after his father, Sir Fairfax Moresby, Admiral of the Fleet. In a theatrical gesture he gave 'some little éclat to the ceremony' by using a capped coconut palm as a flagstaff to raise the Union Jack and claim possession. Lieutenant Francis Hayter wrote a rare account of this ceremony.

> On John emerging from the Bush which he did in a way creditable to any Provincial Stage, we presented arms and the Bugler (who

we had to conceal behind a bush as he was one of the digging party and all covered with mud) sounded the salute ... spoiled by the Marines who, I believe, fired at the wrong time on purpose, because they didn't like being put on the left of the line.

Moments of high comedy never failed to pepper this procession of explorations. On one occasion a Lieutenant Yule escaped murder by dancing along the beach nearly naked, dressed only in his shirt. The warriors were so convulsed with laughter at the sight, he managed to reach the safety of the ship's boat.

The stimulus to explore remained strong among adventurers and geographers, naturalists and ethnologists, not neglecting the joyful and sometimes misguided missionaries who attempted to wrest the islands from the clutches of the Devil. Malaria, earthquakes and cannibalism took a fearful toll of their lives. In the north-west of the country, twenty-five years of evangelism had resulted in more missionary deaths than villagers baptised. The profiteers of the East India Company found little to attract their purses. Their settlement at Restoration Bay in 1793 was soon abandoned. The fabulous plumage of the birds of paradise, pearls and pearl shell, bêche-de-mer and sandalwood became the most important items of trade wherever a European settlement became successful. The British, the Dutch, the French and the Germans, a thousand Hungarians and even a Russian, perhaps the greatest scientific adventurer of them all, Nikolai Miklouho-Maclay, attempted settlements with varying degrees of success.

Colonial flags rose over New Guinea like a flock of doves. The British Imperial Government proclaimed their Protectorate on 6 November 1884 by raising the Union Jack on HMS *Nelson*, one of five men-of-war present in the harbour at Port Moresby. The local people squatting on the deck heard in Motu the ambiguous words that were to cause much future suffering and discontent – 'your lands will be secured to you'. German New Guinea had been annexed three days earlier on the island of Matupit in Neu Pommern (New Britain). On 4 November, Kapitan Schering,

Kommandant of the Korvette *Elizabeth*, took possession of the Bismarck Archipelago by raising the German flag on the island of Mioko in Neu Lauenburg (the Duke of York Islands). Another fluttered in the fetid heat of Finschhafen on 12 November, claiming the north-east mainland of New Guinea (Kaiser Wilhelmsland).

Many eccentric and extraordinary individuals were attracted, and still are, to this destination of the imagination. The formidably inhospitable terrain was explored by a bizarre collection of colonial adventurers, a veritable New Guinean *comédie humaine*. Some exploits were not believed when first reported, but most turned out to be true despite their outrageous detail.

The melodramatic Italian explorer Count Luigi Maria D'Albertis was obsessed with the power of explosives, an authentic pyromaniac, and used every opportunity to set off landmines, petrol, fireworks, rockets with or without dynamite attachments, even Bengal lights which emitted a vivid blue radiance – all to intimidate the warriors in the most flamboyant style. Accordingly, on a May morning in 1876, this theatrical explorer assembled his crew – two Englishmen, two West Indian negroes, a Fijian named Bob, a Chinese cook, a Filipino, a resident of the Sandwich Islands, a New Caledonian, a head-hunter boasting thirty-five prizes to date, and his son acting as a navigator. To defend themselves and pacify the local people, they loaded nine shotguns, one rifle, four six-chambered revolvers, 2000 small shot cartridges and other ammunition, the usual dynamite, rockets and fireworks, a live sheep, a setter named Dash (later taken by a crocodile) and a seven-foot python to discourage pilfering from the luggage.

This extraordinary group entered the estuary of the Fly River in the Gulf of Papua on the south coast aboard the diminutive steam launch *Neva*, to sail into the interior of New Guinea for the first time. In order to divert himself from the difficulties he encountered, D'Albertis captured specimens of *Paradisaea raggiana* (Count Raggi's Bird of Paradise) and examined phosphorescent centipedes. When under attack from villagers, he forced them into terrified submission by igniting cascades of fireworks and rockets. With enviable detachment he wrote in his book *New Guinea: What I Did and What I Saw* that he admired the beautiful reflections on the water the explosions made between the banks of dark forest. After some 580 miles from the mouth of the river he was forced to turn back, his legs paralysed by the onset of a mysterious illness. Nine war canoes of warriors blocked his path near Kiwai Island. He charged through them with the engine at full steam throttle, Bengal lights ascending into the sky, funnel pouring black smoke whilst he bellowed out an aria from *Don Giovanni*. He died in Rome of mouth cancer in 1901 after amusing himself in a hunting lodge of Papuan design built on stilts in the Pontine Marshes.

German traders had begun to move into the Pacific during the race for colonies and the first trading stations were set up in Apia in Samoa in 1856. The history of exploration in the Bismarck Archipelago, my destination, is less well known. By the 1870s, business was being done in 'savage' New Britain. The German hegemony over the Bismarck Archipelago and Kaiser Wilhelmsland lasted from 1884 to the outbreak of the Great War in 1914. This was a classically ill-fated German colonial adventure, first under the disastrous and punitive Neu Guinea Compagnie and later ruled by the Imperial Government itself despite the fact that Prince Otto von Bismarck was not an enthusiast of colonial adventures.

German New Guinea also attracted its share of fearless explorers. The Austrian Wilhelm Dammköhler spent thirty years travelling through German, Dutch and British New Guinea. He worked on pearling luggers, prospected for gold, and explored the mainland. He was a man with a literary bent as well as a person of some sartorial distinction. In 1898 he had a close shave with a Tugeri head-hunting party. The ferocious Tugeri were among the most feared of all the tribes. They took heads to provide their children with names. They would cover themselves with chalk, set out in their canoes to attack a village and then after grabbing a victim would demand or cajole his name from him. They would then remove the screaming head with a bamboo beheading knife, memorise the name and bequeath it to their newly born.

On this occasion Wilhelm was collecting fresh water, having anchored his cutter, the *Eden*, at the mouth of the Morehead river. As he rowed upstream he carried, in addition to the water containers in the dinghy, a copy of Byron's poetry, two silk shirts, a pair of Russian calf boots and a pair of white duck trousers. After some thirty miles he encountered the Tugeri. They calmed themselves when they mistook him for a missionary. Dammköhler played along with the deception:

> On the following morning, the chief signed to me to read prayers,
> whereupon I opened my Byron and read some stanzas out of that
> . . . I remained with these friendly natives a fortnight, mixing freely
> with them, hunting with them etc.; and I kept up my missionary
> character all the time, reading to them out of my Byron morning
> and evening during my stay.

How Lord Byron would have loved such an incident. Poor Wilhelm finally bled to death after being attacked on a tributary of the Watut river near the present city of Lae. He was skewered like Saint Sebastian with a dozen fiendishly-barbed arrows in the arms, legs and chest. One severed an artery.

For those romantics and eccentrics, missionaries and mercenaries, desperate speculators, searchers after extremes, explorers, adventurers, swindlers, prospectors and a thousand other misfits who fled from so-called 'civilisation', the Black Islands had become a source of mystical and fictional descriptions, ultimately a magnet. New Guinea has always offered the possibility of self-transformation to depressed though imaginative underachievers and individualists. Outsiders unable to accept the prosaic nature of life in the bourgeois society of Europe have always been seduced by New Guinea and its promise of unspeakable adventures.

In the circulating libraries of the time, the public could read of a phantasmagorical world of fabulous creatures like Captain Lawson's deer, endowed with long manes of silken hair, birds that sounded like locomotives, striped cats larger than the Indian tiger, mountains thousands of feet taller than Mount Everest. They read of men with vestigial tails who sat in their huts allowing the tails to protrude through special holes cut in the floor. There were reports of native cavalry that rode striped ponies and women who ate their children as a form of birth control. They read of the web-footed Agaiambu people, who lived in the marshes and swam through the reed beds, had flaring nostrils like a horse, small legs and buttocks, strange muscular protuberances on their scaly inner calves, walked with the 'hopp-ity gait' of a cockatoo on flaccid, straggling toes and whose feet bled when they walked on dry land. They kept pet crocodiles tethered with vines and raised pigs in slings. New Guinea was a domain of impenetrable tropical jungle and gothic phantasms that might well have been imagined by the French naive painter Le Douanier Rousseau on a particularly creative day.

The islands also attracted visitors of genius who had serious academic intentions. Perhaps the most famous of these was the Polish anthropologist Bronisław Malinowski who, at the out-break of the Great War, pioneered new methods of fieldwork in the Trobriand Islands. Melanesia subsequently became a cultural laboratory for European anthropologists and one of the most closely studied of the 'unknown regions' on earth.

In 1906 Australia took responsibility for British New Guinea and the British Protectorate ceased to exist. The new territory was now to be called Papua. The first Lieutenant-Governor of Papua, Sir Hubert Murray, was an empire builder of Olympian accomplishments. He was a character who seemed to have stepped straight out of *Boy's Own* fiction. Born in Sydney in 1861, he stood six foot three, weighed fourteen stone and was amateur heavyweight boxing champion of Great Britain. A row-ing blue at Magdalen College, Oxford, and possibly the finest swordsman in Australia, he had read Classics and achieved a double first.

A career as an Australian barrister beckoned but he abandoned the idea as 'too tedious'. In 1904 the post of Chief Judiciary Officer became available in New Guinea. In need of diversion, he replied to a newspaper advertisement and was offered the post. He adopted a paternalistic style of governorship, promising the Papuans, 'I will not leave you. I will die in Papua.' At the time he was considered progressive but now is considered by indigenous historians as being regrettably colonial. He greatly admired men who exercised self-discipline and refused to open fire on the most threatening of warriors. While travelling the country on his circuit he carried a portable library. His nephew recalls seeing him reading a Greek text in rough weather, seated in a chair lashed to the deck of his small government vessel *Laurabada*, holding it above the waist-high foam to keep it dry.

He wrote a number of excellent books recalling his tours of duty, full of wry observation. He mounted expeditions into the interior and developed a degree of understanding of native customs and languages unusual in colonial administrators of the time. His laconic style keeps one turning the pages. He describes murder in his book *Papua or British New Guinea* published in 1912:

> . . . murder to these outside tribes is not a crime at all; it is some-times a duty, sometimes a necessary part of social etiquette, some-times a relaxation, and always a passion. There is always a pig mixed up in it somewhere . . . *Cherchez le porc.*

He later refers to the reputation of '. . . the Rossel islanders who were quite oblivious to the most ordinary rules of hospitality' and ate 326 Chinese who had been shipwrecked on the island. He informs us that in some villages 'a thief is punished by killing the woman who cooks his food' as this causes great incon-venience to the thief.

Both his wives Sybil and Mildred clearly lacked the sense of humour required to survive the colony, hated every minute of it and left him alone for long periods. Rumours of his mistresses were legion. Government House became the dwelling of a bachelor, full of books, manuscripts, saddles and muddy riding boots on the veranda. In February 1940 he suffered his final illness but refused to be carried off the *Laurabada* at Samarai hospital. 'You can carry me when I'm dead, but not before.' He was seventy-eight and had been in office for thirty-five years. On the forty-first day of mourning, thousands of Papuans came together at the stilt village of Hanuabada in Port Moresby for the funeral feast. There was total silence among the lighted fires and torches except for 'the quiet tapping of a thousand native drums'.

At the outbreak of the Great War, the small Australian Naval and Military Expeditionary Force, derisively called the 'Coconut Lancers' by my uncle, the Gallipoli veteran, released the German Government from further responsibilities in a minor military engagement at Bitapaka near Rabaul. '*No more 'um Kaiser, God Save 'um King*' read the Australian proclamation issued to the bemused villagers.

Much of the German administration was retained. Those Germans who took an oath of neutrality were allowed to return to their properties. There was an abortive move to rename the German colonies the Kitchener Archipelago or even Australnesia. The Australian military administration replaced the more enlightened, or as they saw it, 'soft' German bureaucracy, with a regime of questionable severity. Both Germans and local people were treated with arbitrary and undisciplined brutality. A number of Germans were photographed being flogged in public.

After the war, Sir Hubert Murray advocated measures that would create the combined state of 'Papuasia', comprising Australian-governed Papua and the former German New Guinea. His dream was for it to have an educated and affluent indigenous population. However, after protracted discussions throughout 1920 and threats by the then Australian Prime Minister, Billy Hughes, the League of Nations finally handed the entire former

German possession to Australia in 1921, now to be known as the Mandated Territory of New Guinea. The capital would be Rabaul on the island of New Britain. Papua was to remain under a separate Australian administration with the headquarters in Port Moresby.

Great rivalry came to exist between these two Australian colonies. All German possessions in the Mandated Territory were expropriated – the magnificent German colonial buildings, the immaculate plantations, even wedding photographs. 'More like looting,' some residents thought. Many plantations were sold to Australian ex-servicemen who had little understanding of proper methods, more a romanticised vision of white verandas overlooking tropical lagoons, the plantation worked by armies of cheap labour. A fortune would be guaranteed. The young Errol Flynn ran such a plantation near Kavieng on New Ireland, deftly concealing behind that dazzling smile his complete ignorance of the copra industry. Particular bitterness resulted from the post-war inflation that made the German government's compensation to German planters worthless when it finally arrived. The welfare of the indigenous population was of course ignored.

The influence of the previous military administration remained strong under the civilian mandate, many soldiers becoming government officers. Unfortunately, it signalled a return to some of the worst excesses of the Neu Guinea Compagnie. Punishment was often entirely at the whim of the District Officer or *kiap* ('captain' corrupted into Pidgin). There was little accountability and few criteria for the capricious penalties imposed by these men, referred to by some as 'God's shadow on earth'. Substantial authority was sometimes placed in the hands of inexperienced boys as young as twenty-one. They had come to New Guinea in search of 'adventure', and found themselves in control of enormous tracts of territory and large numbers of the indigenous population. Severe regulations were implemented such as the puritanical 'White Women's Protection Ordinance' which meant that a Papua New Guinean male even smiling at a white woman was fraught with the danger of imprisonment. But scattered among the neurotic and unstable were many outstanding individuals who felt a strong sense of moral obligation to the colony and considered their service a true vocation. In time, university training and a career structure emerged, albeit military in flavour, and much was achieved in the fields of tropical medicine and construction work.

The Australian public were too preoccupied with the aftermath of the war and their own grim future during the Depression to take a close interest in faraway New Guinea affairs. The government of the day had a strategic interest in Papua and the Mandated Territory, ever hopeful of revenue from gold and petroleum. The colony was expected to pay for itself, many Australians feeling a degree of ambiguity about the whole notion of an 'Australian' colony. Local people suffered greatly under the rule of a nation that was struggling with an unclear view of its own national identity. Villagers scarcely understood the nature of the European wars that had carved up their land so barbarously and confused their allegiances. That there was not more violence speaks volumes for the adaptability of the Melanesians to the Australian mandate, those unpredictable successors to the severe certainties of German rule.

The Highlands up to this time had been considered uninhabited. In 1933 an Australian adventurer named Mick Leahy and his young brother Dan, together with the patrol officer, Jim Taylor, flew over the inaccessible Wahgi Valley for the first time in a Junkers aircraft and saw signs of intensive agricultural settlements. This valley was perhaps twenty miles wide and sixty miles long and contained a long meandering river. To discover if it contained gold they decided to explore on foot. The spectacularly-decorated local people wearing the brilliant plumes of Birds of Paradise had never seen white men and regarded them as their returned spirit ancestors descended from the skies. They in turn were impressed by the appreciation shown by the 'wild men' for Italian opera, played on a wind-up gramophone. The power of art effortlessly to cross cultural boundaries was commented upon until a little translating from the indigenous tongue revealed that the sounds emerging from the trumpet reminded the warriors of the screams of women selected for cannibal feast. Mick Leahy made further expeditions, but gold eluded him to the last. He gave an entertaining paper in 1935 in London at an evening meeting of the Royal Geographical Society where he described warriors with dried snakes and finger joints hanging from their ear-lobes.

> A number of them have an eye shot out. We concluded that they
> were peeping round the shield at an inopportune moment . . .

Mick himself had peeped into skull racks shaped like dovecotes. The Fellows awarded Leahy the prestigious Murchison Grant for his explorations. He is remembered as a dashing, romantic explorer who took brilliant photographs and filmed some of the most astounding footage of Stone Age people encountering Europeans for the first time. New Guinea provided a rich vein for what might be termed 'Macho Adventure Writing'. The American author Jack London sailed the *Snark* through Melanesia in 1908. In his strange story *The Red One* published in 1918, he is unashamedly excessive throughout, continuing the general fascination with ear-lobes:

> . . . her sex was advertised by the one article of finery with which
> she was adorned, namely a pig's tail, thrust through a hole in her
> left ear-lobe. So lately had the tail been severed, that its raw
> end still oozed blood that dried upon her shoulder like so much
> candle-droppings.

During the Second World War all the islands of the Bismarck Archipelago fell quickly to the Japanese. The arrival of the Americans with their enormous quantities of 'cargo' had a profound impact on the local population. Even more influential was the equality with which they saw 'black' American servicemen treated by their white comrades. The notion of colonialism became an anathema in the post-war world. Australia was influenced by a United Nations recommendation in 1962 to prepare the country for independence. Russia under Khrushchev derided colonialism on the international stage with cutting invective. To the villagers of Papua and New Guinea, all the foreign powers were equally heartless intruders.

> Every white man the government send to us
> Forces his veins out shouting
> Nearly forces his excreta out of his bottom
> Shouting you bush kanaka[1]

KUMALAU TAWALI, from 'Bush Kanaka Speaks'

Less abusively, but with a fiery idealism, my friend John Kasaip-walova, the poet and student radical, urged the destruction of the colonial yoke:

> Reluctant flame open your volcano
> Take your pulse and your fuel
> Burn burn burn burn
> Burn away my weighty ice
> Burn into my heart a dancing flame.

JOHN KASAIPWALOVA, from 'The Reluctant Flame'

---

[1] This historical word had a Hawaiian origin meaning 'man'. A 'bush kanaka' was a Pacific Islander employed as an indentured labourer in Australia on the sugar plantations in Queensland. It came into more general colonial usage throughout Papua and New Guinea for any labourer. Initially a neutral word, it evolved into a rough form of colonial abuse, much resented by native labourers. It was also used in the islands as a term of what might be called 'affectionate abuse', so beloved of the Australian male when addressing each other or local 'mates', viz., 'Come and have a beer you bloody kanaka!'

In 1975 full independence from Australia was finally achieved. It was gained without bloodshed, revolution or violence. This in itself was a significant tribute to Melanesian tolerance.

In the decade following independence, economic and trade links with overseas countries were successfully established. Papua New Guinea has received over 10 billion dollars in development aid from Australia since independence was achieved in 1975. In recent years, however, the promising legacy of a Westminster-style government has degenerated through cultural hybridisation into a desert of self-interest, nepotism, corruption and violence. The development of *wantokism*,[1] which evolved quite understandably from traditional community and family ties, is deeply at variance with the notion of 'fair' individual enterprise and parliamentary democracy. An egalitarian culture of mutual support dedicated to subsistence survival has clashed with the European individualistic, dividend-driven market economy creating a climate of desperate contradictions. There are no beggars in Papua New Guinea and, unlike many nations, no one would ever starve. The land is rich and fertile, the clan loyalties strong. Yet traditional cultures can no longer supply solutions to the types of internal and external dilemmas posed by the modern world.

Admiral Sir William Goodenough, the chairman of the 1935 meeting of the Royal Geographical Society addressed by Mick Leahy, clearly had a premonition after the paper was read when he commented:

How far and in what direction is the march of man going to interfere with these people? Every possible care should be taken

---

[1] Literally 'one talk' in Pidgin. People who share the same language, loyalties and cultural heritage. They perform shared activities and support each other as they come from the 'same place', village or province. The preferential nature of so-called *wantokism* is a source of much dissatisfaction in modern PNG and the explanation for numerous misplaced criminal accusations of nepotism in both the private and public sectors.

that the people of New Guinea and their country are not exploited in any way.

True values of independence and respect for *kastom* are struggling hard to survive in the hearts and minds of the new generation growing up in the Bismarck Archipelago.

4. Death is Lighter than a Feather

Across the sea,
Corpses in the water,
Across the mountains,
Corpses heaped up in the field,
I shall die only for the Emperor,
I shall never look back.

Japanese Second World War Poem

No roads link Alotau, the provincial capital of Milne Bay Province, with Port Moresby. Boats are infrequent and no longer run to schedule. Fortunately, this has insulated the province from *raskol* activity which blights life in the Highlands. The relative isolation of Alotau meant that flying from Port Moresby was the only feasible alternative.

The 'Islander' light aircraft climbed laboriously out of barren Jacksons airport over the forbidding green of the Owen Stanley Ranges. A gothic landscape unfolded below, the precipitous ranges and valleys resembling the spires and flying buttresses of monumental medieval cathedrals draped in cloaks of thick, tropical vegetation. The gloom and mystery brought to mind the inconceivable hardships endured by nineteenth-century explorers who attempted to probe the interior of New Guinea from Port Moresby. Precipitous paths needed to be cut through thick jungle slowing progress to a mile a day. Sir William MacGregor, first Lieutenant-Governor and intrepid explorer, wrote in his diary of the 'deathlike stillness' that prevailed in the dripping fog that swirled about the moss-covered trees as he climbed Mount Victoria. Gurney airport, which services Alotau, occupies the same position it did in 1942. Charles Gurney was one of the colourful band of aviators who opened up New Guinea in the 1930s and was a squadron leader in the RAAF (Royal Australian Air Force). Below the spinning hub of the propeller, endless plantations of oil-palm unrolled sporadically through breaks in the low cloud.

The terminal was the usual fibro affair with slowly circulating ceiling fans. My luggage was quickly unloaded and packed into a truck. The road into town was severely potholed, with river fords constructed of shattered blocks of concrete. Torrents rushed across them into Milne Bay, threatening to sweep us away. It was hot and uncomfortably humid. Luxuriant vegetation swathed in mist fell like velvet curtains from the ranges lowering into the Coral Sea. The province showed little sign of development.

This formidably rich cultural area is known by Europeans as the Massim. The inhabitants speak Austronesian languages, unlike the majority of Papua New Guineans, who speak one or other of the 300 Papuan languages. Much of the Milne Bay Province consists of islands and the finely-carved artefacts reflect this marine environment. Spectacular ornamented canoes, weapons of ebony and black palm, red spondylus shell necklaces and armshells of tangled beauty, clay pots of abstract shape, and noble war shields are the legacy of the past. Gifted master carvers worked along this coast in the nineteenth century, but the production of such masterpieces has tragically declined in modern times. The culture of the north supports large village communities, whereas in the south, clusters of tiny hamlets nestle along the shore. Their belief in sorcery and magic is strong. Compulsive sweeping keeps the villages free of any personal waste that a sorcerer might fix upon and use in casting an evil spell.

Massim communities are composed of clans ruled by symbolic totems representing an animal, bird, fish or plant. Children take the totem of their mother and it is forbidden for members of the same totem to marry or have sexual relations. In the past the punishment for illicit sex was death. In the almost complete absence of tourism, apart from hermetic groups of divers, the people subsist through fishing and agriculture. Crocodiles and turtles, pigs and megapodes (a small, dark-feathered scrub fowl), bananas and coconuts are being insidiously supplanted by modern supermarket foods, particularly cheap tinned meat like Spam that has scarcely altered since it sustained troops in the Second World War.

The Massim was one of the first areas of New Guinea evangelised by British missionaries in the late nineteenth century. The local people's profound belief in the power of magic posed the greatest challenge to Christian conversion. The villagers continue to believe that supernatural spirits inhabit trees, streams, rocky places and swamps. Belief in Jesus has not removed the fear of sorcerers who kill by projecting fatal diseases and cannibalistic flying witches that become airborne after dark, snapping bones and tearing entrails. More seductively, erotic magic weaves love spells with flowers plucked in secret groves. More optimistically, a happy life after death is guaranteed by a magic that promises three states of Paradise and an afterlife where Hell is sensibly absent.

Alotau developed from the original Second World War American military base, but the capital was transferred officially from Samarai Island only in 1968. This islet, lying off the easternmost point of the mainland, had become inconvenient as it was only accessible by sea and overcrowded with residents. Alotau has the reputation of being the safest town in Papua New Guinea, a reputation jealously guarded by the inhabitants.

I had arrived in the sultriness of mid-afternoon. Hundreds of listless people were sitting under the rain trees in the shade cast by the awnings of prefabricated supermarkets. They were almost motionless in the oven-like conditions and appeared to be in shock, as if a terrorist bomb had recently exploded. Clumps of resentful youths were chewing *buai*,[1] smiling women in colourful cottons suffered tugging children, a covered market baked in the heat. Banana boats skittered across the glittering harbour and rusting Taiwanese trawlers disintegrated at their moorings.

A path crossed a stream, and I climbed some steep stairs through an avenue of trees past the tables of female sellers of *buai*. This green globular betel nut, which is the seed of the Areca palm, is laid out in carefully measured rows and beside each nut lies a small betel-pepper stick, the fruit of the pepper vine. The husk of the Areca seed is removed and the tip of the green pepper spike dipped in lime, now usually contained in an old plastic film container. The mixture is then enthusiastically chewed. In the past the powder was taken from an attractive gourd or the shell of the young coconut, often decorated with intricate burnt-in designs and fitted with a woven or wooden stopper. A beautifully carved ebony spatula with a pig or human carved on the handle would be used to remove the lime. Some spatulas were made from the bones of relatives. The family would then be able to suck the bone when taking lime as an intimate reminder of the departed. Lime powder is no longer made from crushed coral by women in secret locations. Spatulas are seldom used.

Gallons of saliva are generated from the chemical reaction and the resulting vermillion juice is spat in jets like an uncontrollable haemorrhage. The effect is horrifying to witness and mildly narcotic to experience. The villagers say it 'makes them feel strong', '*makim head good fella*'. Certainly it makes people more talkative, but excessive use creates a drugged daze in the chewer. Nuts are often presented to visitors. In the past, if the point of an offered nut faced away from the stranger, it was a secret signal to kill him.

I was overcome by nausea and an atrocious bitterness during

---

[1] Pidgin for 'betel nut'.

my first attempt at 'wearing New Guinea lipstick' as it was popularly known. Gales of laughter accompanied my facial contortions and twitches but much friendliness followed. Captain Cayley-Webster, travelling through New Guinea in the late nineteenth century, referred to betel as 'a veritable *pâté du diable*'. Chewing is paramount in social relations in Papua New Guinea, and the cosmetic clash this creates with modern sanitised life has become a symbolic focus of cultural freedom and *kastom*. Betel is chewed by men and women when working in the gardens, attending feasts and travelling by canoe, when making love and meeting friends. Television advertisements encourage people to stop chewing because of risks of mouth cancer and to present a 'clean image', but for most it is a universal refuge begun early in life, a balm to the rigours of existence in these islands.

I would be staying at Masurina[1] Lodge, run by Chris Abel, grandson of the missionary Charles Abel. I had never met him, but I had read a great deal about his unique regional business. In 1973 Alotau was a tiny place consisting of five trade stores and a post office. Chris established the Alotau Tea Shop with the help of Mila Walo ('Aunty Mila'), one of the outstanding women who emerged from Charles Abel's Kwato Mission. This tea shop was the forerunner of a large local public company called Masurina that Chris Abel established at Milne Bay, with interests ranging from accommodation to fisheries and construction. The local people of Milne Bay are major shareholders in what has become a symbol of the commercial way forward in modern PNG. I wanted to visit Samarai Island and the Kwato Mission and talk to Chris about his controversial forebear.

'The Lodge', as it is known locally, is situated high above the harbour and has the flavour of an early South Seas colonial resort with prefabricated units painted with large blue numbers. Reception and what might be termed a drinks veranda have a

---

[1] 'Masurina' means 'the fruits of an abundant harvest' in the local Suau language.

pleasant colonial atmosphere. A number of fans were ranged along the balustrade to keep the air moving and the mosquitoes at bay. My room overlooked the last thrust of the Owen Stanley Ranges, a line of jagged peaks heading down to the sea. Coconut palms crowned the hill above my writing table and from a garden below, a disembodied, unearthly monody sung by a child floated on the breeze. A dog barked in a distant valley. The weather felt unstable, the peaks shrouded in knotted clouds that were cut by the occasional flicker of lightning followed by sombre thunder.

A couple of attractive local girls with engaging smiles were looking after reception and talking quietly in the Tavara language. A figure sat at a table drinking beer and reading. He wore olive-green officer's fatigues as part of his tropical kit, the crown of his Australian Akubra hat covered with a colourful woollen cap from the Highlands. A furled racing umbrella was propped against the arm of the chair. Some artefacts and a slim volume entitled *Betel-Chewing Equipment of East New Guinea* lay on the bamboo table. Clear, grey eyes and a welcoming face framed by a well-trimmed beard greeted me as he lowered his clip file.

'Come and sit down. Get yourself a beer.'

It was a relief to relax near regular puffs of air from the fan. Carrying my luggage the short distance to my cabin had sent the sweat streaming down my face. Any movement in this sweltering heat apart from drinking seemed excessive. I bought an ice-cold beer and sat down in a cane armchair.

'Who do you work for?' he asked directly. The pressing need to speak to a European faintly betrayed itself.

'No one actually. Just wandering the islands.'

'Really? A wanderer is pretty unusual round here. I work for AusAID – Biomedical Engineer checking equipment – at the hospital.' He would be the first of many aid workers I would meet on my journey.

'So, what's the state of the hospital equipment in PNG?' I asked, unsure whether I wanted to hear the answer. Assembling my own travelling medical kit had taken weeks of thought and terror, as the list of possible ghastly diseases and the range of conflicting advice grew.

'Dire, absolutely dire. The hospital in Alotau though is actually quite good with excellent staff.' There was disappointment in his voice, overlaid with an almost convincing pragmatic realism.

'What sort of problems do they have?'

'Well, the main problem is lack of maintenance. The cultural mentality is so different. They think sterilising only requires the instruments to be washed in Omo.'

I felt that the constant struggle with cultural 'otherness' had made him almost unnaturally phlegmatic. He smiled wryly.

'Is it the same all over the country?'

'The Highlands are worse of course. I saw an ambulance in Mendi drenched in blood. I thought, "God, it's bloody violent. Even the ambulances are blood-soaked!" Actually, it was betel juice from people spitting on it. Looked just like blood! But spitting on an ambulance?'

I smiled but my feigned bravado concerning health matters was ebbing away. We sat in silence, the fans whirring and the occasional tortured dog screaming in agony.

'The tribes up there are spearing each other again. They love fighting and drinking. Some died recently after downing a hellish cocktail of coconut juice, methylated spirits and turpentine. It's reverting to pre-colonial days.'

An unmistakable tone of angry disillusionment and ruined hopes marked his voice. So many aid workers begin with high ideals that fade in the face of indigenous resistance to change. The benefits of being rushed headlong into a technological paradise from the Stone Age are not immediately obvious to men still profoundly involved with their elemental natures.

'Don't you ever worry you might be targeted?'

'Sometimes, but I *am* related to a Napoleonic general!'

Despite the off-hand smile, an expression of cultivated stoicism hardened in his eyes. An easy man to underestimate.

'Is that so.' I looked away.

I must admit to being sceptical of Napoleonic references in this part of the world. I had heard many such claims while travelling through Polynesia in my younger days. Ravings mostly. The South Pacific attracts extraordinary characters often beset by cosmic visions.

Heavy tropical rain had begun to fall on the iron roof and the storm channels were brimming with water. Night was quickly closing in as the fans hummed lazily. Village girls carried platters of food into the dining room. I rolled down my sleeves – a precaution at dusk in this malarial area. The female *Anopheles* emerges to strike at close of day. Small, silent and deadly.

'But you're not French are you?'

'No, English, actually. Born in Surabaya in Java.'

'So who was the French general?'

'General Alexandre Mocquery. He attended the Military School at Fontainebleau. Around 1806, I think it was.'

He chuckled in the way that those moved by the memory of illustrious relatives often do – a mixture of respect combined with a feeling of comparative inadequacy. The silhouetted coconut palms began to dissolve in sheets of water.

'How did he die?'

'Fever in Algeria.'

We both fell silent and looked out into the opaque, watery atmosphere, listening to the muffled clatter of a tropical deluge on the broad leaves and thought of Europe. The ghosts of a hundred misguided adventurers and metaphysical questers seemed all around us.

'Shall we go in for dinner?' he said at last.

Sele and two other men were standing by the roadside as the Toyota Hilux four-wheel-drive whizzed past. They were bending over, staring up and listening to a rattle in the suspension of the truck.

'It's not serious,' one said.

'It'll get us there,' said the other with finality.

'Do you really think so?' I said.

The vehicle had turned around and was steaming back down the road toward us, sitting high on its suspension. Again they bent over with ears cocked.

'The brakes were all right last week,' Sele said.

'I went down to East Cape last month in it,' another said.

'Oh, come on! Let's go!' I said. The Toyota roared past once again sounding pretty rough.

Sele, myself and two laughing island girls, Rachel and Marie, loaded up with provisions and plenty of chilled water. Our excursion to East Cape, the most easterly point of Papua New Guinea, would take most of the day. The atrocious road was full of the usual potholes requiring the skills of a rally driver to negotiate. We would need to cross some fifteen rivers and streams swollen by unseasonable rain. Some had warning signs of treacherously deep water: 'Jesus Loves Careful Drivers – Take the Right Side'. Love messages with hearts and arrows had been picked out in white shells on the river beds.

The area toward East Cape is relatively unspoilt, and we passed the immaculate hamlets, villages and family communities of the Tavara people that have been erected at the very edge of the water. Clear, swept areas of sand have been carved from the dense tropical jungle to accommodate the thatched-roof huts erected on stilts with diapered walls of palm leaf. Smaller detached huts nearby serve as kitchens. Bedding of patterned sleeping mats and pillows was laid out in the sun to air. A Milne Bay woman stood at the window waving and smiling through the brightly coloured washing hanging on the line. Beautiful children squealed with intense pleasure as the family pig blundered about the yard accompanied by a wretched dog with its scrawny pups. The road caused us to be thrown about inside the cabin like rag dolls.

'Where were you born, Sele?' I asked. The girls craned forward, bumping my shoulder and listening intensely to my words, collapsing in fits of giggles if I caught their eye.

'On Logea Island, near Samarai.'

'Really? I hope to go there. I want to visit the Kwato Mission.'
Coincidentally, we were passing a church, one of many along
this road. Pale blue walls with a simple black cross. As it was
Sunday, a large congregation had filled the building. The huge
windows were thrown open and hymns were being sung with a
passionate enthusiasm that saturated the tropical groves.

'They are good people!'

Sele had a stained ivory smile and seemed illuminated from
within by his Christianity. A good man. The girls had never been
to East Cape before and were in a state of high excitement,
chattering and giggling interminably.

'Are you still at school, Rachel?' I glanced over my shoulder
at the pair bouncing in the back.

'No!' they chorused, '*Mipela iwok lon Lodge, insait lon
kisen.*'[1] Few children go on to secondary school. We bumped
along, the springs often bottoming out in the potholes.

'Is there much violence around Milne Bay, Sele?'

'No. It's peaceful here. A ship came in from Lae with many
raskols last month. It was in the harbour. Many crimes happened
but we got rid of it pretty quick. We don't want such things here
in Alotau.'

He seemed proud of taking the moral high ground and clearly
wanted me to judge Milne Bay and the islands as far superior to
the rest of the country.

Picturesque family groups were sitting on the beach in the
shade of flowering pink and white frangipani trees, talking,
laughing and looking out to sea. Many elegant canoes with outriggers were drawn up on the shore under rosewoods. The hulls
had faded to a delicate pale blue or jade green. Groups of children
happily played with models fitted with sails. These childish rep-
licas were to be the only sails I saw whilst in Papua New Guinea.

The puncture we got from the brutal coral road was only to
be expected. Sele showed not the slightest exasperation, treating

---

[1] 'We work at the Lodge in the kitchen.'

it more as a slight inconvenience than a drama. The spare wheel was loosely chained to the vehicle and the change was accomplished in record time. The girls were extremely helpful, as I attempted to be, but the humidity and the searing sun made physical effort an exhausting task for a *dimdim*.[1]

This road had been a quagmire in 1942 when the airbase at Gili Gili was being hewn out of the jungle to defend Milne Bay against a landing by Japanese Marines. The local people had watched spellbound as gelignite was placed in holes drilled in the base of coconut palms and the detonation propelled them vertically into the air. Our situation reminded me of photographs I had seen of bogged trucks and ruptured tanks that had skidded into ditches during the battle.

During the Second World War, Milne Bay possessed great strategic significance as it guarded the sea lanes to Australia and the eastern approaches to Port Moresby. By mid-1942, the area had become enormously important to General MacArthur in his campaign against the seemingly invincible Japanese. The Imperial Army had conquered the Bismarck Archipelago and was poised to strike at Australia. Pearl Harbor, Singapore, Malaya, the Philippines and the Dutch East Indies had already been consumed by the forces of the Rising Sun. A remarkable victory was achieved here by the Australians and the Americans against malaria, typhus, bombs, scorpions, mosquitoes, rats, falling coconuts, crawling insects, green dye bleeding from their uniforms, disease, forbidding terrain, incessant rain, clinging mud and a fanatical enemy. This largely forgotten battle marked the extraordinary first defeat of the Japanese on land and halted any further advance in the Pacific, west or south.

The local villagers played a significant role in the victory and suffered terribly at the hands of the Japanese. Many village men were tied to coconut palms with signal wire and bayoneted in

---

[1] Originally a Milne Bay word long used for white men, probably meaning 'stranger from across the sea'.

the chest or anus. A girl of fourteen was staked out, stripped naked, a bamboo stake driven through her chest, her breasts cut off and placed on the ground beside her. Poor food supplies meant the Japanese even turned to cannibalism. Australians took their own violent turn, and carried out summary executions of villagers they suspected of collaboration. For the local people it was a foreign war of which they understood nothing and cared less. Despite the atrocities, their loyal support of the Australians led to them being known as the 'fuzzy wuzzy angels'.

The perpetration of unspeakable tortures by the Japanese has an explanation of sorts. The private in the Japanese army was treated as a cipher by his officers; animals and weapons were treated better. They became intoxicated in the heat of battle and disregarded discipline. Brutalised soldiers may have been effective against Russian, Chinese or Manchu troops, but permitting emotion to dominate proved fatal in the Pacific War. In the jungle they suffered from malaria, fatigue, poor food and heavy, outdated equipment. A private had no method of relieving the pressure of his pent-up fury. Japanese officers intended their men to hate them. The officer class was driven by elitism and a sense of fanatical loyalty to the Emperor and his Imperial Army. When unsheathing the sacred regimental sword, an officer would bind his mouth with cloth to avoid breathing upon it and as a war journal observes, 'amorously caress the naked blade with white silk'.

'On the way back I'll show you where the Japanese landed,' Sele said. 'Local people told their spies the wrong place!'

The sudden silence that followed once the truck had stopped wrapped us in a cloak of birdsong, the laughter of children, tiny waves rapidly lapping on the shore and cicadas racketing in the tropical heat. War seemed a distant memory and little appeared to have changed for millennia. Milne Bay is one of the least disturbed areas in the whole country.

East Cape has glaring white coral beaches, a decaying schooner hauled up under the palms and a granite Methodist Mission Memorial baking in the sun. The small village of Bilubilu is nearby. Across the Goschen Strait the looming bulk of Normanby Island seemed to deserve its reputation for sorcery and cannibalism. Sele and the girls unpacked our lunch and I sat with my back against a gnarled tree at the edge of the turquoise sea and ate my sandwich, watched carefully by a group of shy children who put their fingers in their mouths and tugged at their clothes.

The cobalt waters that swirl up between the vastness of the Coral and Solomon Seas are diamond clear and support an unparalleled profusion of marine life. Many forms are still to be classified by biologists. The reef drop-off is perfect for snorkelling. More screams of laughter as I climbed into my diving gear – lycra suit, gloves, booties, fins, mask and snorkel. Coral cuts become infected in seconds in these warm waters, so rich are they in bacteria. There are the added attractions of fire coral that cause long blisters when touched, lionfish with beautiful but treacherous spines, cone shells that shoot poisonous darts, stinging hydroids, the occasional shark. I was taking no chances. Children swim constantly with no protection but I never saw an adult Melanesian swimming for pleasure.

As I slowly headed out to sea, superb tropical fish and a kaleidoscope of soft corals were laid out beneath me like a living carpet. Visibility in these glassy waters can be as much as fifty metres. Tiny electric-blue fish formed constellations around isolated outcrops of rock; rainbow fish swam lethargically away to shelter beneath the coral shelves; butterfly fish abstractly painted in swathes of luminous purple and chrome yellow shot into crevices; black and wild-green specimens with long, pointed mouths ignored me completely; gossamer-thin angel fish flowed in the crystal current like fabric; fantastic lacy scorpion fish mimicked plants and defied my most careful observation. The seabed as far as I could see was covered with ultramarine starfish, mauve-tipped clusters of beige coral and enormous brain corals. I remained among these enchanting coral gardens for more than an hour.

Sele seemed pleased that I had enjoyed my swim until I mentioned the skull cave I knew was nearby. In the Massim the dead were buried twice. In the second interment, bones were placed on rock shelves overlooking the ocean, or in dank underground holes near the shore. The cave was five minutes from the village, but no one would agree to show me the place. There remains a great fear of sorcery and witchcraft in Milne Bay. Strange apparitions still manifest themselves at sea in contradiction to mission teaching.

'We don't believe in such things anymore,' Sele said not terribly convincingly.

'Well, then, it doesn't matter if you show me.'

He would not argue and shuffled about looking at the ground, occasionally spitting a jet of crimson, anxious to be off. The girls too had lapsed into silence.

'The first mission school was called Under the Mango Tree. We were happy.'

As he crunched the Toyota into gear a couple of young boys chewing betel nut and dressed in sharp sports-shirts begged for a lift into Alotau. There is no bus service and hardly any transport this far along the Cape. Sele politely asked me if I minded, so naturally I agreed. They gratefully climbed into the tray behind the cab. As we jolted along I pointed to a pretty bush-material hut on stilts standing in the water and asked the girls what it was. Screams of laughter came from the back seat.

'A toilet!' they said after catching their breath.

Soon after leaving we were flagged down by some local Tavara villagers whose banana boat had run out of petrol during their Sunday outing. They also piled into the back. Sele ignored our precariously perched passengers despite the lurching of the vehicle, which threatened at any moment to catapult the whole laughing crew into the palms. Clearly they were accustomed to hanging on for grim death. We stopped at a small beach overhung with rosewood trees, a few rusty spikes poking up through the tide washing the sand.

'This is Wahahuba where the Japanese landed. Wrong place! They came on a raining night and crawled under the huts. We thought they are dogs and pigs looking for dinner.' Sele smiled with strange equanimity.

'Were the people frightened?' One inevitably asks trite questions concerning war.

'Much shouting out. One *meri*[1] quickly grabbed up her covers thinking her baby is there, rushed away, but later she finds her bundle is empty. Terrible.' Sele seemed almost tearful. 'They bayonet people to keep them silent.'

On the showery night of 25 August 1942, a heavy naval bombardment preceded the attack of the Japanese Special Naval Landing Force. The Japanese began the engagement with only a vague idea of Allied strength. Neither side had been trained for living and fighting in the tropical jungle. Their equipment was inappropriate to combat the incessant rain and mud which destroyed their boots and rotted their feet. Some Australian troops who fought as part of Milne Force were young volunteers of the Militia, popularly known as *chocos* (chocolate soldiers) or *koalas* (not to be sent overseas or shot). They had trained sporadically at home during their free time. A great deal of ill-feeling, known as the 'choco smear', was expressed towards the Militia by the professional, battle-hardened troops of the Australian Imperial Force (AIF) who had recently fought bravely in Tobruk in the Middle East. 'Scum, scum, the Militia may kiss my bum,' they would chant derisively.[2]

The Milne Bay Battle was problematical and confused. Troops

---

[1] Pidgin for Papua New Guinean woman.

[2] At the outbreak of the Second World War, Australia maintained three separate armies of volunteer personnel. The Militia were part-time, citizen-force volunteers ineligible for service outside Australia or its colonies. The Second Australian Imperial Force (AIF) was a highly-trained volunteer force eligible for service anywhere overseas. Finally, there was the permanent army made up of a relatively small force of trained volunteers. There was friction between these armies due to the differentiation of combat role and degree of professional training. Many Militia units subsequently distinguished themselves abroad when their theatre of operations was extended.

had suffered from seasickness on the stinking hulks that brought them to New Guinea. The only food was Bully Beef and 'Jungle Juice' distilled from palms. Coconuts fall when the stem swells with moisture, and after rain these missiles killed and injured many men. The troops glowed a livid shade of green from skin treatments and dye bleeding from their wet tropical uniforms. There were no proper maps, only rough sketches of the terrain and their radios were useless. No mosquito nets had been provided. A malignant strain of tertian malaria laid low more than half the fighting force through ignorance of correct preventative measures. Equipment was inadequate. No one knew what was going on.

By dawn of 27 August the Japanese were pinned down by intense strafing by Kittyhawk fighter aircraft of the RAAF (Royal Australian Air Force), some flown by former Spitfire pilots. These 'flying shithouses' (as they were affectionately known) were polished with beeswax for speed, but the rain and ooze made flying conditions an indescribable nightmare. Fighters slid off the runway and collided with the bombers. One of the Japanese officers, Lieutenant Moji, became physically ill at the protracted onslaught and in his diary noted in his oddly mechanistic way that 'the tone of our systems was feverish and abnormal'. However, moments of humour were ever present. Some Japanese soldiers attempted to confuse the Australians by shouting unlikely phrases in English.

'Is that you, Mum?' was rapidly answered by a burst of machine-gun fire.

Another four-wheel-drive had become stuck in the riverbed just in front of us. Sele contemplated the scene of impotent activity for a long time. He suddenly gunned across the torrent, all the while being egged on with shouts of excitement and delight from our precariously-positioned passengers. More picturesque tropical beaches were glimpsed through the palms until we encountered the final memorial which marked the western- and southernmost point of the Japanese advance. Some eighty-three unknown Japanese Marines, who made a suicidal charge against impossible odds, lie buried here. The Japanese military maxim, 'Duty is weightier than a mountain while death is lighter than a feather', seemed to possess an even deeper significance in this theatre of war. Soldiers would feign death, lying open-mouthed among the fallen, and then the 'corpse' would suddenly spring to life and shoot an Australian or American in the back. Numbers of dead are uncertain owing to the large quantity of body parts – legs, arms, hands and heads – that were left hanging sickeningly in the trees after the explosion of bombs and shells.

We arrived back at the lodge as dusk was falling. A late afternoon storm was gathering in the mountains. 'General' Mocquery was seated on the veranda in his customary position near the fans talking to a tall, fair-haired man whose complexion and features betrayed all the signs of having spent many years in a tropical climate. He was wearing shorts and his bare legs carried a number of small plasters covering insect bites. Slightly damp, thinning hair accentuated his faintly feverish appearance. Mocquery in full tropical fatigues gestured for me to come over.

'. . . no dental treatment available at all,' he concluded and glanced up.

'Good trip to East Cape?'

'Marvellous! Went swimming. The water's so clear!' I felt elated.

'I'm Chris Abel.' The fair-haired man smiled briefly.

'Ah! I've been waiting to meet you.'

'You must be the writer fellow.' His voice betrayed unusual caution. An engaging yet slightly defensive attitude revealed itself in his English accent. We shook hands and I flowed into a bamboo chair.

We discussed his childhood on Kwato Island and his grand-father, Charles Abel, the famous and controversial missionary.

Chris had spent some twelve years in Popondetta as an Agri-cultural Extension and Development Bank Officer. During an election it was discovered that many of the villagers were unable to read the ballot papers, so he invented what they called 'the whisper vote'. The locals would whisper their choice in his ear, and he would mark their ballot paper accordingly.

Large drops of rain began thudding onto the roof with increasing velocity. A mysterious figure carrying an ancient Gladstone bag wandered onto the veranda. He was wearing a beige linen suit, maroon-spotted cravat and heavy brogues. His engaging face and sculpted beard achieved a wan smile, but he was way overdressed for the tropics and sweating heavily.

'A Victorian detective looking for the ghost of a missionary,' Chris Abel commented wryly.

The BBC were making a programme about the Reverend James Chalmers, a famous nineteenth-century missionary eaten by cannibals on Goaribari Island in the Gulf of Papua. The next day I saw the optimistic film crew board a decrepit yellow coaster and dissolve offshore in a dark tropical storm. Abel suddenly turned to me.

'And what exactly are *you* doing here?' His eyes hardened and a measure of suspicion crept into his voice.

'Just travelling around the islands and writing about the culture,' I answered carefully.

'A couple came here recently *for a good reason*.' He emphasised the words meaningfully. 'A lad came back with his father who had fought in the Battle of Milne Bay. He's going to write a book about it.'

An atmosphere of unspoken confrontation entered the conversation. He seemed suspicious of writers. Russell Abel, his father, had written an excellent biography of Charles Abel in 1934 called *Forty Years in Dark Papua*. But the latest published biography of the missionary had made the whole family angry. One reviewer reported that the book contained errors, twisted facts and nasty allusions.

'And we gave the writer access to all the private papers.'

Clearly I had uncovered a nest of scorpions. The downpour

blotted out the light and almost stopped conversation. He was forced to shout over the noise. Water was swirling everywhere and the storm drains were overflowing. He raised his hands in a gesture of hopelessness at attempting to talk over the hammering rain. The fans rushed moist air over our faces.

'I'll dig out some books for you to look at. You can set the record straight!'

'I'm going to Samarai and Kwato tomorrow in the Orsiri[1] dinghy.'

'Have a good trip!' he shouted as his slender figure disappeared into the murk.

'What was all that about?' commented Mocquery rhetorically.

'I have absolutely no idea.'

---

[1] A local trading company based on Samarai Island in China Strait.

5. Too Hard a Country for Soft Drinks

The elderly 'whiteskin' standing on the wharf at the Alotau harbour side, casually dressed in check sports shirt and light trousers, was waiting for the *St Joseph* putt putt[1] to tie up. I was waiting for the Orsiri banana boat to finish loading and head off for Samarai. The fresh bread delivery was delayed so I hung about smoking a rough cigarette made from tobacco rolled in newspaper. An albino Melanesian ambled past squinting against the sun, his pink skin shockingly blotched, yellow hair dazzling against the palms. Decrepit trade boats were taking on crew who sat on the stern rails, ejecting jets of scarlet into the water and calling out to their friends in passing trucks. My nose was assaulted by a peculiar mixture of fish, yeast, distillate and copra. Banana boats packed with produce and drums of diesel skated across the harbour towards the islands like hunting water spiders. The sun beat down.

'Good morning!' I was the picture of bonhomie.

'Good morning, my son. Are you visiting Alotau? We don't get many of your sort, oh no.'

I thought this was an extraordinary way to greet a stranger. He had an Irish accent and mottled complexion. All 'whiteskins' who have lived in the tropics for years have this wan appearance. We stood side by side rocking on our heels in a foolish colonial manner,

---

[1] Pidgin for 'launch'.

looking at the colourful activity, glancing from time to time at the oil slick and coconut husks floating in the water below the wharf.

'I'm the Catholic priest in Alotau ... oh yes ... the Catholic priest.' He volunteered in answer to my quizzical look.

'Ah! How long have you been here, Father?'

'Oh yes ... must be getting on for thirty years now, thirty years since I left Ireland. I've stopped counting, I have that.'

'I suppose there have been a lot of changes in your time.'

'Oh yes ... murders and break-ins are increasing all the time, they're always about, they are that. That's right. A boat from Lae brought in a whole criminal element, it did. It's gone now, thank goodness, together with the murdering, thieving boyos we hope ... oh yes ... we do hope that.'

'I heard about that boat.'

'Did you now. You must have your ear to the ground. Oh yes ... they know where you are all the time ... they're watching all the time ... oh yes ... now he locks his door ... yes ... now he's gone out ... yes ... yes ... now he's come back. He's gone inside and locked the door ... He's turning on his light. Now he's having a shower. They know it all and see in the dark ... oh yes ... they see in the dark, they can do that.'

'Has your health stood up over the years, Father?'

'Well, to tell you God's truth, I've had the fever recently ... and it laid me desperate low but I seem to be all right now ... oh yes. Age creeping on now.'

More shuffling and gazing.

'I've read that sorcerers and magic are still about.'

'Oh yes ... the magic and the witchcraft are strong, strong. They might be Christian but that old black magic is still there in them ... it's a terrible ting, terrible ting, terrible, terrible ... Propitiate the spirits of the departed now ... it's a dark existence out here to be sure. It certainly is that.'

The *St Joseph*, freshly painted in yellow ochre, finally tied up at the wharf. The priest waved to some village women dressed in Victorian cotton smocks.

'That's my lot there . . . oh yes . . . I'll have to leave you now. God be with you on your travels . . . yes . . . God be with you,' and he wandered over to his flock.

The bread had arrived while I was chatting and the banana boat prepared to leave. This powerful vessel had twin seventy-five horsepower outboard motors and two plastic garden chairs. We powered out of the harbour and the cool wind brushed our faces and lifted our spirits. The rusting Taiwanese trawlers were soon left far behind as we sped along the south coast of Milne Bay towards East Cape. A young village girl carrying some shopping sat in the seat beside me. She prattled on in Pidgin to the two boys piloting the boat but they said nothing at all to me. I put my feet up – one on a carton containing an electric lawn mower and the other on a carton of several hundred tins of baked beans. The dinghy began to buck as we headed towards the open ocean and I noticed there were no life jackets. Later I was to learn that this omission is quite normal practice. I would have felt decidedly wimpish to have mentioned this in such a 'masculine' society. Be a man and drown or laugh as you are taken by a shark. My panama began to whip around my legs as I held it down out of the wind.

We followed the coast east for only a short time, sailing parallel to the road I had travelled only a couple of days before. The sea became rougher as we turned south towards Samarai and the Coral Sea. I felt a tremendous sense of exhilaration, my face dashed with spray, slicing through the azure water, the dark-green jungle defending the mountainous interior coming up on the right. Coconut palms, transparent green water fringing crystalline beaches, a swirl of smoke from the occasional bush hut. Young, brown, white-breasted sea eagles soared on the up-draughts, their wingspans majestically spread against the vegetation. Fragile outrigger canoes weathering the sea swell were cheered on by the boys.

The currents in China Strait are treacherous. The pattern on the surface of the water changes from seahorses whipped by the wind to smooth powerful eddies of deep blue streaming up from the abyss. The pastel outlines of numerous small islands appeared, jagged peaks lifting from valleys shaped like cauldrons. The sun broke through gunmetal clouds and burnished the sea, biblical rays that appeared to be guiding us to salvation. I realised with surprise I was soaked to the skin.

Captain John Moresby landed on Samarai from HMS *Basilisk* in April 1873 hoping to evade the unwelcome attentions of his 'savage friends'. He settled down to dinner with his officers but they were followed by a hundred fighting men, who squatted quietly on the beach beside the blue water and watched the proceedings with close interest. Moresby offered them a stew made of preserved soup and potatoes, salt pork, curlew and pigeon, which, not altogether surprisingly, disgusted the warriors. The sailors unsuccessfully tried paddling canoes which resulted in capsizals, hilarious moments for all concerned. The warriors opened the officers' shirts and stroked the white skin of their chests in wonderment and appreciation. Captain Moresby wryly named the place Dinner Islet to mark this unusually human and peaceful encounter. The local name of Samarai soon replaced the cannibalistic associations of the former.

The island appeared a deserted ghost town at first sight. The former provincial headquarters, which is an older settlement than Port Moresby, had clearly seen better days. I climbed up onto the Orsiri trading wharf to take my bearings. Ruined warehouses lined the neglected International Wharves site, warehouses gaping like skulls set on a rack, the empty interiors propped up by partitions of broken bone. Planks and beams jutted out like shattered teeth. Clumps of resentful youths were loitering around the general store and glared at me without a smile, but the women and children greeted me with friendly waves.

'*Apinun!*'[1]

'*Apinun!*'

---

[1] 'Good afternoon!'

The mown grass and coral streets (there is only one rarely used motor vehicle on Samarai) seemed like sections of an abandoned filmset. I walked between the abandoned shells of two buildings in which some boys were shouting and playing football. I hoped I was heading towards the Kinanale Guesthouse, run by Wallace Andrew, the grandson of a cannibal. Some attractive colonial houses were ranged around the perimeter of a waterlogged football field. In the sultry heat I leant exhausted against an electricity pole near a memorial obelisk before heading for a small beach in the distance. Rain trees and old flamboyants offered cooling shade.

'Hello there!' The voice came from a porch at the top of a flight of steps to my left. 'Come up and have a drink!'

A tall, smiling 'whiteskin' wearing shorts, his legs covered in the ubiquitous small plasters, beckoned me in.

'I'm looking for the guesthouse run by Wallace Andrew. Do you know where it is, by any chance?'

'It's just there!' and he pointed to a large house partly covered in old wooden scaffolding. I had thought this structure was an abandoned building project.

'Wallace is about somewhere, but come up for a minute.' He disappeared inside.

I went into the sitting room and collapsed into a chair while he brought some iced water from the fridge. Furniture was clearly hard to come by on Samarai and the room had the feel of a temporary arrangement that had drifted out of control into permanence. For no good reason I imagined I could see mosquitoes everywhere. Probably the beginnings of tropical madness.

'Hi, there. I'm Ian Poole, Manager of Orsiri Trading.' I instinctively felt he was open and lacking in the customary consuming demons, a rare quality in the tropics, although it was a slightly mad challenge maintaining a business in this remote spot. Australians generally manage extremes with equanimity and dry humour.

'I'm just travelling the islands. Chris Abel from Masurina told me to look you up. I've heard you have an encyclopaedic knowledge of the place.'

'Well, it's certainly interesting. Wallace knows a lot more about the local villagers and Kwato Mission, of course. He was born there.'

'I'm surprised there are so many old buildings left. Didn't the Australians carry out a scorched-earth policy to stop the Japanese?'

'Yes. Unnecessary, though. The Japs buzzed the island a few times in flying boats and dropped a couple of bombs but that was it. The Aussie Administration Unit set fire to all the commercial and government places including the famous hotels. Tragic loss. You can still see the few survivors around the football pitch.'

Missionaries were the first Europeans to establish themselves in this part of Eastern Papua New Guinea. By 1878 the Reverend Samuel MacFarlane had made Samarai a head station for the London Missionary Society, but the LMS soon exchanged land on Samarai for the island of Kwato a short distance away. At the turn of the century Samarai had no wharf or jetty and goods were off-loaded from trading vessels into canoes. It was a government station, a port of entry and a 'gazetted penal district'. Local labour was forcibly recruited here for the plantations. Copra and gold-mining dominated life, and this tropical paradise became a more important centre than Port Moresby. The town, if it could be described as such, consisted of 'The Residency' (a bungalow built on the only hill for the first Commissioner General, Sir Peter Scratchley, appointed by the Imperial Government in 1884 and dead from malaria within three months), the woven-grass Sub-collector's House, the gaol (with the liquor bond store in the roof), the cemetery and Customs House, and two small stores plus a few sheds. There were no hotels or guesthouses in the early days of the 1880s, the traders living mainly on their vessels and the gold-diggers camped out in their tents.

The Europeans besieged Whitten Brothers' premises since alcohol was dispensed from their store. The empties were hurled onto the sand from the roofed balcony. A village boy collected the bottles the next morning and counted them. Whitten then divided the number of bottles by the number of men drinking and so accounts were democratically settled. According to the entertaining reminiscences of Charles Monkton in his book *Some Experiences of a New Guinea Magistrate* published in 1921, men dressed mainly in striped 'pyjamas' or more festively in 'turkey-red twill, worn petticoat fashion with a cotton vest'. He describes picturesque ruffians roaming the palm-fringed shore. One incorrigible known as 'Nicholas the Greek', after pursuing an absconder through impenetrable jungle, returned with only his head in a bag. When questioned about the missing body he laconically commented, 'Here's your man. I couldn't bring the lot of him, so I'll only take a hundred [pounds].' Monckton also describes 'O'Reagan the Rager', who was 'never sober, never washed, slept in his clothes, and at all times diffused an odour of stale drink and fermenting humanity'. The spectacular sunsets and moonlit tropical nights of Dinner Island had formed a cinematic backdrop to all types of mysterious schooners, yawls, ketches, cutters and luggers with eloquent names such as *Mizpah*, *Ada*, *Hornet*, *Curlew* and *Pearl*.

'Life was pretty primitive here in those early days, I suppose.' I wanted to draw Poole out. He obliged me at length.

'Simple pleasures as now, I reckon. The malarial swamp was filled in by the prisoners to make a cricket field. Local people were forbidden to wear shirts in case they spread disease. They believed "the fever" came from the miasma that rose from the stagnant water. *Mal aria* means bad air, I think. Samarai was a deathtrap. The cemetery was always full. In fact, the fatal swamp occupied the football pitch right in front of your guesthouse. Sheep were imported from Australia and used to graze there until they were needed for meat. The residents, I think there were about a hundred and twenty in the early part of the century, played tennis, cricket and the children went swimming. Simple pleasures as I said.'

'Sounds idyllic.'

'There was always the demon drink. That's a story in itself. The first hotel was called The Golden Fleece. One large room and a veranda. It was built of palm with a thatched roof. No doors or windows. Guests were expected to bring their own sheets, knives, forks and plates and sleep on the floor. Drunks would stumble in at night in hobnailed boots and fall over each other cursing and swearing.'

'But I thought Samarai was famous for the glamour of its hotels!'

'True, but that was later in the 1920s. There were a couple of better hotels – the largest one had two storeys and was called The Cosmopolitan. Another called The Samarai, was at one time run by a real merry widow named Flora Gofton. Missionaries had to drag the drunks into church. One called "Cheers!" during the consecration.' I had to smile at this.

'And the hospital?'

'Oh, they built two hospitals and two schools – one each for Europeans and Papuans. Water was segregated as well. "Pride of race", they called it. On moonlit nights people would go around the island by launch singing. Everybody loved the place, although it was pretty wild with drunken miners and labour recruiters.'

'Women must've had a pretty rough time.'

'Some sad stories there. Many just upped sticks and left. There was a Swede named Nielsen who worked his butt off and made a few quid. Then he married a pretty Australian girl who was a bit footloose, you know the sort. She finally pissed off and went south. Every three weeks he would paddle his dinghy out through the mangroves to meet the steamer. Dreamed she would be on it. He always dressed up to the nines to meet her, immaculate – tan boots, clean shirt, tie and white duck trousers. She always disappointed him and never arrived. He would return to Samarai cursing all women and get drunk as a sponge that night.'

'What about married women?'

'Spent all their time looking after ill children. Helluva life. I've got part of a letter here somewhere that will give you an idea.' He fished around in a hefty file by his armchair and extracted a dog-eared photocopy. I noticed he kept scratching his legs and bare arms as did most expatriates I met in Papua New Guinea.

'This one was written by Nell Turner, married to an officer of some sort. It's from The Residency, used to be up on the hill, dated January 1909. She writes: "Alf is not at all well tho' he is gaining weight this last month . . . Kate is a lot bigger than mother – gets bigger and fatter after each baby. Mollie had convulsions on New Year Eve and took over two hours to come out of it, was quite stupid for a couple of days after, she seems quite recovered now. Munrowd had a fit a week after Mollie, but is well again. Jean had a dose of fever but is on the mend."'

'Sounds drastic to me. Matter of fact my own wife is in Australia at the moment. It's tough for women here.' He carefully placed the letter back into his archive. 'I hope to write a history of Samarai one day. No time of course.'

I suggested we find Wallace, so we went over to the guesthouse. I called out but there was no reply. The large room was sparsely furnished and seedy, like an old people's home. A meal was laid out on a table under white gauze. It was dim despite the fierce sun blazing down outside.

'*Yu yah! kamap pinis!*[1] I had gone down to the wharf to meet you!'

A voice came from the gloomy interior at the back of the house. A patriarchal Melanesian in an immaculate white shirt emerged from a corridor limping slightly. His grey hair was carefully groomed, teeth mauled by betel, warm eyes that expressed

---

[1] 'There you are!'

The Genoese Count Luigi Maria D'Albertis (1841–1901) was one of the most colourful explorers to investigate the interior of New Guinea. He explored 580 miles up the River Fly in the diminutive steam launch *Neva (below)*, launching fireworks to frighten the warriors. Among his picturesque crew was the brilliant engineer, Lawrence Hargrave, one of the pioneers of Australian aviation. The white-lipped python (*Leiopython albertisii*) is named after him.

Above The influential missionary Charles W. Abel of Kwato (1862–1930) with three close friends, all former cannibals. Note the bow tie and cricketing blazer – an heroic attempt in the torrid climate to maintain high standards. Born in Bloomsbury, London, Abel believed passionately in the moral power of cricket and assessed his charges both spiritually and at the crease.

Right The master carver Mutuaga (1860–1920) was born into the *maliboi* or flying fox clan and lived on the mainland opposite Suau Island, now in Milne Bay Province. He produced superb carved ebony lime spatulas used in betel chewing. The missionary Charles Abel appreciated the exceptional carvers of the South Massim region and commissioned human figures from Mutuaga for sale to visitors. This squatting figure on an unfinished base is in the Chris Abel Collection, Alotau.

The landing place of the charismatic missionary Charles Abel, who moved to Kwato Island in China Strait in August 1891. The memorial is on the edge of his cricket ground, land originally reclaimed from a pestiferous swamp. Kwato commands one of the most beautiful marine vistas in the entire Pacific.

Tedworth House, Wiltshire. This Palladian mansion was the boyhood home of C.T. (Charlie) Studd, the most brilliant all-round cricketer of his day. Educated at Eton and Cambridge, in 1884 he gave up a fabulous inherited fortune from Indian tea to become a missionary in China (inset: C.T. Studd in his later years in the heart of Africa).

J. E. K.   C. T.   G. B.
Captains of Cambridge University, 1882–3–4

THE CAMBRIDGE SEVEN
C. T. Studd, M. Beauchamp, S. P. Smith
A. T. Polhill-Turner, D. E. Hoste, C. H. Polhill-Turner, W. W. Cassels

The 'set of Studds' - J.E.K. (Kynaston), G.B. (George) and the most celebrated, C.T. (Charlie) Studd - were in the 1877 Eton XI together and all became Captains of the Cambridge XI, a unique family achievement. C.T. brought together a famous group of athletes and aristocrats who were known as the 'Cambridge Seven' and became missionaries in China.

The Reverend James Chalmers ('*Tamate*'), 1841–1901, was an adventurous Scottish clergyman from Argyllshire. He arrived in New Guinea in 1877 but, in an uncharacteristic lapse of concentration, was eaten by cannibals on the island of Goaribari. He was described by Robert Louis Stevenson as 'an heroic card…a big, stout, wildish-looking man as restless as a volcano and as subject to eruptions'.

a mixture of love and disappointment. One of his hands had been amputated midway down the forearm.

'Got your letter. I'm Wallace Andrew.'

Ian left us. Wallace immediately sat down at a bare table and began to play a game of patience with a limp deck of cards. It was as though he needed to erect a barrier to communication as a safeguard. Clearly he had spent years of his life playing this game in lonely isolation. The skill with which he shuffled the deck and deftly dealt and gathered the cards in with the stump of his arm fascinated me.

'I've come to see Samarai and Kwato.' I pulled up a worn chair.

'Ah, it's so beautiful there. We'll go together, you and I, to Kwato. Many people used to come, but there are few visitors now.' The cards flopped softly onto the table. He scarcely noticed if the game 'came out' and took even less interest. Time seemed to have come to a shuddering halt. I realized with alarm that nothing was actually going to happen in the next five minutes, the next hour, for the rest of the afternoon, for my entire life if I stayed on the island for long enough. My own arrival was the main event of the week. I needed to slow down to Melanesian time. It was quiet in that room and baking hot. The ceiling fan motors had probably burnt out long ago.

'Dinah will show you up to your room. Then come down and have lunch,' he suddenly said.

A petite village woman with a beautiful smile gestured for me to follow her up what was almost a grand staircase. The central carpet had long since disappeared, but the unpainted wooden strip in the centre was a ghostly reminder of some past attempt at luxury. We took the right flight of the staircase and passed through two bare rooms with flaking paint. Broken lampshades, mattresses and lumber lay abandoned on the floor in a corner. Her bare feet noiselessly brushed the cracked lino. A long veranda opened off a landing, but the bleached scaffolding hid any view. She pushed open the door to No. 8, a large room furnished

with a double bed, a single bed, a dressing table, a wardrobe and a fan. It was clean and comfortable with screened windows against mosquitoes. French doors opened onto the veranda. I looked through the maze of planks over the former swamp to the few colonial buildings that had survived the destruction of the war.

'You share the bathroom and toilet,' she said in excellent English and showed me the most basic of conveniences. I noticed a sign in red letters under a sheet of discoloured plastic on the wall of the shower: 'For hot water pull string.'

Dinah smiled again and disappeared. A corridor led out to what I thought was a rear entrance, but I found that the stairs had been removed and a twenty-foot drop into empty space yawned below. In a shed I could see a wrecked dinghy. I wandered back into the stifling room and sat on the bed. Glancing up I caught sight of my reflection in the glass. A crumpled traveller, sweating heavily, weighed down with notebooks and maps, wearing a sand-coloured colonial shirt, a planter panama and blue suede boots. Overdressed for the occasion I thought. I noticed there were no locks to the door of room No. 8 as I went down to lunch.

'Where were you born, Wallace?' I had poured myself some livid green cordial and was helping myself from a platter of reef fish and bananas.

'On Logea Island, near Kwato.'

'Really? Some people feel that Kwato was where the nation of Papua New Guinea began.'

'Certainly it was.'

The legendary island mission station of Kwato gives rise to strong passions and controversial opinions. Many of the most distinguished people in public life in the national government attended this mission school. Wallace often lapsed in and out of pidgin which confused me on occasion. Fervent Christianity was obvious in every sentence.

'The people of Milne Bay and the islands *wantim* Word of

God, very much they *wantim*. Charles Abel tried to make a new Papuan society that was Christian and educated for working. He taught us boat-building and metalwork. Mainly discipline and concentration he knock it in their heads. Young people don't want these things now.' He dealt the cards to himself all the time he spoke, cultivating chance. Dinah was clearing away the remains of the lunch.

'Respect for custom certainly seems to be passing away.'

'Gone. Gone now. Young people are too lazy to keep *kastom* alive. Prayer is the answer to all problems, Michael. You must pray. Even when they stole my television and wrecked my boat I prayed.' He began to hum a hymn tune I vaguely recognised.

'Did you get it back?'

'No. God didn't want me to watch any more television. It was a sign. I read and write more now.' His fatalism appeared to be the final tremors of a departing soul.

'Is anyone else staying here, Wallace?'

'Yes. Two government ministers. The Prime Minister has stayed here. The High Commissioner in London, Sir Kina Bona, he stayed here. They all know me. We had hundreds at a celebration not long ago. I built a dancing and picnic area beside the guesthouse. Did you see it? That was before the Englishman betrayed me.' He had stopped the mindless card-dealing and actually looked at me, animated yet with traces of anger.

'Who was that? What did he do?'

'Not now. You go out and look at our beautiful island. We'll talk tonight.'

'Fine. I'll have a look at your dancing area.'

'Ah, yes. Do that. Not many come here now, but in the future we'll once more have many people . . .' His voice trailed away as if he had lost confidence in the remainder of the sentence.

'I tried to have a shower but there was no water.'

'No, that's right. You must tell us first and then we will turn on the pump. Guests usually shower after meals.'

'I see. Well, I'm going now. See you later on.'

'All right. Will you come up to the hospital with me sometime? I need some more tablets. Arthritis they say it is.' He drifted in and out of this world, bolstering himself with prayer and medicine. I dragged the heavy fly-door open and walked out into the fiery furnace of Samarai.

A coral path shaded by coconut palms and pines encircles the island. It is known as Campbell's Walk after the Resident Magistrate who constructed it in the early 1900s using local labour from the prison. The transparency of the porcelain-blue water is transformed to a deeper cobalt as it reaches down China Strait and out towards the impenetrable mainland and pearly lips of Milne Bay. Small coves with upturned canoes invite fishermen to dream on the rocks. Fibro shacks nestle into the sides of slight hillocks waiting to be consumed by the exuberant palms, bananas, ferns, frangipani and hibiscus. The sound of an electric train instinctively caused me to search the horizon until common sense prevailed and I realised it was the sighing of the pines on the island of Sariba. Two women were cooking over open fires in a kitchen hut adjacent to a narrow beach where a gleaming new dinghy with outboard motor was knocking on the tide. The picturesque schooners and yawls, sails bellying in the wind, carving like swift blades through the currents, have long since disappeared.

I had been walking for only fifteen minutes and was already half way around the island. One of the most distinguished visitors to Samarai was the anthropologist Bronisław Malinowski. He was delayed here in November 1917 while waiting for the cutter-rigged launch *Ithaca*, which would take him to the Trobriand Islands. His favourite occupation on this walk was to read Swinburne and write his private diary in Polish. It revealed him as a man who had embarked upon a profoundly personal quest.

Malinowski was born in Kraków, the capital of the Austro-Hungarian province of Galicia in partitioned Poland in 1884.

The family would spend part of each year in the Tatra mountains in the Polish summer capital, Zakopane, a resort which became a haven for artists and intellectuals. Around 1904 he read Sir James Frazer's *The Golden Bough*, a vast collection of myths and magic from around the world, which inspired him to become an anthropologist and writer. Despite frail health he studied mathematics and physics, being awarded the highest academic honours in the Hapsburg Empire – *sub auspiciis Imperatoris*[1] – when he graduated in philosophy from the Jagiellonian University of Kraków in 1908. The deputy to Emperor Franz Josef personally presented him with a gold-and-diamond cluster ring at an opulent award ceremony in Kraków. In 1910 he moved to London where he began anthropological studies as a research student at the London School of Economics. He met and corresponded with eminent anthropologists at Cambridge, became a staunch Anglophile and is generally considered to have created the modern subject of British social anthropology.

By 1914 he was in Australia at the outbreak of the Great War. Although an Austrian subject and technically an 'enemy', he was given financial support to proceed with his work in New Guinea, first at Port Moresby and then on the island of Mailu. He also made two long trips to the Trobriand Islands from 1915 to 1918 where he followed the example set by the great Russian pioneering 'ethnologist', Nikolai Miklouho-Maclay, and formally introduced the concept of extended contact and methodological fieldwork into anthropology. The complex rituals of yam cultivation, the *kula* trading ring and, most notoriously, the liberated sexual practices of the Trobriand islanders, gave rise to a series of remarkable publications. His works became seminal studies of their kind, controlled, objective, classical and charming

---

[1] This Latin phrase means 'Under the Emperor's Seal'. It was the highest academic honour in the Hapsburg Empire and only one or two were awarded in any one year. The recipient was given a jewelled gold ring carrying the Emperor's seal which was conferred at a grand ceremony by a representative of the Emperor. Malinowski lost his ring.

accounts of remote peoples. But his private diaries, written mainly in Polish,[1] reveal a more complex figure, a man riven by doubt and boredom, a hypochondriac besieged by dreams and fantasies, a puritan wrestling lecherous demons nightly under the mosquito net. At a particularly low point he wrote, 'On the whole my feelings towards the natives are decidedly leaning towards, "*Exterminate the brutes.*"'[2] They reveal a man who had embarked on a painful journey of self-revelation. This contradictory character read novels of contained passion such as *Vilette*, and *Tess of the d'Urbervilles*, classical French works such as *Phedre* and the rhapsodic *Lettres Persanes*, even Thackeray's *Vanity Fair* in the midst of a Trobriand pagan paradise. On bad days, unable to work, he would leaf through the naughty caricatures in old copies of the French magazine, *La Vie Parisienne*.

Constructively sublimating his eroticism was difficult. He became infatuated with the owner of The Samarai Hotel, the soon-to-be war widow Flora Gofton, and accused himself of libidinous thoughts:

> . . . on the one hand I write sincere passionate letters to Rozia [his fiancée Elsie Masson], and at the same time am thinking of dirty things à la Casanova.

---

[1] On the inside front cover of the black notebook he inscribed in blue-grey ink: 'A diary in the strict sense of the term,' and immediately beneath: 'Day by day without exception I shall record the events of my life in chronological order. Every day an account of the preceding: a mirror of the events, a moral evaluation, location of the mainsprings of my life, a plan for the next day.' And beneath that: 'The over-all plan depends above all on my state of health. At present, if I am strong enough, I must devote myself to my work, to being faithful to my fiancée, and to the goal of adding depth to my life as well as to my work.' The first entry, on page one, is 'Samarai 10.11.17' (quoted from Michael W. Young's as yet unpublished biography of Malinowski).

[2] Italics written in English and taken from Joseph Conrad's *Heart of Darkness*.

He felt he was betraying the 'sacramental love' of this nurse from Melbourne. In his mind he undressed and fondled the wife of the island doctor, calculating how long it would take him to persuade her into bed. He punished himself with work and exercise. Urging himself to 'stop chasing skirts', he cultivated a solitary passion for making tortoiseshell combs for Elsie, spending hours at this odd task, accusing himself at one point of 'turtleshell mania'. His controversial and explicit work *The Sexual Life of Savages in Northwestern Melanesia* was published in 1929 with a preface by the sexologist Havelock Ellis. Not altogether surprisingly, it celebrates the magic of pagan love free of Christian guilt.

Despite his scientific training he had a creative temperament, beset by the demons of sensual temptation and metaphysical alienation:

> ...I have got the tendency to morbid exaggeration ... There is a craving in me for the abnormal, the sensational, the queer...

In his diary he wrote descriptions of the Samarai landscape that possess the intensity and heightened feelings of German Expressionist painting:

> The evening before: the poisonous verdigris of Sariba lies in the sea, the colours of blazing or phosphorescent magenta with here and there pools of cold blue reflecting pink clouds and the electric green or Saxe-blue sky ... the hills shimmering with deep purples and intense cobalt of copper ore ... clouds blazing with intense oranges, ochres and pinks.

He wrestled with language and culture like his compatriot and friend Joseph Conrad, a similarly-displaced Pole. Some commentators spoke of him, rather inaccurately, as the Conrad of anthropology. This he never was, possessing more of the Nabokovian 'precision of poetry and intuition of science', qualities the great enchanter impressed on his literature students at Cornell. But the prose of the diary, written partly in the Trobriand Islands, does share similar unsettling qualities to those we find in Conrad's tale *Heart of Darkness* set in the Congo. Both writers pressure language to its limits. Malinowski was seeking a cultural truth, the resolution of an identity crisis.

'Bronio' felt that islands symbolised the imprisonment of existence, yet, 'At Samarai I felt at home, *en pays de connaissance* [in the world of knowledge],' he writes. His fastidious nature was repelled by what he considered the inhospitable, drunken and wretched representatives of European humanity he found on Samarai, how they contrasted so depressingly with the natural beauty surrounding him. The 'part-civilised' local villagers he found there offended him equally. He would compulsively circle the island on this path like a caged panther. He sometimes felt he was merely exchanging the prison of self for the prison of cultural research. Yet Malinowski redefined the role of the ethnologist, his work expressing a romantic love for non-European cultures. Here was a man driven to seek his own philosophical nirvana. He subjected his own psyche to as close an objective scrutiny as the Trobriand villagers he studied. One of the great journeys of the modern European mind began on this tiny island of Samarai in 1914.

Campbell's Walk continued along the tropic shore, past rocks defaced by graffiti at the most easterly point but giving way to views of enigmatic Logea island shaped like an oriental hat, and chains of islets married to the sea by golden rings of sand. Clumps of hibiscus with miniature flowers suspended like drops of blood grew abundantly. Arching over the water, a strangely-shaped frangipani tree emerged from a wall, branches loaded with yellow and cream blossom, a canoe silently passing beneath. Almost hidden among the palms were the police station and the modest hospital, a moloch for patients in the early colonial days.

The heat was monstrous. I envied the village boys swimming among the tropical fish in the crystal water. I leant on a broken rail at the end of the wharf.

'Those two are my sons,' a voice behind me said.

I turned to see a tall, elderly 'whiteskin' in a baseball cap. The ubiquitous shorts and plasters decorated his legs, flip-flops on leathery feet, but it was the fathomless melancholy in his eyes that struck me. They were the bloodshot eyes of a hounded man given over to alcohol and grief. The lower lids drooped to catch his many tears.

'Oh! They seem pretty happy. Wish I was a bit younger.' I smiled pleasantly.

'Well, I buried my wife a few days ago in the Trobriands. On Kitava. Died of cancer.'

'I'm terribly sorry.'

'We took her by boat. That's mine over there. The *Ladua*. She was built on Rossel Island in the Louisiades.' I glanced over at an attractive, wooden trade boat painted ochre and grey.

'Why didn't you bury her here on Samarai?'

'I can see you haven't been here long! Her soul must go to Tuma, the erotic paradise in the Trobriands where departed spirits dwell. It's a place where you stay beautiful and there's no old age. You might call it Heaven. You can hear the spirits crying there at night. It's the mirror of the world.' People so rarely speak in this way I lapsed into silence for a time, turning over these poetic images.

'Is sorcery still strong?'

'It absolutely rules the lives of everyone living on Samarai, particularly the women! Everything is explained by sorcery, particularly losses at sea – people taken by sharks or crocodiles. Dinghies often sink in the savage currents. We lost six drowned over there a few weeks ago, and four over here the other day. They try to take the boats as far as Port Moresby!'

I reflected grimly that all my travels through these infested waters would be without a life jacket or radio.

'There's a launch pad for *yoyova* just along the path. Did you see it?'

'No. What are *yoyova*? Strange word.'

'They're the flying witches. They spread destruction and flame from their . . . well, you know! They can change shape into birds or flying-foxes, even appear like a falling star or fire-fly.'

'But they aren't real, surely. What's this launch pad look like?'

'A frangipani tree sticks out from the wall over the sea. It's an odd shape. You can't miss it.'

'Yes! I did see it.'

'They represent the malevolent magic of women, my boy. You must've experienced that. They're real women all right. Some have sex with *tauva'u*, those malicious beings who bring epidemics.'

'Yes, but what do they do exactly? How do they catch you?'

'They pounce from a high place and rip out your entrails, eyes and tongue. They snap your bones then they devour the rest of your corpse.'

Those inflamed, leaden eyes might well have witnessed such ghoulish instincts in action.

'When does this happen?'

'Usually when there's a storm. If you smell shit when you're fishing out at sea, watch out! Then the flying witches will attack by the squadron! They stink!' he laughed out loud, but without conviction.

'They're objects of real terror to these people. It's no joke because their powers are inherited, carried in their belly. Sometimes the spirit leaves them when they're asleep and goes marauding.'

I tried to imagine a world where such beings were an everyday part of consciousness. The books of J. R. R. Tolkien approached such a phantasmagoria, but his cruelty was of a different order.

'Sorcerers don't have the power they did in the old days!' and he looked dreamily out to sea.

The boys had begun diving for pebbles and shells.

'Have you lived here long?'

'Only about sixty years. I came here when I was five.'

'Sixty years on tiny Samarai! I think I would have gone mad.'

'Well, I did. I live over there now. On Ebuma.'

He pointed to a perfect tropical islet surrounded by glittering sand lying a couple of miles offshore.

'It belongs to the Prime Minister. I'm just the caretaker. My name's Ernie, by the way.'

I introduced myself . . . 'Ebuma looks like everybody's dream island.'

'Why don't you come over for a few weeks? We could talk. I could show you the fishing rats. They come down to the shore at night and dangle their tails in the water as bait. Small crabs catch hold and they whip it quickly round to their mouth and fasten onto the crab. Munch, munch!'

'You're having me on, Ernie!'

'There are strange things around here all right. If you see a swordfish leaping out of the water and going crazy, they have a borer in the brain.'

'What the hell is that?'

'A sort of parasite drills into the sword and works its way up the shaft into the brain. The fish goes mad.'

'What a place! Look, Ernie, I'm going to Kwato tomorrow, so I might call in on the way back.'

'It's up to you.'

I could not quite leave this mine of information without a last question.

'A lot of interesting people came through Samarai, didn't they? Malinowski, for instance.'

'Oh, him. You know he used to take opium while he was on the Trobriands? Probably did here too. Hancock, was it Hancock? Can't remember. Anyway, old Hancock told me about it. He was a little boy when the great man came here. Malinowski criticised him as being a spoilt brat in the famous diary. Payback.' He chuckled.

'Come over and see me for a couple of months. We can just eat and drink there . . . on the beach. I'm working on another boat at the slipway just over there. You could help. Want to earn a few *kina*?'[1]

A boy of about ten was running about behaving strangely, banging the walls of the warehouse with his head, laughing manically and fighting off a group of teenage tormentors. He seemed to have no control over his muscles and flopped about like a rag doll.

'He's a bit simple,' Ernie answered my enquiring glance. Boys were leaping into the water in an endless circle.

'May see you tomorrow then. Come over for a month.' He wandered away towards the *Ladua*.

I was sitting on the edge of the wharf when suddenly I was struck from behind by the flailing fists of the disabled boy. He seemed to have gone completely mad and was making the constricted sounds many damaged people make. It became quite painful and I began to slip towards the water. Some of the local boys rescued me and used the incident to give him another beating. He disappeared round the corner of a shed squealing like an animal.

I crumpled in the shade against the gnarled bole of an ancient rosewood growing near a tiny beach. Fragile canoes with delicately lashed outriggers were drawn up amidst scattered coconuts stranded by the tide, the fibre of the husks trembling in the breeze. Women were leaving the market with bundles and launching their canoes to paddle to nearby hamlets. I drifted off to sleep under a blazing copper sky only to be woken by a lizard crawling down my neck. A mangy dog began timidly to sniff my boot as dusk softly enfolded the island in the wings of a giant moth.

The grass-paved street that led back to the guesthouse was dusted with pink and white frangipani rouged by the last light, the pink trunks of coconut palms leaning over a darkening sea. The lonely

---

[1] The currency in use in Papua New Guinea. At current rates (2002), 1 kina equals about 25 pence.

bell from the Anglican church marked the hour. Wallace was seated on his usual perch by the table playing patience.

'So, you're back. I thought you'd been eaten!' and he laughed wickedly.

'No, but I almost ended up under the wharf. That disabled boy tried to push me in.'

'Oh, he's harmless, a sweet child really. Tomorrow you go to Kwato. The pastor is coming in the morning to take you over.'

'Great. Look, I've bought a few beers, Wallace. Let's have one and you can tell me about that Englishman.'

'Oh, him! Not much to tell.' He grimaced as if I had prodded a painful injury.

'He came to Samarai like they all do for a few days, but stayed on. He decided we could restore the guesthouse and went back to England to get the money. Work began and the scaffolding was put up. I built a dancing area in the traditional style.'

'Yes, I saw it. Beautiful local carving on the posts and boards under the roof.'

'Beautiful, yes. But I only used it once. He was an alcoholic and ran up debts everywhere.'

'I had an Irish business partner like that. He drank all the profits.'

'Well, then he left, disappeared into thin air taking what was left of the money with him. I had huge bills to pay. Electricity, workmen. It broke me.' I began to understand why the whole building was surrounded by old, bleached scaffolding. Time had stopped for Wallace as it had for Mrs Havisham.

'Did you tell the police?' The moment I asked the question I realised it was ridiculous out here.

'The police? They aren't interested in things like that.'

'Yes, but . . .'

'He's being sought in London. He's thought to be in Thailand.' The story had the faded, melancholic glamour of a sepia print.

'He sounds a typical predator. Islands attract them.'

Dinah was preparing dinner and the table had been set with

more bright green cordial and chicken, taro and pineapple. I poured a glass of pure, chilled rainwater.

'Is there any local music?' Mercifully, I had not heard any recorded music in the villages of East Cape, just distant *garamut*[1] drumming. The silence was inspiring.

'Oh, yes! Music! That's what I should be playing during the dinner.'

Depression had caused him to forget the past pleasures of hosting guests. Standards had slipped following the Englishman's betrayal years ago.

'My daughter is a singer. She's made lots of tapes. I'll get the recorder.'

He returned with an old recorder that had suffered the ravages of high humidity. The volume was either deafening or scarcely a whisper. Only one speaker was working. The songs were commercial and South Pacific in flavour, but professionally produced. The voice was very musical.

'She calls herself "Salima", which in Suau language means "canoe float".'

'Does she live on Samarai?'

'No. She's in Port Moresby now. All the young people leave the island to find work. My wife left too.'

Again he appeared to be wrestling with terrible dejection and unseen demons. The light went out of his eyes. He returned to playing patience. I battled with the volume control, not wishing to destroy the silence of the island night.

I had almost finished dinner when the two government officials returned from their seminar. They nodded towards me and padded upstairs. I had noticed with surprise their tiny travelling cases on the chairs of their open rooms. They rapidly caught up to my stage of dinner and introduced themselves as Napoleon and

---

[1] The *garamut* is a type of slit drum commonly used in the islands. The carrying power of this simple instrument is extraordinary in the still air of tropical nights.

Noah. One was from the Sepik, the other from Morobe Province.

'What's the subject of your seminar to the councillors?'

'Standing Orders. We need to explain the basis of the West-minster parliamentary procedure.'

'My goodness, that must be quite a task.'

They glanced at each other suspiciously, sensing criticism.

'They're intelligent men. Serious men. The problem is just one of language. You know there are over eight hundred languages in our country. Explaining the concepts behind the English procedure is most difficult. Old English is a strange language for us. Standing Orders are supposed to make parliamentary business easier but in our culture ... more difficult ... some concepts mean nothing to these people even in Tok Pisin. Independence came before we understood how the system worked.'

They both looked dark and fierce with an almost excessive masculinity, as if it was my fault, then they smiled. Such extremes.

'We're having a party to celebrate the end of our mission tomorrow night at the Women's House on the hill. You're invited. And you, too, of course, Wallace!'

He was gathering in the flaccid cards as he thanked them, pleasure struggling up to the surface. The officials rose quite suddenly from the table and headed off to the evening session at the hall. Another hand of cards fluttered down. Wallace turned to me.

'They always stay here, the ministers. Soon I will redecorate the entire hotel.'

He looked around the flyblown walls, the stump of his arm more than symbolic over the cards.

'I plan a stylish refurbishment here. God will bring the cruise ships. Thousands of tourists will visit Samarai. You're just the first of a great wave.'

I switched off his daughter's music. Mass tourism on the scale of Fiji or Vanuatu is an impossible, even undesirable dream on Samarai. The situation seemed ineffably melancholic.

'I'm sure you're right. Well, I think I might go to bed, Wallace. Could you switch on the water pump?'

'The pastor will be here in the morning. Everything will be fine.' His voice trailed away as I climbed the bare stairs.

The air in the bedroom was hot and thick. Garish streetlamps lit the window covered by a thin curtain printed with a tropical landscape hung upside down. I switched on the fan and went for a shower. Huge cockroaches crawled up from the drain but fled as the water fell. I pulled the string that promised hot water but with no result. A blessed coolness bathed me, the effect remaining for a full two minutes. I was slightly worried about being unable to lock the door and decided to sleep with my passport and wallet under my pillow.

I had felt insecure about my personal safety and possessions ever since my arrival in Papua New Guinea. There is something in the air that combines with the menacing expression in the male Melanesian face that is unsettling to a European. The dark and brooding sensibility of the men in particular, creates an ever-present feeling of threat. I felt my presence was tolerated but deeply resented. Smiles shielded a deeper animosity; an ancient impenetrable psyche lay behind those dark eyes. I was not wanted here, the past was resented and there was jealousy of my imagined riches. Covetous glances settled on my belongings. Serious health risks could not be avoided. So came upon me the first temptation to abandon the whole enterprise and return to Sydney. This was to become a common feeling I was forced to fight. Only the idyllic beauty of the islands, the complex cultures and the occasional warm personality kept me travelling. Wallace was a truly good man, but what had it brought him? Theft, vandalism and betrayal. I lay on the bed and stared at the fly-spotted ceiling. The lonely Anglican bell marked the passage of European time. A solitary bird was singing, a species that sings after sunset for the entire night.

Whispers below my window woke me. I could see some youths had clustered around the marble obelisk and were looking up at my window and pointing. I remembered the Catholic priest at

Alotau. 'They know where you are, if you're asleep, he hasn't locked his door ... oh yes.' They wandered away at length and the memorial was bathed in moonlight.

The story of how this obelisk came to be erected is one of the legendary tales of this Province. It began with the cannibalistic murder in 1901 of one of the first missionaries to come to Eastern New Guinea, the Scotsman, the Reverend James Chalmers. He was a friend of Robert Louis Stevenson who described him as 'an heroic card ... a big, stout, wildish-looking man as restless as a volcano and as subject to eruptions'. He was as much an explorer and adventurer as a missionary. The title of his book *Work and Adventure in New Guinea* (1885) describes his attitude to missionary activity succinctly. On one occasion tracing a journey on a map in a village hut, he noticed that drops of liquid had begun to fall from a bulky package lodged in the roof. Grandmother's remains were being dried by her grandson. In many parts of the country the corpse was not buried immediately after death but retained by the family, placed on a platform outside the hut, perhaps smoked and stored or the remains given to the children to play with. In this way the relatives clung to the spirit of the dead for some time after the passing of the body. 'It quite spoiled our dinner,' Chalmers laconically commented later.

His book is full of bizarre cultural descriptions. One of the most celebrated is that of the 'man-catcher'. This was a hoop of rattan cane attached to a bamboo pole that concealed a spike. The hoop was slipped over the head or body of the fleeing victim and then suddenly jerked tight. The spike would penetrate the base of the skull or spine, neatly severing the spinal cord. Ernie had told me during our talk on the wharf that Chalmers carried a Bible under one arm and a shotgun under the other as the instruments of conversion. Certainly not your average missionary, more an aggressive soldier of Christ unwittingly preparing the ground for the arrival of the colonial service.

The charismatic Chalmers was known as 'Tamate' by the

people of Rarotonga. He was a fine figure of a Victorian gentleman and possessed a head as noble as that of the composer Brahms. Both his formidable wives succumbed to malaria. He writes of having to exhibit his chest to the warriors on numerous occasions each day. One friendly chief offered his wife a piece of human breast at a feast, declaring it a highly-prized delicacy. Chalmers wryly observed that this was the end of his chest exhibitions in that part of the country.

In 1901 the London Missionary Society schooner *Niue* set sail along the coast of the Gulf of Papua from Daru. It anchored off the ironically named Risk Point on Goaribari Island near the mouth of the River Omati. This area was well known as one of the most dangerous parts of New Guinea, an area of torrid mudflats and swamp crawling with tiny crabs and fierce cannibals. Early on the morning of 8 April some warriors with faces and shaven heads painted scarlet, their eyes ringed in black, paddled out to the vessel in a fleet of canoes and persuaded a landing party to come ashore. The unarmed Chalmers and his young and inexperienced assistant Oliver Tomkins, together with ten mission students from Kiwai Island and a tribal chief, landed from the whaleboat in a creek close to the village of Dopima. Chalmers had attempted to convince Tomkins to stay on board but the intrepid youth would have none of it. The warriors trembled and giggled with excitement, their cassowary plumes and long tails of grass swishing and shivering in anticipation. The Europeans entered the enormous *dubu* or men's longhouse, all six hundred feet of it, and greeted the occupants. The air of the long, gloomy tube was thick with suffocating smoke and heavy with acrid odours. Rows of enemy skulls by the hundreds were arranged on shelves and racks, some fixed to macabre carved figures hanging from the roof.

The visitors were immediately struck from behind with stone clubs, and fell senseless to the floor. Tomkins managed to escape as far as the beach but was brought down with spears. This was the signal for a general massacre. Chalmers was stabbed with a cassowary dagger and his head was immediately cut off. Tomkins and the rest of the party of young mission boys suffered the same fate. The bodies were cut up and the pieces given to the women to cook. The flesh was mixed with sago to produce a monstrous stew and eaten the same day. The heads were divided among various individuals and quickly concealed from view. Ironically, the party who had expected to return to the schooner for breakfast had unexpectedly become breakfast. The *Niue* meanwhile had been boarded by a canoe raiding party and looted. The Captain managed to get under way and brought the grisly news of the slaughter to the wretched settlement of Daru.

After twenty-five years working among the 'skull-hunters', it is surprising that Chalmers allowed himself to be fooled. He was famous for possessing an infallible instinct for reading primitive moods and knowing when to leave. The precise reason for the butchery is unknown but there is speculation that he insisted on visiting in the middle of a ceremony that was forbidden to outsiders.

That this was an unprovoked cannibal murder rather than a revenge killing was clear. A punitive expedition was mounted three weeks later from Port Moresby. When the Government steam-yacht *Merrie England* (a most versatile vessel that reportedly could 'go anywhere and do anything') finally left Goaribari, some twenty-four warriors lay dead, many wounded and all the sacred men's longhouses on fire. But the heads of Chalmers and Oliver had not yet been recovered.

A year or so later, a young lawyer, Christopher Robinson, was appointed Chief Justice and was acting as Governor of the Possession. He decided to go to Goaribari in one of the pretty gilded cabins of the steamer, retrieve the heads and capture the murderers for trial. He had learned that in the matter of identification of skulls, those that had artificial noses attached were from people who had died from natural causes; those skulls without noses had been killed, the noses bitten off by the killers. As fate had it, the party he assembled were chronically inexperienced in dealing with villagers or had only recently arrived in New Guinea.

In April 1903 the *Merrie England* once more anchored off the cannibal shores of Goaribari. Some of the highly excitable local people were enticed aboard from their canoes with trinkets and trade goods. The murderers were known to be among them. The 'grand plan' was that the constabulary would grab them upon a given signal. The plan went horribly wrong. Wild fights erupted all over the deck. The red-painted warriors remaining in the canoes attacked the ship with arrows which drew rifle fire from anyone on board who could lift a weapon. Nearly all lost control in the ensuing panic and blazed away at everything that moved on the water. One, a letter copyist, collapsed in a fit of shrieking hysterics at the sight of a man being shot. An unknown number of the inhabitants of Goaribari were killed.

The facts of the case were instantly sensationalised and exaggerated by an Australian press starved for scandal. The missionary from Kwato, Charles Abel, demanded a Royal Commission to investigate the circumstances of the reprisal raid. Robinson was vilified with sulphuric slander and offered up for immolation. The innocent steamer *Merrie England* was absurdly compared to the infamous Australian 'black-birder',[1] the slaving brig *Karl*, owned by the Irish physician, Dr James Murray. Robinson was summoned to Sydney and a junior magistrate appointed in his place as Governor. Like Timon of Athens he was now abandoned by all his false friends. He took the only course open to a gentleman of honour in those days. While the occupants of Government House in Port Moresby were peacefully sleeping, he wrote his account of the incident, accepting full responsibility for the actions at Goaribari. He then took his revolver, walked out to

[1] A 'black-birder' was a slave ship or the captain of one that forcibly abducted men from island villages for labour on the Queensland or Melanesian plantations. Dr Murray and his brig, the *Karl*, were notorious in South Pacific waters for sensational cruelty and murder in their pursuit of profit through slavery.

the base of the flagstaff in the moonlight and blew out his brains over the withered grass. He was thirty-two.

The marble obelisk, ghostly in the silver moonlight below my window at Samarai, commemorates this sad saga. Part of the inscription reads:

> His aim was to make New Guinea a good country for white men. This stone was set up by the men of New Guinea in recognition of the services of a man, who was as well meaning as he was unfortunate, and as kindly as he was courageous.

The monument is now considered to be politically incorrect and the plan is to tear it down.

6. 'Mr Hallows Plays No Cricket. He's Leaving on the Next Boat.'

CHARLES ABEL

Letter from Kwato Mission to
his sons studying at Cambridge

My early morning walk around Samarai unveiled the islands of China Strait floating on glass, filtered through magenta gauze. Perfect silence reigned apart from the occasional fish breaking through the mirrored surface of the sea. The air was still and almost cool, the grass streets lightly scented with frangipani blossom. Scavenging dogs held their tails between their legs and cringed away from a lone fisherman heading towards the beach. Local children were screaming with joy as they entered the school in a crocodile line. I watched until lessons began; their enthusiasm and sense of mischief was electric. Papua New Guinea is a republic of children.

Wallace did not look too happy at breakfast. 'I don't sleep well these days. And then there is my arthritis. You will come down to the hospital with me, won't you?'

'Of course, but aren't we going to Kwato today?'

'Yes. I asked the pastor to come before lunch. His assistant will get a dinghy for you. But I won't be coming. I've got accounts and letters to settle.'

I concealed my disappointment.

'I suppose you don't want the flying witches to get you!'

He suddenly became grave and serious. 'If you fear the witches they have power over you. If not, they can't touch you. I don't fear them, Mr Michael.' His tone indicated I had overstepped an invisible line.

'Strange lights appear above the water at night and witches can *kaikai* you. On moonlit nights on Kwato, the spirit of Charles Abel appears. Ask the pastor about it.'

I changed the subject slightly. 'Did your grandfather ever meet the Reverend James Chalmers?'

'I'm not sure, but I do know after Chalmers was killed and eaten they tried to cook his boots! Ha! Ha! What do you think of that? Yes, the cannibals thought they were part of his feet. They boiled them for days but never could get them tender enough!'

'Then, he must've known Charles Abel.'

'Yes, we knew all the Abels. I was born on Kwato remember. I knew his son Cecil Abel. He was a good man. At Wagawaga Charles would walk up and down the beach at night praying in the moonlight. He was in strong communion with God.'

The cards were produced and the flight from reality began again. The government ministers came down for breakfast, faces transformed by friendly smiles. They ate quickly and headed off for the final day of the seminar.

'Don't forget the party tonight, Michael.'

'Certainly not.'

'We'll send a councillor down to guide you.'

I ate a few more slices of sweet pineapple and perfumed paw-paw. The sun was up and the room heating slowly. The torpor of the day had already begun to set in.

The early European settlement of New Guinea is the scarcely credible story of competing missionary teachers of various denominations carving up the country – Lutherans, Roman Catholics, Evangelicals of the London Missionary Society (LMS), Methodists, Barmen of the Rhenish Mission, Anglicans, Pietists,

Baptists, Protestants and Seventh Day Adventists. On first contact the indigenous islanders were described by both colonisers and missionaries as indolent, mendacious, 'intractable little cannibals', loathsome and depraved, filthy, 'truculent mannikins', sensual, lazy, and, *in summa*, 'hopeless little degenerates' – to list some of the more insulting labels assembled by the Governor, Sir Hubert Murray, in his book *Papua or British New Guinea*. The primary objective of missionaries was to save the endangered souls of the 'natives' from perdition and render them European in the shortest possible time.

The first LMS station had been established in 1871 on islands in the Torres Strait by the Reverend Samuel MacFarlane. By 1877 two local missionaries from the Loyalty Islands had established a presence on Logea Island in China Strait close to Samarai. Here the vividly-painted inhabitants impaled skulls on spears at the prows of their outrigger canoes. This was a region of enthusiastic cannibals, and the missionaries commanded respect often by force of character alone. They were also respected for their powerful possessions, superior technology and items they wished to trade. The evolution of what could be called 'Oceanic Christianity' was a slow process.

The following year Samarai was 'purchased' by MacFarlane from the local people as the LMS head station for 3s 6d. Parts of the island were cleared, houses and vegetable gardens established. The LMS flag of the dove and olive branch now flew alongside the Union Jack. Mission stations inevitably became part of the colonial structure. Through education of the mind and spirit, missionaries innocently prepared local people to accept European values, oiling the wheels of understanding during the imposition of European colonial bureaucracies. But there was a terrible price to be paid in loss of life.

Cultural observations taxed the Victorian mind and weird snobberies were noted with horror by the missionaries. The cannibals in the Milne Bay region considered themselves above their neighbours in the nearby D'Entrecasteaux Group, deploring that on those islands they ate *every part* of the human body, including the hair which they cooked in blood and gorged upon. Others from the coast who boiled their victims considered the manner in which the inland people negligently threw the body on the fire to roast, rather 'bad form'. Despite Conradian gasps of, 'The horror! The horror!', the East End (as Eastern New Guinea and the adjacent islands were then locally known) became the centre of LMS activity. Surprisingly, these 'pagans' were most anxious for the Word of God and indicated they wanted teachers to live among them. By 1891 the head station had been moved to the nearby island of Kwato.

The indigenous pastor of Kwato Mission, Naba Bore, arrived far earlier than expected. He wore a striped tropical shirt and had an attractive manner. My arrival in this remote place had clearly caused curiosity. Courtesy and interest flowed freely. Yet during our conversation occasional shadows disturbed his equanimity like wisps of cloud crossing the sun. We talked about religion and cultural change. Many missionaries to the islands were trained indigenous pastors who understood the local customs well and were galvanised with the religious fervour of the newly converted.

'People are losing their religion and returning to the past, particularly the young ones. They are so lazy! But it's a problem all over the world.'

'Yes, I know, but there are places where this is not happening. I lived in Poland for a few years, and they're building huge new parish churches there. Let me tell you a story of faith. I visited a village in the south of the country, near a town called Łańcut where there is a magnificent castle. The parish priest wanted to build a new church, as the old one was too small. Now, imagine this. The villagers baked the bricks for the church in their own ovens after cooking their food each evening. Then they gave them to the priest and built the church at night by the light of candles.'

'Yes, it reminds me of Kwato,' said the pastor. 'When Charles Abel came here in 1891 everyone helped. I'll show you everything this afternoon.'

'But what was he like as a man?'

'He was the best type of Englishman, not an intellectual or mystic. We called him "Taubada", a word in our language meaning "leader" or "chief". You know, he spoke Suau fluently and translated the New Testament. He believed in action and not words. Abel had a big heart and believed in work, sport and healthy fun. Work was fun for him. He had a zest for it!' The pastor was leaning forward with great intensity.

Wallace took over.

'One of the first things he did was to build a cricket pitch and teach us to play. His Kwato XI used to play the white Samarai XI every week. They were good. They even played in Moresby.' Wallace had not brought out his solitaire pack and was transformed with enthusiasm.

'Alaedi, the Kwato googly bowler, once had Frank Laver, the Australian Test batsman, out for a duck! What do you think of that?' I confessed to being mightily impressed.

'He thought cricket would build us up. Cricket and religion woven together. Ministers had to be able to preach *and* bat well or you were out!'

'So you really think cricket helped you?' I remembered my own terror of the slip machine at school, Brother Andrew fiendishly throwing in ball after ball, cassock flying, my hands stinging.

'Cricket built a team spirit and taught you how to be fair, how to lose well and how to keep going no matter what. Character and morality, that's what Abel taught. Young people don't want this any more. They just want money and to do anything they want. Terrible, terrible.' I found this veering from elation to dark despair rather wearing. Wallace continued the darker mood.

'There's crime now, not cricket, on Samarai. *Raskols* ruined my dinghy and sank the motor in the sea!'

'The Kwato Mission is finished. There's no money to keep it going.' They looked dejectedly at the floor for a long time.

'After lunch you can go,' Wallace said.

The pastor accompanied me to the wharf and we sat in the

shade of a vast casuarina to wait for the dinghy. The tiny Samarai market was operating in slow motion in the heat. A woman and her baby lay peacefully asleep on the ground between the huge roots. A dog tormented by mange sat in the sun and scratched itself ferociously until its bare flesh was pinker than ever. Small cancers seemed to be hanging from its stomach, ribs pushing against the flaking skin. It collapsed with hunger next to the outrigger of a canoe on the beach. The sea shimmered transparent green, fluffy clouds hung motionless above the scattered islands. The searing heat of the sand soon forced the animal to heave itself up with enormous effort and slough away under a cool hut, from which refuge it was violently ejected.

'The dinghy will be here soon,' said the pastor.

'There's plenty of time,' I replied.

Time. Of course Europeans consider time as a commodity to be partitioned and rationed. For Melanesians time is fluid and organic, a part of their existence and not outside of it. Days are measured in suns and the waning of moons. Beneath coral trees in scarlet bloom we talked about youth unemployment in Papua New Guinea, the conflicts between religious belief and *kastom*, the lack of established standards of behaviour for young people. He was shocked to discover that we shared so many problems, even in the so-called 'rich countries'. But the dinghy did not come.

'The dinghy should be here. Ah! Here's Dinah. She'll know what's happened.'

Dinah looked flustered when she arrived.

'He's shot off! The boy went at eleven thirty before we had lunch!'

'How could this have happened? We must arrange another dinghy. Dinah, go and fix it.'

She left again and did not return for at least an hour. Many people were publicly consulted about the problem around the market. Villagers wandered over and were informed of the treachery of this lad. They were sympathetic and clicked their tongues. Confusion over the payment for petrol turned out to be at the root of it. Finding dinghies in the islands was driving me mad, a European with a schedule. Dinah and the pastor exchanged places and he went off to arrange something with the schoolmaster.

'I used to be a teacher. I taught on Logea, over there.' She pointed to an island quite close by.

'Yes, I used to teach in London.'

Another dog seemed to be chewing the bitumen off the road near the wharf, savagely trying to extract something from the blistering tarmac, nose squashed, teeth bared. I asked why dogs are treated so cruelly in the country.

'It's true. A vet was invited to the island once and was so shocked at their condition that twenty-eight dogs were shot, stones tied around their necks and their bodies dumped in China Strait for the sharks.'

I lapsed into silence and hoped the dinghy would arrive soon. The afternoon wore on. A fierce-looking local man in a red shirt who might have some petrol, who might have a dinghy, might possibly have taken us, vaguely approached, then drifted away and picked his teeth. Dinghies are a complicated business for a *dimdim*. Clouds fleetingly crossed the sun.

'We can't go with those young boys by the supermarket. They've been drinking all afternoon.'

Women were leaving the market and packing their frail canoes, lifting skirts over strong legs, pushing off from the beach and paddling to other islands with power and elegance. A feeling of unease and of matters slipping out of control began to erode my confidence. The pastor returned and we all sat there stricken. He saw the ruin of his country in this incident.

'This is the problem with us Papua New Guineans, we cannot organise things properly. Cannot follow through. It's a difficult country. I'm so sorry. If we don't go soon it'll be dark.'

It was as though the mismanagement of time had become a form of moral weakness, a source of failure. Suddenly, a messenger!

'The schoolmaster's dinghy will take us when it returns from Logea!'

We were soon skimming across the five minutes of water towards Kwato. The wharf was ruined, with many planks missing. The boys would wait with the dinghy while we explored the island.

Pastor Naba Bore was full of detailed information.

'That's his cricket ground. There was a swamp in that place with a serpent lying at the bottom. The missionary Mr Walker swam the swamp and covered in slime, climbed the hills. Malaria was everywhere and it took them four years to fill it in. The pitch is used for drying pandanus leaves now.' A tone of routine nostalgia crept into the pastor's voice.

The pitch was covered with strips of bleached pandanus leaves, the lumpy ground dotted with heaps of dried grass and tunnelled with crab holes. In building this cricket ground Abel may have been influenced by the most brilliant cricketer of his day, C. T. Studd. He was a schoolboy idol, good-looking, high-spirited, a household name, who gave up a fabulous inherited fortune in Indian tea in 1884 to become a missionary in China. He was a cricketing blue at Trinity College, Cambridge, and regarded by W. G. Grace as the finest all-rounder in England. Six other young aristocrats joined him, ranging from the stroke oar of the Cambridge boat to a baronet's son.

They had all been converted by the sermons of the American evangelists Dwight L. Moody and Ira Sankey, who had also influenced Abel. These evangelists inveighed against

... the hollow, drifting life with feeble mundane ambitions – utterly selfish, giving no service, making no sacrifice, tasting the moment, gliding feebly down the stream of time to the roaring cataract of death.

In 1882 Studd had taken the entire England Test team to hear Moody. The 'Cambridge Seven' as they were known, created extraordinary public interest at the time, even rivalling the concern felt for General Gordon at Khartoum. Athleticism, virtue and Christianity were powerfully linked in the Victorian mind in a manner unimaginable today. I fancied I could see a team of ghostly Papuan players in perfect cricketing whites on a rolled lawn, outlined in action against pink and white frangipani. Abel himself played in a bow tie. His notebook describes his mission boys: 'Bele, a clever youth and fine medium bowler. Maru, nice serious lad – very clever behind the sticks.'

'Here you can see the old engineering shop, blacksmith's shop and sawmill.'

The pastor indicated a joyless scene.

The buildings had the same melancholic, decrepit air of those on Samarai. Rusty engines, cogs, shafts and all sorts of implements were slowly settling into the long grass. Kwato had been what was known as an 'industrial mission'. Abel instituted a training programme in boat-building, wood-working and metal-working. As an Evangelical he believed in progressive Christianity. 'We must be generators, not merely receivers,' he wrote to his sons at Cambridge. The Mission owned copra plantations which gave rise to anger among the traders as posing unfair competition. The LMS itself considered any industry by missionaries to be the vile laying-up of Mammon. Abel clearly had a vision of the coming age of business, and realised that even missions needed to be financially viable if they were to survive. He pressed on courageously against all forms of opposition. The mission eventually separated from the mother house and Abel became a respected though controversial figure.

Unlike the Anglican Melanesian Mission on Norfolk Island,[1] where the students were trained to return to their villages, on Kwato they were intended to form an autonomous group through intermarriage. Concentration was focused upon the few, the *elect*

---

[1] Norfolk Island is one of the most remote islands in Oceania, lying one thousand miles east of Australia across the Tasman Sea. It is home to the descendants of the mutiny on the *Bounty* and was one of the most feared penal settlements during the period of convict transportation. This small island has superb Georgian buildings and is covered in tall, endemic pines.

chosen by the Lord. Local children were removed from the debauchery of the Europeans living on Samarai, away from the dangers of sorcery and the 'pagan' sexuality fuelled by lascivious dancing. Victorian missionaries loathed dancing. They saw it as the prelude to sexual excess, itself the prelude to Hell.

Charles Abel had strong principles and convictions. This was the man who tore up Galsworthy's *The Forsyte Saga* as decadent trash and threw it into the sea. The ritualised homosexual practices involving semen and sago that he witnessed during the initiation of young boys appalled him. In parts of what is now Papua New Guinea and Irian Jaya, 'boy-inseminating rites' were the main means of transforming a boy into a strong, masculine and fearless warrior by an older male or 'mentor'. This temporary period would lead to customary marriage and heterosexual behaviour. According to the anthropologist Gilbert Herdt, 'homosexuality is the royal road to heterosexuality' in many of these cultures. Missionary activity has all but eliminated even the memory of such 'sinful' practices.[1]

---

[1] The anthropologist Gilbert H. Herdt states, 'We are dealing, in Melanesia, with one of the most complex and intricate sexual systems in culture that has ever evolved.' If one encounters the word 'semen' in anthropological indexes concerning Papuan cultures, an extraordinary array of functions and symbolism is revealed in different parts of the country. Far beyond its erotic and procreational value (about the limit of Western meaningful significance), semen defines relationships, can be considered as a commodity and scarce resource, can be equated with different varieties of food such as pandanus nuts, may be mixed with various foods to provide love magic and festive meals, may be gathered from group intercourse and fetishised, considered in the same way as breast milk to 'suckle' and 'strengthen' boys, may be a source of political power, used in initiation rites of men's secret societies, required for medicinal purposes (e.g., smeared on bamboo poles which are then 'exploded' with fire in Kimam Papuan villages attacked by epidemics). Smeared on the body, it is thought to render warriors invisible in time of war. Semen is considered the essence of life by the Marind-Anim culture. For a detailed anthropological examination of this complex subject, see *Ritualised Homosexuality in Melanesia*, ed., Gilbert H. Herdt (University of California Press, 1993).

Abel felt his mission was to save the Papuan race from extinction and he believed grave dangers accompanied any return to their village origins. A sequestered generation of outstanding Papuan leaders was moulded here, trained in the skills that would be required when the country became independent. Many occupy influential positions in government today. It is hard to criticise this policy given the circumstances of the time. Present-day PNG leaders floundering in the current miasma of corruption could learn from the moral strength, high standards and virtue inculcated in those fortunate enough to have been taken to Kwato.

A wrecked trade boat, the remains of the once-proud boat-building tradition, had been hauled up on the grass and lay rotting against a palm tree. We climbed a path originally hewn by prisoners, through an overgrown, once magnificent and extensive English landscape garden. Orange trees, roses, rhododendrons, oleanders and twenty-eight varieties of hibiscus once grew here in abundance with tropical plants and other shrubs imported from Australia. The entire summit of the hill had been removed by prison labour to provide a flat area for the house, the earth used to fill the fever swamp below. It took three years.

'Beatrice, his wife, made a beautiful home here. General Mac-Arthur stayed during the war. Nothing was burnt on Kwato. Not like Samarai.'

The pastor made a broad sweeping gesture over a cleared area of grass, conjuring with words solid buildings and a vibrant family life. Beatrice was an intelligent and thoughtful girl who had mastered a variety of Victorian social accomplishments and excelled at the ladies' academy she had attended in Brussels. She trod reflectively in the footsteps of Wellington at Waterloo and spoke excellent French. This polished, highly moral young woman of twenty-three arrived in 'savage' New Guinea as a new bride in 1892. Her leadership qualities and refinement were emulated throughout the islands and allowed the Kwato Mission to run with graceful precision. Women from the Massim area (the present-day Milne Bay Province) are independent by nature and moulded in self-confidence by the matrilineal society in which they live. Beatrice trained many female students in teaching, nursing and domestic science to succeed her as competent female missionaries. Kwato itself became rather like a finishing school for young women.

'It had broad verandas and panelled walls, a marble fireplace, beautiful furniture, carpets, silver candlesticks, yes, and even a piano. Charles Abel would never remove his jacket, even in the hottest weather and even wore spats in the tropical heat. Everything had to be correct!'

'Yes, there was a piano. She would play Beethoven or Chopin. It was a beautiful marriage. They loved the dark-eyed Papuan babies. Mrs Abel looked after the children while their mothers learned sewing. There was even a Kwato fashion. They made simple tunics for the women, not those terrible Mother Hubbard[1] things.' Dinah echoed the devotional tone.

'Yes, we used to sing part songs to the organ on moonlit nights. It was beautiful.'

I glanced up at the exquisite views through palms and swathes of scarlet and yellow hibiscus, down China Strait, over the jewels of Ebuma and Samarai and on to the distant mainland.

'Look behind you.' The pastor broke into my daydream of Victorian family life transferred to the tropics.

An enormous church built in traditional Massim style crowned an elevated grassy plateau. The shingled roof had the acute pitched shape of the *dubu* or men's longhouse, supported on brick columns with large open spaces between them. The congregation would be in direct communion with nature during the service. The eaves, gables and posts were decorated with fine

---

[1] The term 'Mother Hubbard' originates in the early illustrations to the nursery rhyme. Mother Hubbards were long, shapeless gowns that missionaries imposed upon native women to completely conceal the 'lascivious', bare-breasted body. It also prevented native women from engaging in the sexually suggestive 'pagan' dancing that so terrified Victorian evangelicals.

Massim carving and the internal wooden bracing of the roof was of great complexity. The setting sun streamed past coconut palms long-shadowing the lawn. Logea and the gauze-green islands scattered out to the horizon breathed gently in the goldening light.

'This is the Memorial House of Prayer,' Pastor Naba Bore informed me. 'One of the finest churches in Papua New Guinea, the first to be built of stone.'

The pastor was becoming excited at my evident interest in his domain. He guided me around the church showing me various commemorative brass plaques. The atmosphere of an Arcadian paradise was potent on Kwato and a feeling of uncanny transcendence took hold of me.

'The original floor was covered in crushed coral. Abel was a master of improvisation. Baked taro and coconut milk were used instead of bread and wine as the elements of the sacrifice. But look, time's passing! We must go down to the hospital.'

'The hospital? There was a hospital on Kwato?'

'The best in the area. Much better than the one on Samarai. I live up there.' He pointed to a dilapidated house above the church. He waved to his entire family who were sitting on a long bench. I met his wife, elderly mother and beautiful daughter. Everyone smiled and said they did not want to ever leave Kwato, but at the same time looked inexplicably crestfallen. Naba Bore explained.

'Living is difficult here now. Everything's gone. No electricity or telephone. No doctor or school. No shop. Nothing. It will cost too much to rebuild the house. It's falling down. Beautiful view, yes, so beautiful, but we must leave soon. I'm the last pastor.'

We descended the one-kilometre winding track to the beach, constructed by schoolchildren from the Kwato school. Charles Abel had planted a cleared area with numerous plants and shrubs brought from Sydney. Fragments of the garden were still clearly visible, although overgrown with weeds.

'Let me tell you a funny story about Abel, the doctor,' the pastor said. 'A sailor had come into the hospital with his hand blown off by an explosion of dynamite. Careless fishing, probably. Abel and the magistrate Armit got their instruments together – a saw, quills from a goose for draining the wound and some sharp knives. Unfortunately, the patient woke up from the chloroform anaesthetic halfway through the operation. He reached for his pipe and smoking hard, took swigs of rum and brandy while watching the rest of the interesting events right to the end! Those were the days!'

By 1929 when the hospital opened, Abel had already broken away from the London Missionary Society. On a fund-raising trip to America his zeal and persuasive powers had inspired the American evangelicals to form the New Guinea Evangelization Society under its first president, Samuel R. Boggs. The laws of New York State empowered the society to receive legacies, and money was collected for the construction of the Kwato hospital. An alternative to the loathsome and filthy native hospital on Samarai was desperately needed. The white hospital there was quite respectable, naturally. The Governor, Sir Hubert Murray, would die in it. The Abel house had been full of local cases of malaria, dengue fever, measles and pneumonia. Beatrice was exhausted through nursing them.

The hospital was constructed in the shape of a cross using American imported timber which resisted tropical rot. The structure was still surprisingly sound, though obviously neglected, and squatters had moved into the once-gleaming corridors and spotless wards. In the past the long verandas and large windows had calmed fevered brows with cool breezes and therapeutic views of China Strait.

'When the white man left, everything fell apart,' Pastor Naba Bore said as we ascended the tropical path, sprinkled with emerald beetles and white orchids. He repeated this many times on the strenuous climb and descent to the jetty, shaking his head and clicking his tongue against his teeth.

'All gone away,' he said, accompanied by a sigh, 'when the white man left.'

That such an uplifting concept as the Kwato Mission, a place of such serenity and peace, should have fallen into ruins is indeed a tragedy of our distracted times. The mission was essentially closed down after the declaration of independence in 1975 when financial assistance stopped. Abel had died in a freak car accident in England in 1930, and his universally beloved Beatrice died on Kwato in 1939. His sons continued the vision, building a deep comradeship between local people and Europeans. The gifted and courageous Cecil Abel sailed two mission-built boats complete with crews to the naval authorities in Port Moresby as a contribution to the war effort in the 1940s. Many of his missionaries died during the conflict, some beheaded by the Japanese. He was a devoted enthusiast of native customs and wrote many historical papers for the University of Papua New Guinea. True to the implicit political nature of missionary work, he was also one of the founders of the Pangu Party[1] and wrote the Preamble to the Constitution of the Country. He died in 1994. The Kwato Mission was now condemned to death through cultural change and lack of religious fervour. Few were trained outside the Abel family to maintain the spirit. The light went out.

'Whenever the Abels returned from a trip at night, torches and bonfires would be lit along these shores and on both sides of China Strait. Blasts would be blown on the conch shells. The drumming from the villages was intense.'

The pastor stood above the dinghy on the wharf and made another sweeping gesture in the half-light towards Logea Island and beyond, depicting the history with undimmed passion. He was not returning to Samarai with us but returning to his house

---

[1] In the 1968 elections for the second House of Assembly, the Pangu Party was one of six parties that contested the election. Its largely unrealised policies advocated rural development, welfare and opposed foreign investment. It was the largest and most influential party in the move towards independence in 1975.

on the hill. He seemed weighed down with sadness. I suppose I would be one of his last visitors from faraway London.

'Thank you for coming,' he said simply and turned away.

The boys gunned the dinghy in a wide exhibitionist arc. The sun was setting over the cricket pitch and the rough memorial stone that commemorated Charles Abel's first landing on Kwato. The penumbra spread across the mirror of the lagoon and filled the space with deep shadows. Pastor Naba Bore has died since I began writing this book. Perhaps he had a premonition which gave rise to his palpable melancholy that afternoon.

7. Constitutional Crisis in Makamaka

'If I hadn't gone to Kwato I would've eaten you!' I joked as Wallace greeted me from his perch above the cards. 'We were so delayed! What happened to the arrangements?'

'Those boys just shot off! I can't explain it.' He was annoyed.

The government officials were wandering about in high spirits. The seminar was over.

'Don't forget the party tonight, Mike. We'll send someone down to escort you in the dark.'

'What time?'

'About seven. Have a shower and you can come up with Wallace and Dinah.'

'Fine. I heard you're leaving on the government launch early in the morning for Alotau. Can I hitch a lift?'

'Yes, of course. We leave at six. Quite early.' The offer was a great relief as I did not want another struggle to obtain a dinghy. The sometimes treacherous China Strait needs a skilled boatman and the government pilot would at least be reliable and sober.

That evening we threaded our way gingerly up some steps, through the dark gardens to the women's guesthouse. Wallace was struggling a little with his arthritis. Indistinct shapes flitted about with the insects. The tangled path opened out onto a terrace where a large group of local men and delegates from other island groups were seated in a circle. There was hardly any light. Figures lying in the grass and leaning against palms slowly materialised as my eyes accustomed themselves to the gloom. Napoleon introduced me and I sat down. The atmosphere was serious and respectful despite the plastic bucket crammed with bottles of South Pacific lager in the centre of the group. The other minister was discussing Standing Orders.

'You should be operational now. Remember, Standing Orders are to speed up business in the council and limit useless talk. You must be courteous and not make disorder with much abusive language.'

Sage nodding of heads all around the bucket. More bottles were passed.

'Tell us again of the difference between "adjournment", "postponement", and "recess". These seem to be very different ideas in my language.' The questioner had an intelligent face, a large moustache and a halo of tightly-curled hair.

The minister from Morobe Province adopted a grave mien and the group fell silent.

'A *recess* is a period of holiday between sessions of a committee or parliament. If an enquiry or meeting *adjourns*, it is stopped for a short time. *Postponement* means a meeting is put off to another time, later.' He carefully emphasised and enunciated each word.

'So they all mean the same thing! And we don't have any money for a holiday!' Everyone was listening intently, although smiling.

'No, you must use the correct word for the situation. Some words can be used where others cannot.'

'I don't understand.' Murmurs of agreement came from the dim recesses of the garden.

'Let me give you a clear example. Take the word "poke" or *puspusim* in Tok Pisin. Sometimes if you are sick, down there,' everybody laughed and commented bawdily as he pointed towards his genitals, 'if you are sick down there, the doctor, because he is a big man, a professional man, a good man, will not ask, "How often do you poke your girlfriend?" or "How often do you *puspusim*?" He will ask, "How often do you have *sexual intercourse*?" You see he is using the correct, professional, not crude language, for the situation of visiting the respectful doctor.'

He seemed quite pleased at this clever masculine example to demonstrate the idea of register in language and looked over to me for approval.

'Very good!' I said.

'But in my language there is only one word for fucking. What's the difference?' The man with the moustache was intensely emotional, almost desperate for clarification.

'Have another drink, bro. You will feel the difference soon enough! After many committees!'

'What words can't we use in our council meetings?' asked a man in a battered red T-shirt.

'Well, you cannot call someone a dog.'

'But what if he is a dog? I know many dogs in the council!'

'Well, you cannot use unparliamentary expressions like that. Also you can't call a man a rat, or a coward, or a criminal or a pig.'

'I'd never call a man a pig. I like my pigs too much. But there are many rats in government, and criminals. If that's what they are, that's what I'll call them.'

'If you do it, big trouble.' The minister was not smiling.

I reflected that although this conversation was not without humour, the difficulties of translating the terminology and the concepts of Westminster parliamentary procedure into the hundreds of obscure languages spoken in the islands of Eastern Papua New Guinea must be almost insurmountable. Few Standing Orders have been passed since 1832 and many are couched in Victorian English. Some were passed in the seventeenth century and even a native Englishman would find them difficult to comprehend. Cultural differences with England, and also among themselves, greatly exacerbate the linguistic problems. But the councillors approached their governmental tasks with great dedication and I could not help but respect them. However, they looked worried and studied the tops of their bottles.

'Never let people see that you *fear*. If they realise you fear, you're finished!' Napoleon interjected vehemently. I remembered Wallace saying something similar about the flying witches. Without warning, as if at a secret signal, they all stood up, put down their beers and went into dinner. Rather confused, I carried mine with me.

The ladies who ran the guesthouse were standing in the corridor that led into a large room where tables groaned under vast quantities of food. They greeted me with velvet voices that were soft and gentle. Such a contrast to the harsh masculinity I had recently left. My 'new friends' were now garishly lit by fluorescent tubes. Many were dressed in faded T-shirts, tattered shorts and had bare feet or wore thongs, while the more senior made an effort with crisp sports shirts, trousers and thongs. Despite their rough exterior, these men showed a curious fastidiousness by not carrying their beer into the meal. I was the only 'whiteskin' in the company of about eighty islanders and the only person uncivilised enough to bring his alcohol with him.

The agenda for the evening events was announced. Various long speeches were made full of emphasis upon duty and honour, listened to in stony silence. A welcome was given by the manager of the Women's Association Guesthouse. She hoped, with gentle irony, that we would enjoy this modest meal 'prepared at such short notice'. A particularly long Grace was said during which all the Christian virtues were extolled. The minister then introduced me as an honoured guest. In the middle of all this high seriousness, he indicated that I was to serve myself first. The pressure of almost a hundred pairs of Melanesian eyes was intense. My selection was precisely noted as I took a portion of each dish. I felt they were sensitive to the slightest gesture that might shame them in any way. There was an audible sigh when I sat down and happily ate the first mouthful. The ministers then served themselves, and the remainder of the delegates followed.

The food was a delicious selection of local dishes, meat and nuts, fish and taro, pineapple and chicken with coconut sauce. Orange cordial or water was drunk with the food. How courteous these poor, carelessly-dressed government officials were compared with the glamorously attired but often intensely rude, inebriated elites at Palace of Westminster gatherings in London.

'The government representatives from the Engineer Group *ol ilaik mekim wanpela singsing tumbuna*[1] to thank you,' the Master of Ceremonies enthusiastically shouted.

The Engineer Group is a cluster of small islands south-east of Samarai. The ten islanders began a spirited song in close harmony at high volume. One of the group wearing a flowered shirt, open to the waist over a powerfully-muscled chest, accompanied the singing with a dance of wild movement and intense feeling. He illustrated the words, limbs fully outstretched grimacing and smiling with great animation. The singers stood at the end of the long banqueting table while the spectators were ranged along the walls. The dancer made his way along the line of diners gesturing in a frantic fashion. The song became more and more like a savage ceremony. An elemental fierceness began to bleed into the edges of the performance. The seminar on Standing Orders had become the thinnest of veneers. Nothing could have been further from the behaviour of local councillors in Croydon who also use Westminster procedures.

Another serious speech by the Senior Adviser followed this display.

'You now have a mandate to govern and act responsibly towards your wives and children and your dogs and your pigs. Tomorrow morning, Napoleon and I must leave early before any of you are awake. Our English friend here will accompany us. We are heading back to Alotau where we have a heavy responsibility. There is a constitutional crisis in Makamaka. We must attend

---

[1] 'The government representatives from the Engineer Group would like to sing a traditional song to thank you!'

for the good of the country. You may soon have your own constitutional crisis here. They are everywhere, my brothers. Now you have a mandate to deal with them.'

Wallace looked distinguished in a white dress shirt with his handless arm nonchalantly resting on the back of a white plastic chair. I felt privileged that I had been invited to this closed meeting. In my attempts to socialise, if I made the effort to smile broadly, the intimidating faces of the Melanesian men blossomed and dramatically metamorphosed into extreme friendliness. Everyone showed great interest in my travels and was evidently pleased I had chosen the islands. A great shout went up.

'A hymn from the men of the Engineer Group *pinisim olgeta samting*!'[1]

They were more reluctant on this occasion but soon another forceful rendition of a typical 'Oceanic Christian' hymn lifted the rafters. And another, and another.

The ministers abruptly and formally closed the meeting and the people began to disperse. We headed back to the Kinanale in pitch darkness to have another drink and more talk. A number of local people were clearly waiting for our arrival. Wallace began to 'interview' me for their benefit and it became obvious I was expected to give some type of sermon or address.

'Michael, tell us about your visit to Kwato. How you felt about it.' Again I was being lured into talking about Christian belief, moral behaviour and desperately searching my soul to be honest. The witty, world-weary cynicism I engaged in at Notting Hill parties would not suffice here. These people were earnest Christians and expected me to uplift them in some way. I genuinely praised Kwato and lamented its loss. The silver-haired ladies and some of the young men nodded and smiled in agreement. They seemed fascinated by my language, in thrall to my opinions. It was very strange.

---

[1] 'A hymn from the men of the Engineer Group to finish!'

'Michael, tell them about the Polish Catholics and the bricks.'

I repeated the story. They seemed extremely impressed by this demonstration of faith and clicked their teeth in approbation.

'Benny Hinn came here,' one old lady said.

Benny Hinn was a thundering American evangelist of the old school who performed faith-healing on stage. The handicapped person would be brought forward, Benny Hinn would lay his hand on their head and they would collapse in a heap amid alleluias, miraculously cured. This evangelist with his superbly coiffed hair was extremely popular in Papua New Guinea, many villagers watching him on satellite television. I felt I needed to make amends for my levity and told another Polish story.

'On 1 November, All Souls Day, the graves are swept and tidied, flowers placed and candles lit on every one. At the end of each grave a seat is constructed so the relatives can sit and commune silently with the departed. The most famous people of the country are remembered – musicians, scientists, artists and poets. At night the candles burning in every village create a wonderful atmosphere in the countryside as you go from one to another. Oases of light and love. I think it's remarkable to remember our loved ones like this.' They whispered their appreciation.

By this time the room had slowly filled with people, although the younger ones had drifted away to the television. I suddenly began to feel like a missionary myself with a captive audience and was covered with confusion. Clearly, the charisma of the white preacher still has an effect on these people. I was unable to deal with the situation and excused myself, pleading an early start in the morning. I could feel the disappointment behind my back as I climbed the bare staircase. Christianity is a living force in Milne Bay Province. Little has changed since the late 1880s when the teacher Ipunesco exclaimed to Mr W. G. Lawes, the senior missionary for the London Missionary Society in New Guinea, 'My word, the people here too much want him Word of God.' But they also believe in flying witches.

*   *   *

Wallace was sad I was leaving.

'The first meeting was good. But the second will be better. If someone comes back it means something.'

He had already settled down to a game of patience after breakfast. He waved and looked back at the softly flapping cards. I grabbed my rucksack, slid open the insect door and walked past the deserted skeletons of buildings towards the wharf. They were loading cartons of official papers into the dinghy as I climbed down the rickety wooden ladder. No seats of course, no life jackets, just the bottom boards, so I sat at the back against a plastic container of petrol. The two officials climbed down with me. We were ready to leave Samarai.

The fibreglass dinghy leaps noisily into the silence of the China Strait. At dawn, the dim shapes of the islands begin to emerge from the tropical night and a soft breeze caresses my cheeks. My two companions possess noble profiles in their own land crossing their own sea. The shoreline with its curtains of complex vegetation, leaning coconut palms and fringes of beach is home to the early hunting sea eagles, brown and white, gliding against the deep green. The first rays of the sun rapier across the glassy surface.

'Ah! The heat of this terrible sun!' exclaims the officer from the Highlands with the fearful intonation of a white explorer crossing the desert. He pushes his baseball cap to one side as a shield. Suddenly, we are slowing and rounding a point, heading towards a wharf on the mainland opposite Samarai. Mangroves come down to the shore here, their roots a set of rib cages ready to swallow any intruder. Wretched wooden boats lie on their sides, cabins dissolving in the mud, propellers pointing to the sky. The rusting hulk of a freighter is disintegrating with elaborate slowness. We approach beneath two figures standing on the wharf. They look down at us through some missing planks. This wharf is known as the Belasana Slipway.

'Is Captain Moses there?' asks Napoleon.

Silence follows. We get nothing positive out of them despite long discussions about the precise location of a particular house.

We drift under the wharf and the voices echo as they shout down to us. One points to another part of the shore and indicates Captain Moses may be there. Our skipper guns the light craft out of the shadows towards a dwelling standing among coconut palms. The water is extremely shallow, perhaps two inches over the reef, and we must punt the dinghy towards a narrow beach. Figures appear in the landscape and lethargically approach. I have begun to grasp that the daughter of the minister promised to meet her father this morning but has failed to appear. He conceals his annoyance not altogether successfully.

The wife of a villager points us towards yet another small estuary near the slipway. More punting and after a short burst of power we enter a far more sinister place. Threatening mangroves close in on either side. Heavy weed fouls the propeller and the water turns magenta. We punt towards stinking black mud flats and a pitiful hut clinging to the shore.

'Careful!' the skipper whispers as we begin punting towards an outrigger canoe. '*Pukpuk*[1] are watching us.'

Some ripples appear in the still water surrounding the mesh of roots and form into currents which slowly dissipate.

'People on the hill are watching, too.' Islanders are preternaturally observant.

An atmosphere of danger had unexpectedly penetrated our return to the provincial capital. More ripples and a commotion on the surface. Shouts float across the mire. A man, a woman and toddler come down to the shore and begin to wade towards us through the foul mud. The air is unnaturally thick and motionless. The man lifts his toddler clear of any danger and puts her, struggling wildly, into an outrigger canoe moored nearby. A discussion in Suau takes place with the pilot. Napoleon turns to me, the passive observer of this fruitless search.

'She was here but has gone somewhere. They don't know where.'

Impatience has taken over the mood in the dinghy as we labour

---

[1] Pidgin for 'crocodiles'.

against the weed so as to punt out of this ghastly place. The air is full of insects and menace as the sun begins to strengthen. We sit on the floor in heavy silence staring at the horizon. They reluctantly decide to give up the search.

Now we must drive ever harder and more alarmingly into the waves to make up for lost time. The currents are strong and the banana boat flimsy, but the pilot standing in the stern, hand tightly gripping the throttle, reads the waves skilfully. Occasionally, we are drenched by a bow wave. We hug the shore and I can see with great definition the dense weave of the jungle, crystalline beaches rising from pale green water, bush huts in clearings beneath the swooping eagles. A shoal of flying fish skips from the sea in a glittering arc and shoots past our boat, then slows and cascades like foil onto the surface of our wake. Dolphins with graceful bodies and rigid fins follow our hectic dash, cavorting in the current with superb skill. The scene is primeval, timeless; we are travelling through any century but our own. Suddenly there is a curious flash of light from within the jungle curtain.

'A glass is shining!' shouts our pilot.

We instantly turn landward in a fierce arc and speed towards the shore. A villager and his family emerge from a hut glued to the wall of foliage. The man steps forward into the water waving. The officials ask the pilot to tell him in Suau that this is the Magistrate's boat and to ask if it is an emergency. Much shouting as we rock on the tide just offshore. No, not an emergency, just wanting a lift into Alotau. Sorry, you must wait for another boat, this is an official mission. The man is perfectly happy and rejoins the family. They all wave as we spin in another demented arc and roar away towards Milne Bay.

The rest of the journey is uneventful. In the harbour the activity of the town takes over with sundry yells, heaving of merchandise, plumes of diesel fumes, jets of crimson betel juice. The officials bid me a perfunctory farewell, avoiding my eyes and getting into a Toyota Hilux. Were they embarrassed by the whimsical daughter or was our time together simply finished?

The remainder of my day was spent in a fruitless search for a charter to the D'Entrecasteaux Islands. Chris Abel tried to be helpful but there are now so few boats on scheduled runs you could be hanging around the wharves for days, even weeks, perhaps months looking for a suitable trade boat. There was never a guarantee of a return trip, and one could be left stranded. I called in at the Masurina shipping offices on the harbour without success. The difficulty of arranging anything to a timetable had begun to depress me, but Chris compensated in an unexpected way. He showed me some carved figures by a former master carver from this region called Mutuaga that were languishing in his office. The carvers of the Massim are arguably the finest in Papua New Guinea and little is known of them outside the international salerooms of Sotheby's and Christie's.

The art of Africa and Oceania had a great impact on modern artists such as Picasso, Alberto Giacometti and Max Ernst. There has been a severe decline in traditional carved artworks in Oceania this century, but collectors continue to roam the islands in search of rare treasures from the past. Original ethnic carvings, masks and ritual objects have increased in value and are much sought after by the great auction houses. Collecting was characteristic of the first Europeans to visit New Guinea in the late nineteenth century and in many regions this activity moulded the way in which they were perceived by the local inhabitants. Some villagers believed every white man was a collector, and they were often associated with the recruitment of native labour. Artworks were exchanged for steel axes, beads, bottles and red calico cloth. Among ethnologists there was a more macabre commerce in human skulls, often of tribal ancestors. Hundreds of skulls and bones entered European collections. Charlatans in this traffic abounded. One dealer bought up scores of skulls and grew orchids in them, using them as flower pots. These exotic blooms were later fraudulently sold to Europe as 'Devil Orchids', said to be worshipped by the Papuans in depraved cannibal rites.

Mutuaga was born into the *maliboi* or flying fox clan around

Replica of *Endeavour* at Trial Bay, New South Wales. The ship was constructed from 1988 to 1994 in Fremantle, Western Australia, at a cost of AU$17 million. The author experienced something of the astonishing achievements of the old explorers while sailing this tall ship from Southport to Sydney before setting out for Papua New Guinea.

Samarai Island in China Strait, former gateway to British New Guinea, Papua and the Mandated Territory, was considered the 'jewel of the Pacific'. Looking out to the beach at Deka Deka.

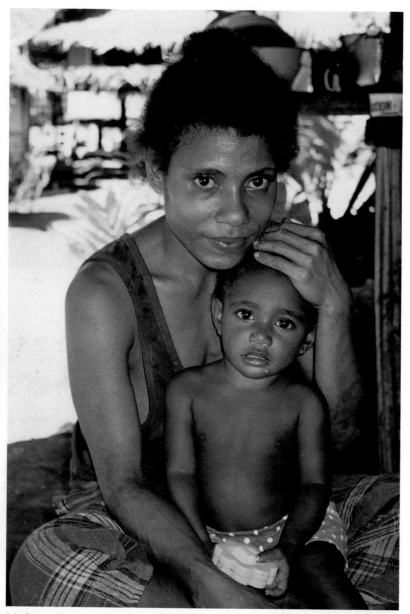

Mother and son at Bilbil near Madang, famous for its earthenware pots. The great Russian explorer Nikolai Miklouho-Maclay spent considerable time on Bilbil island sailing *Balangut Vang* canoes with his legendary friend Kain, a local master mariner. He felt Bilbil was the perfect tropical island.

Children from Ohu or 'Butterfly Village' performing the song *Mormor* near Madang. The joy in life expressed by Papua New Guinean children is extraordinarily intense considering their comparative lack of material advantages.

Demas Kavavu, Paramount Chief and former Premier of New Ireland, holding his personal *malagan*. Note the rare and beautiful *kapkap*, symbol of chiefly status, hung around his neck. Village of Fatmilak, New Ireland Province.

The sorcerer from New Hanover consuming broken beer bottles at the opening of the General Store on Tsoi Island, New Ireland Province. His activities were intended to draw evil from the store, thus ensuring its commercial success. Great bodily heat is required to function effectively, so he may chew ginger and drink salt water.

Slaughtered pigs at the Panafiluva *malagan*, New Ireland Province. Note the white band of lime from forehead to snout indicating an elevation in status when the pig is slaughtered at the sacred place of the *malagan* ceremonial.

1860 and lived in the village of Dagodagoisu on the mainland opposite Suau Island. Many outstanding carvers came from this region, but he was the most eminent. Charles Abel appreciated the artistic beauty of the local carving and took a serious interest in the local customs of the people. Mutuaga was in his thirties when they met and was not a Christian. Abel became his patron and commissioned him to carve some distinctive human figures mounted on bases for sale to the mission's visitors. Such figures were unique, not normally carved in the Massim, and may have been associated with protective magic. He produced superb ebony lime spatulas with carved handles in the shape of stylised human figures or pigs. Walking sticks, house boards and special handles for the ceremonial presentation of shell money also emerged under his hands.

Chris showed me a distinctive carving of a Mutuaga human figure on a base.

'They're very fine, don't you agree? Mutuaga paddled his canoe loaded with such carvings ninety kilometres around the coast to Kwato.'

The quality and exotic distinctiveness of the pieces were unmistakable. They certainly possessed the 'Voice of Silence' that André Malraux felt was common to all great works of art, irrespective of their cultural origins.

'We're going to build a Massim Cultural and Recreational Centre in Alotau with a section devoted to Mutuaga. Preserving the indigenous culture is vital to this area.'

'When did he die, Chris?'

'Around 1920.'

'The finest carvers are from around here, the South Massim. Have you seen the boards around the courtyard up at the lodge? Excellent local work I commissioned.'

He drifted into an outer office to deal with some shipping business so I slipped away to the bustling harbourside. The singular mixture of bush huts and modern developments makes Alotau one of the most pleasant towns in the country. I hoped to head next day for Losuia, 'capital' of the Trobriand Islands, and so began the long trudge up the hill past the hospital to the lodge to pack.

The road to the airport had dried out since I last travelled it, so instead of drowning, I was choked by clouds of fine dust that blocked the air conditioning on the Toyota. Huge lumps of coral were scattered about and potholes still full of deep water threatened to overturn us. The vast oil-palm plantations were no longer hidden by mist and rain, and a monotonous monoculture was revealed, as tedious as it is depressing.

Prominently displayed on the desk in the terminal was a large, almost illegible sign that read 'Flight Check-In Closed'. This meant trouble. The small Milne Bay Air (MBA) Islander aircraft had been directed to fly to Rabaraba first and would return to collect us and then fly on to Losuia. A delay of some four hours in the hotbox of Alotau terminal was anticipated. I settled down with the *New Guinea Diaries* of Nikolai Miklouho-Maclay, some dried apricots, cashews, dates and water and hoped someone would turn on the fans.

A couple of hours had passed when I was joined by Keith, a bluff Australian in shorts, huge boots and a battered canvas hat. He was also waiting for a plane. Expats tend to gravitate towards each other at remote air terminals.

'Waitin' for a flight, mate?'

'Yes. To the Trobes.'

'Well, ya know what they say about MBA. The "maybe airline" or "married but not available". Ya get used to it mate.' He laughed without resignation. 'Travelling through are ya?'

'Yes. I'm visiting the islands. Beautiful and no violence.' I always felt obliged to explain the reason for my visit, as I did not wish to be drawn into interminable discussions about the breakdown of law and order in the Highlands.

'Milne Bay is a great province for tourists. I've been here for years. Love the place.'

'What are you doing here?'

'AusAID. Small Scale Mining adviser.'

'Ah! Another AusAID bloke. So where are you off to?'

'Misima. The gold mine there is being run down and due to close quite soon.'

'Mining gets a pretty bad press.'

'It's not too bad.' A shadow of disillusionment edged with optimism crept into his voice.

'Well, the Ok Tedi mine is an environmental disaster and Bougainville copper is completely wrecked,' I said forcefully, suddenly angry and feeling for some reason like thumping the ecological drum.

'That's why I'm in small-scale mining, mate! Small is beautiful. I want to help the people help themselves so they can own and run their own mines. They share everything in this society. The villagers should get the money, not just the bloody transnationals and those elite government sods in Moresby. There are more "Sirs" in this country than you can poke a stick at! It's the country of "Sirs"!' An attractive contrast of rough manners and high ideals was something I often encountered among aid workers.

'I heard about a gold mine on Tabar in New Ireland opening up. It'll destroy the culture of the area, won't it?' He swiftly became impatient, took off his hat and wiped his damp forehead. I was about to receive the gunfire of his philosophy.

'Look, mate, so many young aid workers come out to New Guinea with bloody odd ideas about helping the people develop. I've been here twenty-five years and I try to tell them that many of these local people don't *want* change or development. Aid workers want to change everything overnight. Like most people, the villagers want more money without doing much more work for it. They like life the way it's been for generations. But others want to work hard and change, live a different life. They're the ones we can help. Advise and don't impose.'

His lecture was cut short by one of his team, an islander asking about the loading manifest. The advisor adopted a teacherly role, the wraith of colonialism entered the proceedings, yet he remained warm and helpful.

'Now ya know ya should've checked there was a separate waybill for this equipment. You must have the yellow form before the green one, then keep the pink one. Now go off and do it properly.' He stretched his massive boots and heavy socks out in front of him and noisily exhaled with frustration. He waited for me to say something he could disagree with, so I obliged him.

'Independence certainly was a good thing.' I sounded forthright.

'I'm not so sure, mate. The people weren't ready to maintain the structures, poorly prepared. The Australians left too early.' I had heard this view from many different quarters. He seemed to veer from real concern to cynicism concerning future progress.

'Like some water?' I offered a bottle of the local brand, imaginatively called H2O.

'No thanks. Never touch the stuff.'

'How do your family like it here?'

'They're in Cairns in northern Queensland, mate, and happy to be there too. I'm going back next week. Ah, here's the plane!' He seemed relieved to escape my questions. A small turbo prop landed and spun along the runway in front of the oil-palm plantation.

'Does it go to Misima and then on to Losuia?'

'No mate. Yours'll be much smaller. Well, good luck!' We shook hands and he clumped away.

An Air Niugini[1] jetliner from Moresby landed soon after my mining engineer and his voluminous cargo had departed. There must have been some VIPs on board because it was welcomed by a group of 'warriors' dressed in nylon grass skirts, listlessly waving spears and clubs. The heads of their silent drums were covered with black rubbish bags fastened with elastic bands.

---

[1] Air Niugini is the national airline.

They were wearing orange make-up and shuffled about dispiritedly which only added to their generally impotent appearance. A group of American divers frantically snapped away at the 'local colour'. The loitering youths I had seen on Ela Beach in Moresby were infinitely more intimidating than any of these fraudulent 'skull-hunters of Alotau'.

I soon returned to my book and nuts but not for long. I must attract expatriates or perhaps they are simply desperate to speak to a visitor, convince him of their courage in the face of overwhelming odds, tell him of unimaginable horrors, in need of a shoulder to cry on, a beer to weep into, a temporary refuge from tropical torpor. Terminals are full of them.

'Where are you off to, mate?' An obese man wearing a baseball cap and a hunted, haunted expression sat down and wiped the sweat from his streaming forehead. Heat and panic radiated from him in suffocating waves as he leaned forward to fix my attention. His eyes started from their sockets in a frantic need to communicate.

'Trobriand Islands. I've just come back from Samarai and Kwato.'

'I'm collecting some guests.' He paused for a minute or so and then suddenly erupted. 'And I'm facing ruin!' His eyes began to well up with tears. This impulsive confession was so unexpected I sat stunned. His hotel in Alotau had gone into receivership and the bank was about to seize his assets. His marriage had fallen apart. He began coughing uncontrollably and had probably been drinking. This was a drowning man. I began to wonder what the people standing nearby thought of his extraordinary *cri de coeur*. I put my hand on his heavy shoulder and discovered he was trembling. It was humid and stuffy in the terminal. The German colonists called his emotional stress *tropenkoller*.[1] He seemed to relax a little.

---

[1] German word associated with tropical illnesses such as malaria, but it came to mean a type of 'tropical madness' or 'frenzy' that might affect white people after an extended period in the tropics.

'Where are you staying in the Trobes?' His trainers squeaked on the linoleum.

'At The Argonauts' Guesthouse.'

'You can't stay there! Now look, when you arrive you must contact . . .' I must confess I had stopped listening. The man was clearly on the verge of a nervous breakdown. But the news that grim accommodation awaited me was a shock. Suddenly there was an announcement.

'Would Mr Moran please come to the information desk. Mr Moran.'

It took me a few seconds to react. I apprehensively approached the desk.

'Yes?'

'Mr Moran, I'm afraid we have to offload you from the flight. The pilot said you are the heaviest with luggage. The plane is too heavy to fly without crashing. Could you identify your bags please?'

I must have looked confused because he repeated his statement. Too much drama for one day. So it was back to Masurina Lodge for another night, having spent five fruitless hours in a fetid oven. Papua New Guinea is not called 'the land of the unexpected' for nothing. The main problem was that the next flight to Losuia was not for another five days.

Later that night, as I idly watched a malarial mosquito slowly dying in the sink, I wondered if I would ever escape Milne Bay. The phone lines to the Trobriand Islands were down again. In the end, I decided to change my plans entirely, return to Port Moresby early next day and fly from there to Madang, a town close to the birthplace of the German New Guinea Empire. All plans seem to begin and end in Port Moresby however hard you tried to avoid it.

Rain was coming down in solid sheets as I prepared to leave for the airport at dawn. The storm drains were gushing and it was hard to make oneself heard over the deluge. Milne Bay has two hundred inches of rain annually. By an unconnected coincidence, 'General' Mocquery had also been offloaded from Air Niugini and never made it to Wabag. The satellite dish had been vandalised again and he was setting off into the unknown. The only other passenger, a forestry official, struck out with exaggerated violence at any insect that approached his face. Malaria had begun seriously to frighten him after several severe bouts of fever. He was wearing the ubiquitous jungle shirt, shorts, boots and rolled socks. A moth-eaten canvas hat was pushed back on his head. We were talking about the lack of roads and boats, how reliant one becomes on air travel.

'Gettin' around is a work of art in itself. Up in the jungle the trees are growin' behind you as fast as you can chop 'em down!' He wildly slapped his own face. He looked closely at his empty palm. I told him about my experience of being offloaded. He grabbed at his nose, carefully inspecting his fingertips.

'I know a better one. On one flight they pumped out fifteen minutes' worth of safety fuel to accommodate an extra passenger. How about that for regulations!'

The dust bowl to the airport had become a raging torrent. The bus began to slip sideways as we crossed one raging ford. At the airport, another group of American divers was returning from a live-aboard dive boat. Milne Bay offers some of the best scuba diving in the world. Errol Flynn physiques gingerly carried tiny models of outrigger canoes; a couple wore matching his and hers flowered trousers; enough diving cargo to start a cult.

8. 'O Maklai, O Maklai!'

or

The Archipelago of Contented Men

On the fine morning of 17 October 1870 His Imperial Highness
Grand Duke Constantine Nikolayevich was in St Petersburg
inspecting the ships soon to be leaving for the Pacific at the naval
base at Kronstadt. He was welcomed aboard a rakish steamer
moored in the harbour. After inspecting the vessel, the General-
Admiral and President of the Imperial Russian Geographical
Society was shown into the cabin occupied by an anaemic,
bearded young man with flashing blue eyes and reddish-brown
hair. The Grand Duke had authorised the naturalist to take passage on the steam corvette *Vityaz* (*Knight*) that was joining the
squadron of the fleet in the Far East. En route he would be landed
on the shores of New Guinea.

The young scientist had sufficiently valued his conversation
with the Grand Duke to record it in his journal.

'Ah! Nikolai Nikolayevich! Can I be of the slightest assistance
in your endeavours?'

'Your Highness, everything I could wish for has been done. I
would like to express my profound gratitude for all the assistance
rendered me in my enterprise.'

'Of course, of course. Is there anything at all you find yourself
in particular need of?'

'You are aware, Your Highness, that the aim of my journey

to New Guinea is scientific exploration of this little-known island. I cannot say how long I shall be there or if I shall return, fever and native customs being what they are. I have taken the precaution of putting several copper cylinders in my luggage for manuscripts, my diaries and notes. I would be very grateful if you could send a Russian man-of-war to visit my hut so the cylinders might be dug up and sent to the Imperial Russian Geographical Society if it appears I am no longer alive.'

The Grand Duke listened carefully to this formal request and replied.

'I pray that God will protect you. Be assured I shall forget neither yourself nor the cylinders, my dear Nikolai Nikolayevich.' He took the young explorer's hand and pressed it firmly.

Known as 'the Caledonian Cossack', Nikolai Miklouho-Maclay bore a name devised from his father's rather undistinguished Russian Cossack surname and the questionable Scottish military ancestry of his grandmother. His mother was of German and Polish origin. Her father was an officer in the Russian Army who fought against Napoleon in the patriotic war of 1812 and had been physician to successive Russian tsars and Polish kings. His paternal grandfather, Semyon Miklouho, was a Cossack in the service of the Empress Catherine II, and in 1788 distinguished himself at the battle of Ochakov against the Turks under Count Grigory Potyomkin.

Into this aristocratic atmosphere of smoke and mirrors, Maclay was born in 1846 and educated partly at home. He studied for short periods at Leipzig and Heidelberg, finally electing to study medicine at the University of Jena, at that time a hotbed of discussion concerning the evolutionary ideas of Charles Darwin. Dubbed 'Baron' Miklouho-Maclay (this unusual title is believed to be inherited from his mother's side – the Empress Ekaterina gave this title to 'foreigners' who had distinguished themselves in service to her), he shared rooms with Prince Alexander Alexandrovich Meshchersky, a member of the great but wayward Russian aristocratic family. Under the influence of his brilliant zoology professor, Ernst Heinrich Haeckel, who had had numerous conversations with Darwin ('a venerable sage of ancient Greece, a Socrates or an Aristotle'), he became interested in comparative anatomy.

Maclay travelled a great deal at his mother's expense studying fish brains. On an expedition to the Red Sea, he avoided the dangers posed by fanatical Muslims by adopting theatrical disguise and being photographed in Moroccan and Syrian dress. On board a dhow it was discovered with horror that he was an 'infidel', although he spoke fluent Arabic. He had fallen into a malarial delirium and had begun to rave in Russian. Upon awaking, he kept his adversaries at bay by threatening them with a microscope which they mistook for a revolver. He was constantly in debt and often short of funds to mount his scientific expeditions.

In a bitter and desperate attempt to obtain some financial sponsorship for a projected voyage to the Pacific, Maclay began to play off the Imperial Russian Geographical Society against the British Royal Geographical Society. After protracted negotiations and ruthless letters, begging for 'such trash as money', his mother was forced once again to make up the difference between the small subsidy offered by the Russians and his sizeable requirements. Sir Roderick Murchison, the Russophile President of the Royal Geographical Society in London, had promised much but was now dying. His successor had begun playing the 'Great Game' and warning of Russian threats to India. Maclay was on his own with a pocketful of 'ridiculous kopecks'.

The departure of the *Vityaz* was further delayed by the Franco-Prussian war and Maclay spent his last weeks in Russia at the Palace of Gatchina as a guest of the family of Tsar Alexander II. The fabulous ostentation did not move him unduly as he strode through the royal parks planning his audacious excursion from New Guinea to the Sea of Okhotsk. The Imperial Geographical Society wanted him to concentrate on meteorology and physical geography, worried that he might dally over his beloved swim bladders, sponges and corals or, even worse, attempt to discover the vanished continent of Lemuria, thought to lie beneath the Indian Ocean.

Maclay now joined the ranks of those whose imaginations were overwhelmed by fantasies of New Guinea. He became obsessed with the anthropological comparison of Papuans to other 'tribes of the Pacific Ocean'. Unknown to the Society, he was secretly looking for *Homo primigenius* to explain the origin of man. He felt New Guinea might provide him with promising material. During his time there, many native men were brought to his remote hut so he could draw what he considered to be their vestigial tails (referred to more scientifically as a 'cutaneous polypus').[1] He had already taken an active interest in a now extinct hairless Aboriginal tribe in Queensland. It was not until 8 November, after many delays, that the *Vityaz* finally sailed under the command of Captain Nazimov, together with over one hundred cases of Maclay's possessions including a library containing volumes by Goethe, Kant, his favourite pessimistic philosopher Schopenhauer, Byron and Molière, as well as scientific texts. The journey to the fabled land would take him almost a year.

They made a landing on 20 September 1871 on the shores of Astrolabe Bay. The ship anchored near a small promontory, and Maclay explored a village near the coast. He refused to bear arms, only colourful gifts, and from the outset established friendly relations with the local people. This non-confrontational approach distinguished him from most explorers of the time, and led to his name being cherished throughout history along that coast. The following day was the birthday of His Imperial Highness Grand Duke Constantine Nikolayevich. The ship was dressed and a salute of guns prepared. Maclay decided to remain with the villagers during the salute, as he anticipated the effect the explosions would have on them.

---

[1] A drawing of a 'little tail' is in the Macleay Museum at the University of Sydney.

While waiting, he chose the site for his hut. He selected a spot some ten minutes' walk from the village near a stream and a grove of trees. 'To impose my presence upon them I considered tactless,' he wrote in his diary with characteristic sensitivity. The guns, as expected, created panic among the villagers with much cowering, stopping of ears, falling to the ground, writhings and rushings about. Maclay could not stop laughing which actually proved the best antidote to their fear. The bay where the *Vityaz* lay at anchor he named Port Constantine in honour of the Grand Duke. He decided to stay at the spot, characteristically against the advice of the surgeon who thought the coast 'shrieked of malaria'. The surgeon's judgement was subsequently vindicated.

Four days were expended constructing his *palazzo* from native materials. The Artillery Officer, Lieutenant Chirikov, suggested preparing some mines and placing them in a defensive circle around the hut. The warriors observed the many wires and plungers with bemused interest. At first, Maclay's pale skin convinced them he had come from the moon. His dwelling was packed to the palm roof with books, goods and basic provisions. A small boat given to him by Captain Nazimov was hauled up on the beach. At 4.00 in the morning of 27 September he wrote his last letters to his family and Grand Duke Constantine, bade farewell to the commander of the *Vityaz* and rowed ashore. Earlier he had ordered his Swedish servant Ohlsen to lower the Russian flag over his hut as the warship departed, but found him pathetically sobbing on the steps. It was to be an omen for his future craven behaviour in the face of adversity. His young Polynesian servant, simply named 'Boy', was preparing breakfast. As the corvette steamed away he wrote in his diary, '. . . henceforth I was left to myself – in future everything depended on my energy, will and labour'.

So began the first period of extended anthropological fieldwork in European history. More was to follow, most notably by Bronisław ('Bronio') Malinowski. It was my intention to explore Madang and the Rai Coast (as it is now named, a local corruption of Maclay) and the villages Maclay frequented. The site of his hut at Garagassi had been visited thirty years before by the Russian marine research vessel the R/V *Vityaz* and a small memorial erected. I wanted to make this demanding journey as a minor tribute to the 'Moon Man', a strikingly courageous individual and one of the few early New Guinea explorers who treated the 'native races' with even the slightest degree of humanity.

Madang, situated on a fine harbour on the north-east coast, is arguably the most beautiful town in Papua New Guinea. A council sign greets you at the entrance to the town with the self-effacing and charming philosophy: STRIVING FOR BEAUTY. The people are extraordinarily friendly and willing to talk. Backed by the most precipitous mountains in the country, the Adelbert, Bismarck and Schrader Ranges to the north-west and the Finisterre Ranges to the south, it is the gateway to the Highlands and the mighty Sepik and Ramu rivers. The area has three active volcanoes and some forty-five islands lie offshore.

During the period of the German Empire, arcades of coconut palms led to the town and fine colonial hotels displayed broad verandas to catch the breeze. On postcards depicting the *Strand-promenade nach Friedrich Wilhelmshafen*, the Germans wore white tropical suits and the local people bright red bandanas. My room at The Madang Resort Hotel, rather different to the hut 'enjoyed' by Maclay in the previous century, overlooks Dallman Passage. Beaches break through the dense vegetation like golden tongues. The lawns of Ragetta Mission on nearby Kranket Island sweep down in a civilised gesture to the shore. He named the islands along this coast the 'Archipelago of Contented Men' and was 'given' the island of Urembo by a local chief.

A sound like steam escaping from a locomotive jolted me out of a reverie as I was unpacking my rucksack. The incredible noise announced the arrival of a pair of male Blyth's hornbills that live in the extensive gardens feeding on red berries. Their short wings produce an astonishingly loud sound and their appearance is distinctive, even intimidating – large birds with huge horn-coloured beaks, orange breasts and black plumage. A few fallow deer were grazing on the grass. At dusk, flocks of squealing flying foxes crossed the town and roosted noisily in the trees. The air had the softness of velour as the sun folded behind the palms and canoes glided across the passage to the outer islands. Night is a quickly-thrown cloak near the equator. Once the tropics infects your blood, it enslaves you like a terminal illness.

I arranged my guides immediately, as in PNG you must allow time for unexpected delays and multiple changes of plan, particularly if it is an unusual request. This turned out to be an eminently sensible idea. A long conversation with Roy Bakarum, one of my guides, revealed he was from the village of Bongu where Maclay had spent a great deal of time. As a passionate admirer of the great Russian, he was enthusiastic about my plan to visit Garagassi as he had never visited the landing site himself. It was many years since any European had been there. The journey by road would require careful planning, as there were deep rivers to ford and the road surface was uncertain. My other guide was a Highlander with a big smile known as 'Busy Bee' which describes his character perfectly. He was generous with his fund of local information.

My first morning in Madang was spent in primeval surroundings. I had launched out of bed at 5.00 a.m. and was headed for a section of pristine forest protected from logging, gardening, burning, and hunting by a *bigman* with vision, Kiatik Batet. The Didipa clan have resisted all attempts at logging their sacred land. For these people the forest is imbued with spiritual force, haunted by demons, bush spirits and presences. It is a landscape that has witnessed their social history. To destroy it would strike at the root of life itself.

An old man, dressed in a torn shirt and speaking no English, led me through a riot of rainforest vegetation, ferns, bananas, lianas, strangler fig, rosewood and walnut, ebony and tulip. The luxuriance of untouched forest in any latitude is always astounding. The dilemma facing villagers in Papua New Guinea is that personal cash is becoming increasingly necessary as society evolves out of the subsistence economy. Precious woods are a valuable cash commodity, but logging companies are only interested in a limited number of species. For the clan, the whole living fabric, every plant and every tree, is permeated with significance. The rights to land are owned by kin-groups and not individuals, a feature of the culture that causes severe trauma in many forestry decisions. The clan has no legal identity in European law, so it is mainly individuals or *bigmen* with an enterprise spirit who benefit from logging contracts. They are expected to distribute their new wealth, but many do not. This fractures the traditional egalitarian social order based on sharing and leads to general disharmony. Within the subsistence family, the sexual division of labour that has persisted for centuries is dislocated by these individual initiatives, which leads to further social problems.

Busy Bee rescued me from the ferocious mosquitoes and we drove to an immense wall of vegetation glowing in the newly-risen sun. He said we would see many species of birds flying against the lofty curtains of thick growth draped over a rock-face extending at least a kilometre along the roadside. The formation looked like a castle rampart, a fragment of an ancient civilisation. High on a cable a dollar bird, bluish brown with red beak and legs, sat as immobile as a sculpture in the heat. Sulphur-crested cockatoos flashed their brilliant white against the deep green, calling angrily with ear-piercing screeches. A Lesser Bird of Paradise clung to a fruiting vine in the middle storey of the escarpment. The body of this magnificent specimen was maroon-brown with long plumes of yellow gold trailing behind. In the early morning and mid-afternoon, adult males visit the *lek*, an area of bare earth in forest glades where the white-breasted females congregate. Such display areas can be used by groups of male birds for many years. A dominant male will copulate with numerous submissive females, at first vibrating in ecstasy and then dancing ostentatiously before them, flaunting his plumes. My bird, although not at a *lek*, was engaging in all manner of flamboyant acrobatics. A flash of gold and he was gone.

During the German hegemony from 1884 to 1914, a certain Herr Peterson, Medical Assistant to the Neu Guinea Compagnie, was collecting plumes when he was murdered by tribesmen inland of Madang. In the violent reprisal that followed, five local men were killed and the village of Bemari was burnt to the ground. Smuggling the birds subsequently became a fatally glamorous occupation for adventurers, the plumes exerting an almost supernatural spell over them. Errol Flynn seriously considered hunting and smuggling the species during his peripatetic career in New Guinea.

The spectacular plumes of the forty-three species of birds of paradise play an important part in the lives of Highlanders, providing brilliant decoration at festivals. They also trade in the dried skins of adult birds. When the first skins arrived in Europe in the fourteenth century, the wonder they evoked caused them to be regarded as 'birds of the gods'. As native people had prepared the skins without the feet, it was thought the birds lived by floating in the air, never alighting until death, feeding upon nectar and dew. The first Western naturalist to see a bird of paradise in a forest habitat was the Frenchman Réné Lesson, sailing the Pacific aboard the *Coquille*. In 1824 he too encountered a Lesser Bird of Paradise and wrote of this experience:

> The gun remained idle in my hand for I was too astonished to shoot ... It was like a meteor whose body, cutting through the air, leaves a trail of light.

But it was the work of the great English naturalist Alfred Russel Wallace that drew the attention of the world to 'the most beautiful and most wonderful of living things' in his superb book, *The Malay Archipelago*, published in 1869. Charles Darwin used them in his studies of evolution. 'Plume fever' took hold of the milliners of Paris and Edwardian London where paradise feathers became the rage of the fashion world before the Great War. Indian maharajahs and the kings of Nepal passionately sought these 'treasures' for use in their jewelled *aigrettes*. The illegal trade in smuggled skins blossomed. As light as a feather? From 1890 to 1929 some fifty thousand tons of feathers were shipped to France. Now, plumes cannot leave Papua New Guinea legally unless for a scientific purpose. I felt privileged to have seen one bird if only for a few brief minutes.

We climbed back into the Toyota and pushed on down towards a broad, shallow, rather featureless river, the banks cleared of vegetation.

'Perhaps I can tell *you* something for a change, Busy Bee! Do you know why this river was called Gogol?'

'No, I've never thought about it.' He gave a brief laugh.

'Miklouho-Maclay named it after the famous Russian writer, Nikolai Gogol, who lived in the nineteenth century. He had a very long nose which gave him no end of trouble.'

He chuckled as he threw the truck into the corner after crossing the rattling bridge.

'It interfered with his love life. Actually, he wrote a story called "The Nose" about a nose that leaves its owner and wanders alone around a city in Russia.'

'I should read that. About the love life, I mean.' His voice did not sound particularly convincing.

'I had two wives.' Busy Bee was clearly in a confiding mood.

'Lucky man! One after the other?' We were passing the castle ramparts once again.

'Not really. At the same time. They spoke different languages and couldn't understand each other. It ended badly. I left one and the other left me!'

'Yes, I've been through all that sort of thing. When I . . .' He cut me off.

'Women! They're all sisters but live in different houses!' He laughed and we continued on in silence for a while. We visited a sulphur stream which flows from an ancient cave containing fossilised and petrified corals. To my amazement, tortoises survive in the malodorous waters. A light dusting of sulphur is deposited on their shells and skin like fine talc. They were being fed by beautiful children with dark eyes. The exotic flowers dusted over the ground and growing beside the stream gave the scene the atmosphere of a prehistoric Eden. We soon arrived at another village.

'This is Siar Island.'

Many pretty houses of split bamboo stood on stilts by the water. Some remained identical to sketches by Maclay, bush-material roofs almost touching the ground. A friendly young man was squeezing coconut milk outside his front door. He put the coconut flesh in a cloth and twisted it with a wooden spindle at one end, the other being attached to a pole driven into the ground. The milk was collected below in a plastic bowl. We exchanged pleasantries and then he said, 'The Duke of Edinburgh came here. He said it was just like England.'

'Really? He said that?' I glanced over the lagoon to the huts and tropical palms. I understood the irony of the remark, but I am not sure my friend did.

'Yes. And he went for a swim just over there.' He gestured to a beach where some children were lying under the roof of an open pavilion.

In the nineteenth century the British had given Siar a white ensign to 'protect' it against a German invasion. The villagers carefully preserved this 'totem' in the men's house, endowing it with magical powers. It was destroyed in the Second World War when the Japanese attacked Madang. The village was bombed and the men's house destroyed together with the flag. When Prince Charles visited Madang some years ago he heard this story and decided to replace it. The new flag now lies in a glass case in a back room of the Lutheran church, a faded and moth-eaten reminder of imperial days. The building has a red tin steeple and a severely cracked lintel resulting from an earthquake.

*   *   *

'Pastoral work in Germany was simply too boring!'

Father Wolfgang ('Wolfie') Eisenstadt from the Catholic Divine Word Mission in Madang was answering my question as to why he had come to Papua New Guinea from his home town of Dresden.

'There were three possibilities: Philippines, PNG or Africa. I got PNG and so here I am.'

He was as you might imagine the classic East German missionary to be – a heavily-built man in his fifties wearing sandals with woollen socks, a shovel beard and carrying a briefcase. He had an enthusiastic, almost bombastic manner that veiled a generous heart of great earnestness and simple justice. He had spent most of the day at the ordination of the new Polish Bishop of Madang, Bishop William Kurtz, originally from Silesia. I have been astounded at the number of Poles attracted to this country. Two Polish priests with Nikon cameras told me there were well over a hundred Poles doing missionary work in the mountains behind Madang. Wolfie had a mission station on the Sepik river.

'You missed some ceremonial dancing from Chimbu this morning at the inauguration!'

'I wish I'd known. Does the Catholic Church approve of local culture?'

'Sure! But all those old-timer ideas are gone now! As long as it is not directly against the Christian religion, the Catholic Church does permit things. In some places, even dancing at the consecration. But not everything is permitted. Not those initiation ceremonies, you know the ones when an armband is put on and the boy goes off into the bush to have sex with some girls! Ha! Ha! We will not have that!'

'What's life like for you here, father?'

'Sometimes terrible, sometimes wonderful. One of our sisters was shot in the face recently and also my colleague, a priest, was shot. They expelled some boys from the school for taking drugs.'

'But didn't the police help?'

'The killer was found by the police. They shot his kneecaps

out from behind the legs. They say he is now "settling down" – he cannot move of course! Yes, yes, he is said to be "settling down". Sometimes the police execute criminals or shoot one leg off. "Now he is settling down!" They say that.' A faint smile revealed a wide experience of extreme conditions.

In line with my project of exploring the links between Miklouho-Maclay and the annexation of German New Guinea, I headed up the coast towards Alexishafen, the former Catholic Mission of the Holy Spirit, now a Catholic convent. It was established in 1905 under a Father Limbrock. The harbour is as lyrical as a tropical painting. The clear light gives extraordinary definition to the colours. The landscape is not one of epic grandeur like the volcanic caldera of Rabaul, more the soft, seductive presence of Polynesia. Tall coconut palms of former plantations line sheltered lawn-covered peninsulas, hibiscus flower in oases of scarlet, canoes with curved prows are drawn up on the beach or float on the placid water. Numerous atolls are dotted about the horizon in diminutive green clumps. The mission buildings scattered over extensive areas of clipped grass are impeccably kept. It seemed a deserted heaven.

Alexishafen, to the north of Astrolabe Bay, was Miklouho-Maclay's greatest discovery. He named it Port Grand Duke Alexei after the Tsar's brother Alexei, Grand Admiral of the Russian Navy, a man said to be more interested in women than in ships. His deep and compassionate knowledge of local customs, particularly the vexed question of clan rights to land, led his romantic and noble cast of mind to conceive of a grand development plan known as the Maclay Coast Scheme. He envisioned development of the area as an indigenous Papuan possession under beneficent Russian 'protection'. He imagined Port Grand Duke Alexei as a Russian naval base. The only problem was the increasing presence of the Germans.

Despite the administrative catastrophes further down the coast, the Germans did achieve some fine things at Alexishafen. This mission station became one of the finest achievements of their period in New Guinea. By 1910 a sawmill, a hospital with a doctor and eighteen schools had been established around the beautiful harbour. An excellent road twenty-three kilometres long had been constructed from Friedrich Wilhelmshafen (Madang). The fathers, brothers and sisters trained villagers to be masons, carpenters, fitters and tailors as well as teaching them the German language. A dam was built for the cultivation of paddy rice, hundreds of hectares came under coconuts.

This was not the only mission along the coast. The Protestant Neuendettelsau Mission at Sattelburg near Finschhafen and the Rhenish Mission Society also selflessly gave lives in the service of Christianity, education and health. All the mission stations paid an appalling price. The cemetery at Alexishafen is a veritable war grave, a forest of black crosses with white lettering, standing in serried ranks on concrete plinths under the baking sun. It is a melancholy scene, this last resting place of selfless German missionaries wrenched from life by malaria, blackwater fever, murder and war. These were true soldiers of Christ fallen in the battle to save souls. Young nuns were clearing the rank grass from around the graves of local people in the terrific heat, over-looked by a crucifix and statue of Our Lady beneath which marble plaques displayed the names of hundreds more who died in this pestiferous spot.

We drove further up the North Coast Road to picturesque Malolo Plantation, past groves of vanilla to a volcanic beach of black sand glittering with specks of gold. Busy Bee lived up to his name and began clearing some of the gardens of weeds and then built a fire from coconut husks on the beach. I was hoping to see the volcanic cone of Karkar Island which is almost two thousand metres high. It was wreathed in slow-moving cloud, so we sat on the beach by the fire talking and drinking beer, waiting for it to clear.

An early landing had been made on this island in 1700 by the flamboyant English buccaneer explorer, map-maker, navigator, and pioneer of modern travellers, William Dampier. Dampier was one of the first men to land on the Australian continent in 1688, when the English pirate ship *Cygnet* was driven onto the Western Australian coast near King Sound. He was a cousin of Jonathan Swift and parts of *Gulliver's Travels* were based on the remarkably erudite and sensational journals he kept of his exploits, published in 1697 and entitled *A New Voyage Around the World*. In 1699 the Admiralty outfitted him with a scarcely seaworthy ship, HMS *Roebuck*, for further exploration of *Terra Australis incognita*. The voyage across the Indian Ocean and along the Western Australian coastline was fraught with difficulties – shortage of water, a leaking ship and mutinous crew. Eventually forced to abandon his coastal route, he sailed north to Timor where the ship was reprovisioned. He sighted New Guinea on the first day of 1700 and explored the Bismarck Archipelago, passing through the straits that bear his name, finally sailing along the north coast to Batavia, the headquarters of the Dutch East India Company in the Spice Islands (present day Jakarta). Alexander Selkirk (the model for Daniel Defoe's Robinson Crusoe) went privateering on the Spanish Main with William Dampier, who subsequently helped rescue him in 1709 from the island of Juan Fernandez, four hundred miles west of Valparaiso in Chile.

Maclay had admired the seashell and tortoise-shell ornaments of the Karkar warriors with their black-smeared faces, apparently accompanied everywhere by fat dogs on short legs. He attended to their infected wounds, cleaning out fly-blown pus and hundreds of grubs, bathing and bandaging the legs of young boys. These acts of kindness have never been forgotten.

Time expanded to stillness as a lone fisherman took up his spot on a palm trunk jutting over the sea. I had brought with me the now-rare Sentinella edition of Miklouho-Maclay's *New Guinea Diaries* which I carried with me everywhere I went, and which everyone wanted to read. There is a voracious appetite for history in Papua New Guinea that seems to have been completely neglected by the educational system. In 1883 Maclay had been taken on his third visit to the Maclay Coast by the corvette *Skobeliev* under the command of Admiral Kopitov. He notes in his typical detached style:

23 March

Weighed anchor about 6 o'clock; about 8 we passed through Izumrud Strait between New Guinea and Karkar Island. At the south-western extremity of the latter we noticed several sailing pirogues and after three hours I was convinced that these same pirogues had been drawn up on the beach at Cape Croiselles – this served as a proof for me of a continual interchange between the natives of Karkar and the inhabitants of the mainland.

I thought I was going to Garagassi this morning but there was an inevitable change of plan. After a scenic tour of the harbour I ended up on Big Pig Island (as opposed to its sister Little Pig Island), diving with a group of young Christian visionaries. This uninhabited atoll is some way off Madang Harbour and used to be where lepers came to pass their final days. I saw many species of fish here I have never seen before. Large shoals follow the current like pieces of drifting silk dappled by quick sunlight. I saw my first 'long tom' – a rapier-thin, transparent fish with an amazingly elongated head and mouth. Unforgettable coral gardens sheltered panda clownfish and the ever-present clusters of electric blues. I explored the interior of the island, a claustrophobic tangle of sprouting coconuts, the buttressed roots of massive figs, pandanus palms with tall aerial roots, rubbery ferns, fan palms. Millions of crawling insects of astonishing size and agility were devouring the jungle floor. The shallow beach was littered with massive pieces of driftwood, bizarre shapes that can occupy a fanciful imagination for hours.

As we sped back towards Madang in the launch I examined my fellow passengers. The five young American missionaries boasted a pure vision and alarmingly good health. A doctor, a pilot, a nurse, a teacher and his pregnant wife who seemed terrified of being in the country. They formed a type of 'mission-ary strike package', a cell of one of the numerous sects now swamping Papua New Guinea. The potential they offer villagers for theological confusion is large, the contrast with their potential flock great indeed. They were the absolute antithesis of conven-tional South Seas missionaries. The nurse was tall, sexy, and very blonde, wearing tight, bum-revealing shorts and expensive trainers; her boyfriend, the pilot, was obviously a gym fanatic, judging from his extensive muscular development. All were bronzed and had the smiling, scrubbed look of affluent born-again Christians. All were slightly tense with the pressure of their vocation. They said they were on a 'reconnaissance assignment' to set up a mission station either in Madang or the Highlands sometime in 2003. 'We're taking them the Word of Christ,' the nurse murmured and the pregnant wife clasped her hands over her stomach and turned her gaze inward. I wondered how a year in the jungle would affect their disturbing air of perfection.

Madang is built on a peninsula and laid out with pretty lakes, parks and peaceful inlets. One of the most beautiful roads in the country, known as Coronation Drive, skirts the Prussian blue waters of Astrolabe Bay on the south-eastern side of the town. The road follows a line of coconut palms bordering a rocky shore. The immaculate golf course opposite is relieved by ponds covered with magenta and pink water lilies, adding to the gentle colonial atmosphere. The imperious Finisterre Ranges rear in the misty distance. The centre of the town is fairly ugly, due mainly to litter and a seeming lack of civic pride. A rally of the 'People's Labour Party' seemed faintly threatening. The Old German Cem-etery on a mound near the market is haunted by alcoholics and dope fiends. Only one of the headstones was intact. Of particular insult were the Australian war graves that have been disfigured and submerged in refuse. All the grave markers have been brutally torn away.

Papua New Guineans are a people with an infinite capacity for waiting. Groups were lounging lethargically or sleeping against the trunks of what must once have been fine casuarinas and rain trees. The broad avenues have been carelessly lopped. Cheap advertisements blight the town. The locals are the friendliest I have met, and I was constantly being accosted by youths carrying babies or machetes and wanting to talk. One from Manus with a row of crosses tattooed across his forehead had just graduated from teacher's college.

'What subject do you teach?'

'Everything! I teach all the subjects, sir!' he enthused with pride.

Dusk was falling and thousands of raucous flying foxes were seeking their roosts in a scene resembling a Transylvanian nightmare. I headed back to the hotel, hopeful of a new itinerary for the Maclay excursion. There it lay on my bed. Tomorrow would be the day.

9. *Kolonialpolitik* Defeats the Man from the Moon

'No Melanesian ever died of stress!' Busy Bee said as we headed towards Garagassi. I was telling him anxiously that I feared I had contracted malaria in the night.

Roy Bakarum from Bongu was with us. Tall and slim with wrap-around sunglasses, he was one of the dashing and intelligent managers at the hotel. His clan owned land around Bogadjim and he was a knowledgeable enthusiast of Miklouho-Maclay and his work. They had never had a reason to visit the site of his first settlement at Garagassi and were as excited as I was at the prospect. It would be a long and difficult drive, fording rivers and following unmarked tracks through high kunai grass.

We had loaded some provisions but the first stop was the market to buy cigarettes and betel. After much discussion, Roy and Busy Bee decided to buy three cigarettes. A stall selling individual cigarettes forcibly reminded me that local people operate on a dramatically reduced financial scale.

The first village we visited was Bilbil, which in Maclay's time was known as Bilibili Island. The islanders violently resisted the forced labour required by the German coconut plantations, which culminated in an attack on the Friedrich Wilhelmshafen Station[1]

---

[1] 'Station' was used to designate German settlements during the period of their Empire. This settlement or 'Station' was to become the present city of Madang.

in July 1904. This murderous behaviour can scarcely be credited to the same people who joyfully thronged the beach to welcome Maclay, the *tamo russ* or 'man from Russia'. The moon came to be thought of by the local people as 'Russia', another reason apart from his complexion that Maclay became known as the 'Moon Man'. As a punishment for the attack, the German district office compelled them to relocate to this present village opposite their island.

Maclay had been fortunate to see the magnificent *Balangut Vangs* drawn up on the beach, a type of large trading canoe with a square pandanus sail, decorated mast and large hut built on a platform at the crossbars of the outriggers. These craft could carry twenty people, but were unable to tack and could only sail with favourable winds. He made some fine sketches of them. On Bil Bil he was to befriend the legendary Kain, a master sailor. His diaries are filled with lyrical descriptions of sailing expeditions with Kain. Maclay thought it was the perfect tropical island and felt he could easily live there. Always the professional ethnographer, he compiled a vocabulary of the language.

Bilbil remains famous for its pots along the entire Rai coast. The village seems entirely devoted to their production, and I sat in one of the split bamboo huts and watched a woman forming a pot. She held it on her knees and placed her left hand inside holding a rounded stone, while with her right hand she would pat the outside with a flat piece of wood, so determining the thickness and shape. This deceptively simple craft has remained exactly as Maclay described it. Her husband sat behind her preparing tapioca while a beautiful young girl suckled a baby beside me. Children were everywhere, constantly shouted at by the old women of the village. Now that the great *vangs* have disappeared, only small canoes lie beneath the trees. The wind was disappointingly unfavourable for a crossing to Bilbil island. These particular villagers were clearly used to European visitors and had a markedly diffident attitude. The spontaneous friendliness I had become used to was absent here.

The road became more potholed as we drove through

plantations of old German cultivation. Immense Banyan trees, coffee bushes, tobacco, kapok and the remains of teak forests had run wild after ninety years of neglect. We came to the Yawor river and stopped on the pebbled banks to contemplate alternative crossing points. The swirling waters looked deep, but Busy Bee chose a ford quickly and despite water coming halfway up the doors, we crossed safely. Sweating women on the roadside were carrying heavy logs and loaded *bilum*[1] bags in the oppressive heat like pack animals. Their privileged men sauntered by their side, hunting in their light plaited bags for betel-chewing equipment. Flat land supported open swathes of high kunai grass resembling the African Savannah.

Deep in the jungle near the village of Bogadjim, the polished marble gravestone of the murdered Curt von Hagen looked strangely out of place among the hibiscus, coconut palms and tall grass. Dr Hagen was a doctor from Sumatra who worked for the Astrolabe Compagnie, formed to cultivate the 'limitless' tobacco plantations. He rose to become Administrator of the German Protectorate at Stephansort (Bogadjim), the new headquarters of the Neu Guinea Compagnie after it was transferred from the fever hole of Finschhafen in 1892. Tragically, he had been set to leave the colony within weeks. The remains of the narrow-gauge railway imported from Germany, now overgrown by grass, is the sole relic of that extensive and dramatic industrial period.

The Minsieng and Kior rivers were more difficult to ford but after checking the wheels with a machete, we rumbled across in showers of spray. Some rivers had bridges but they were considered unsafe, so we chose to cross on the river bed. Much of the land around Bogadjim belongs to Roy's clan. In the local language *Boga* means 'wild man' and *jim* refers to 'the fruit of a tree'. It is a sprawling settlement.

---

[1] A net bag used for carrying everything from groceries to children. The handle is placed around the forehead and the bag hangs down the back. Much artistry is put into the colour and weave of the design. These bags are seen throughout Papua New Guinea.

The beach at Bogadjim overlooked Astrolabe Bay to Garagassi and Port Constantin, exactly as Maclay sketched it in 1871. Some houses had window shutters of split bamboo in faded reds and blues, others had plaited walls in coloured diapered designs with words woven into them. Roy told me that houses survive for about eight years and are now built on stilts which makes them cooler and healthier. In Maclay's time the palm roofs slanted down, almost touching the ground. A few of these still exist scattered about. Having a space beneath a house was avoided, as it was believed that sorcerers could burn leaves and the rising smoke would carry a fatal spell to the occupants.

Maclay had first seen Bogadjim from the deck of the *Vityaz*, and after settling at Garagassi he made many trips there. He writes of the attempts by the head man Kodi-boro to persuade him to live permanently in the village by offering to build him a hut and parading several nubile girls as possible 'wives'. At that time it was the wealthiest village on Astrolabe Bay. To 'possess' this magical creature, Maclay, the 'man from the moon', would be enormously prestigious.

The site of Maclay's landing at Garagassi is approached through an ocean of head-high kunai grass, past a bay with outrigger canoes in red, white and blue and a wrecked trade steamer. My two guides had never been there, which is not surprising considering the remote location. A long track descends finally into a hollow filled with coconut palms and meanders out to the edge of a cliff. A modest memorial has been set up which reads:

> To the Glorious Memory of Russian Scientist
> N. de Miklouho-Maclay
> Who Landed Here From
> Corvette *Vityaz*
> In the year of 1871
> From Seamen and Scientists
> Of R/V *Vityaz*
> December 1970

There are panoramic views over Astrolabe Bay and the coastal Finisterre Ranges. In this land of constantly-changing designations, Maclay named the highest peaks after Kant and Schopenhauer. Later, Captain Moresby, in a bout of patriotism, renamed them Mount Gladstone and Mount Disraeli. At present, one peak remains unnamed, the other is called Mount Abilala. Maclay speaks of a promontory enclosing a small bay on which he built his hut. I had long conversations concerning the exact position of it with some elderly inhabitants who had gathered. They had decided the original site was now under water and the promontory had disappeared. On his second visit in 1876 Maclay himself noticed changes in the shape of the mountains and the shoreline near his hut. Part of the area had disappeared when Krakatau exploded in 1883. This eruption caused a sinking of the seabed resulting in a severe *tsunami*.[1] The Germans had subsequently harvested all the Kalapulin trees that remained on the coast for boat-building, which left the shoreline unprotected from storms. The resulting erosion finally destroyed the site and the remains of Maclay's hut slipped beneath the waves.

I climbed down to the tiny beach and the clear stream mentioned in his diaries. Some village women were washing clothes, and a pale blue-green canoe had been hauled up under trees. Leaves still hang down from branches nearly touching the water, just as he described. Large coral boulders litter the shore. The area has preserved the serenity described by Maclay, and has retained all the magical characteristics he sketched in the diaries.

The air was hot and still in the nearby coconut grove, and a few young coconuts were collected and sliced open. I lit my pipe much to the interest and amusement of the local villagers, and they inspected my smoking accoutrements with great intensity.

---

[1] A series of large waves caused by major disturbances on the ocean floor. It could be an earthquake, a submarine landslide or submarine volcanic eruption. They affect much of the Pacific region.

The coconut juice was wonderfully refreshing as the diaries were passed around.

Favourite scenes from Maclay's life were recounted by members of the group – the affection felt for his memory remains strong. Many children along the Rai Coast are still named after Maclay or given names with a Russian derivation. Everyone loves the legendary scene when he first visited a local village and was fired upon with arrows and threatened with spears, the warriors jabbing them towards his eyes. With great presence of mind, being unarmed and unable to speak to them, he decided to lay out his mat under a tree and simply go to sleep, lulled by the birdsong and cicadas. This extraordinary behaviour disarmed them at a stroke. My new friends also spoke of his first Papuan friend Tui, who instructed him in the language and customs and constantly warned him of dangers. His descendants still live nearby.

Maclay came to be regarded as a deity, a *tibud*, and the legends multiplied and penetrated the mountains and ran along the coast. Everything he touched, his tobacco and in particular his gun, became imbued with wonder. The word for gun, *tabu*, became part of the language, as he would shout, '*Tabu!*' to stop them touching the firearm. The weapon magically brought down the black palm cockatoo which he mysteriously dissected and drew in his hut. He cut hair samples from the reluctant local people with fearsome 'scissors' and, in a form of exchange, cut locks of his own hair to give them. His servant Ohlsen soon pointed out that the whole of one side of his scalp was almost bald.

On one occasion he prevented a war between the village of Bongu and neighbouring Gorendu. A young boy named Tui had been bitten by a snake and had died. Naturally, the villagers were convinced *onim* (sorcery or evil magic) was at work. Sudden death always has a malevolent reason in Papua New Guinea. The local people had worked themselves into a state of extreme excitement fearing earthquakes and other disasters. Maclay wrote, 'I decided to ban the war.' He predicted misfortune for them if it went ahead. His demi-god status effectively defused the aggression.

When his Polynesian servant 'Boy' died of malaria, he and Ohlsen secretly disposed of the body at sea in the dead of night, weighted with rocks and hidden from the flickering torches of the local fishermen. One macabre aspect of this story was that Maclay, not being able to 'dissect Boy's skull and preserve the brain for research', ultimately 'dissected the larynx with musculature and tongue' to despatch to his teacher, Professor Gegenbaur of Strasbourg. 'A piece of skin from the forehead and scalp with hair went into my collection.' Often brought low with malarial fever himself, Maclay was selfless, not only in attending to the local people but also to his hopeless servant, normally prostrate with fear.

Amid much laughter, I was told that Maclay had visited the village of Gumbu, where he was asked if there were women on the moon and how many wives he had there. He assured his questioners that there were none. Late that night a rustling beside him indicated the presence of a woman on his sleeping plank. He rose quickly from the *barla*[1] saying he had no need of women. This was a source of great wonder. He steadfastly refused to sleep with the local women which added to his charisma as a superman. The stories continued until it was late, many coconuts had been eaten and my pipe tobacco all but exhausted.

Bongu is now the largest village on Astrolabe Bay with a population of five thousand. The village has a terrific visual impact. The huge central ceremonial area baked silently in the heat, bare of any vegetation except a pink frangipani and a couple of large rain trees. The village houses on stilts were scattered over an enormous expanse of swept, compacted sand and coral. A stool was drawn up for me in the shade near an old man preparing betel with a valve from an old truck engine. I was surrounded

---

[1] A village hut consisting of an open sleeping platform with a sago palm roof.

by some eighty adults and children who simply stared at me, watching my every movement with the greatest concentration. Huge brown eyes in tiny faces were transfigured with attention. I chucked one child under the chin which caused screams of delight and they ran in all directions, scattering like billiard balls.

They drifted back and I indicated I would like to see inside one of the houses. Roy had been keeping up a running commentary on the method of sewing together vertical strips of sago palm for the roofs. As I climbed the ladder to the sitting area, there were more screams of amazement. The dark interior of the house, completely lacking in windows, was bare except for the pad-locked door to the sleeping place. I considered for a moment all the paraphernalia collected on my journeys that was packed into my London flat. These people possess so little above what is strictly utilitarian, they must have thought me mad to inspect a dim, empty space. I climbed back down holding onto a con-venient rope and feeling rather foolish. Roy explained some of the more intricate aspects of family life as a child searched through the old man's hair for what looked like wood chips. Old women smoking pipes, their lips stained with betel, shouted comically at the children who drifted back and forth. I have never been the object of such absolute fascination and it is a disconcerting experience. I could only imagine the effect Maclay must have had on such people who had never seen a white man before and who behaved in such unexpected ways.

On his second voyage to New Guinea in 1876 he lived at Bugarlom on a promontory near Bongu. This time he cleverly brought a prefabricated house from Singapore that only required for completion a roof, stilts and the Russian flag. He enclosed the space below the floor to form a makeshift laboratory. Maclay was far better equipped and provisioned on this occasion, includ-ing among his effects Turkey carpets, Bordeaux and champagne. He planted a garden with coconuts, maize, fruit and vegetables; his European farm animals caused amazement and fear. On his first visit, Maclay noticed that the villagers of Bongu had a fashion for smearing themselves in red paint and producing strange carved artworks called *telums*. Fortunately, he sketched some of these statues of ancestor heroes, so characteristic of the Astrolabe Bay area. He continued collecting skulls, lamenting the missing lower jaws. They were considered so precious that the relatives of the deceased preferred to wear them around their necks. It is a mystery to me that the remarkable life of this pioneering scientist is so little celebrated outside of his native Russia.

Clouds were beginning to form over the mountains which gave a sense of urgency to our return before rain flooded the rivers and swept us away or made the roads impassable. On the outskirts of Madang we rounded a bend in road and suddenly a huge crowd of people were spilling out from a festivity. I could see dancing groups. It looked like a large event.

'Roy, let's stop. I'd like to have a look!' They both seemed curiously reticent about the idea, but we parked the Toyota in the milling crowd and carefully locked it.

'It's some sort of political rally. The politician owns a beer garden, I think. Now what sort of political platform is that?'

'A sure-fire success I would think!'

As we approached I could see at least three groups of dancers. A group of men and young boys playing drums and decorated with croton leaves were encircled by bare-breasted women smeared in rust-red paint dancing energetically, chanting, and shaking long, dried palm fronds. The circle expanded and contracted as their magenta grass skirts swished with the movement. All wore spectacular crowns of white feathers. The women were from Yabob, a village between Bilbil and Madang. They were performing a dance known as *Maimai*, which is a poetic expression of the cool wind in the early morning and late evening, a wind that soothes the spirit. Roy stood behind me surveying the crowd in his sunglasses like my personal bodyguard.

Another group, mainly of children, were wearing fantastic

pink-and-white headdresses which looked dazzling with the sun backlighting them through golden bird-of-paradise feathers. Old men sat in the centre of the dancing circle and the children seemed close to exhaustion. A burst of energetic chanting would wind down and then pause, only to renew itself with fresh exuberance. This was the song known as 'Mormor' from Ohu or 'Butterfly' village. It explains their vivid decoration. Yet another group were dancing with huge painted fish affixed to the tops of long poles. I was told that this dance had been purchased from the Aimaru coastal people for thirty pigs.

We left the celebrations to the sound of breaking glass, passed along the picturesque circuit of Coronation Drive to the hotel, and concluded my tribute journey to Nikolai Miklouho-Maclay. Inadvertently, he had prepared the ground for colonial expansion and all his worst fears would be realised, as one exploiting power succeeded another. Yet he remains one of the true spiritual fathers of modern Papua New Guinea. Bronisław Malinowski considered him a revolutionary anthropologist, a 'new type', who had made the radical decision to live among the people he studied so as to become almost one of themselves. Another admirer was Anton Chekov. He was interested in anthropology, and had studied the Udegeian people along the Amur river on the border with Manchuria. He took a keen interest in Miklouho-Maclay and recognised his enduring fame, courage and incorruptibility.

The 'Baron' subsequently had a rich and varied career in Australia, founding an important marine biology station near Camp Cove in Sydney in 1881. He travelled Europe lecturing and attempting to gather support for his ultimately stillborn Maclay Coast Scheme. Only today is this grand concept of the regional independence for indigenous cultural groups and a 'native Great Council' recognised as truly visionary. More patriotically, he wanted to see the establishment of a Russian colony in New Guinea. All his best efforts to prevent German annexation failed and he ended up a crushed spirit; his hair turned grey and his health broke down.

After returning to Russia early in 1886, his company was sought by all the luminaries of the day. Ivan Turgenev became a convivial companion. Leo Tolstoy felt many of his own scientific theories had been proved by Maclay and begged him to publish his diaries as a contribution to 'the science of how people may live with one another'. Grand Duke Nikolai Mikhailovich visited him in St Petersburg during the last stages of his final illness. A multitude of symptoms had ravaged his body, and the rigours of the numerous tropical diseases he had contracted over the years concealed the true cause of death. In his diary he described the terrifying effects of a common 'paroxysm' of malaria.

> In the change from chill to heat I suddenly felt a strange illusion of my senses. I definitely felt that my body was growing, the head enlarged more and more till it reached nearly to the ceiling, the hands became enormous, the fingers became as thick and as big as my arms. I felt, at the same time, a feeling of the enormous heaviness of my expanding body . . . The paroxysm was so violent and the sensation so strange that I will be a long time forgetting it.

Despite being submerged in this frequent hell, he had worked, written voluminously, studied, dissected, observed and cared for the health of the local people. Nikolai Miklouho-Maclay, debilitated by sacrifice and pumped full of morphia, died in his wife's arms with all the dignity he commanded in life, late in the evening of 14 April 1888. He was just forty-two.

A descendant of Tui, Maclay's first Papuan friend (who loudly lamented, 'O Maklai, O Maklai' at each new audacious deed he contrived) asked at the one hundredth anniversary celebration of Miklouho-Maclay's arrival at Garagassi, 'Why did Maclay let the Germans come here?' as if the explorer was somehow able to prevent it. I spent the morning at the Catholic Divine Word University attempting to answer this question. The only air-conditioned room in the building protects the historical books, many of which are rare German volumes. I was shown into this blessedly cool room, the old refrigeration unit dripping water from its condenser onto the mouldy concrete. The staff and students suffered through the heat, but at least history was comfortable and safe. I leafed through fascinating picture books and pages written in thorny Gothic script that chronicled the old German days.

The history of Germany in the South Seas and the rush for territory is a fascinating and largely unknown cul-de-sac of colonialism. The European presence in the South Pacific had increased in the late nineteenth century, largely as a result of changes in the method of production of coconut oil, used at that time in a variety of industries, particularly cosmetics. From 1850 German traders had been arriving in ever-increasing numbers, and Samoa was used as a base to expand German influence in New Guinea. The German banker Adolph von Hansemann had assisted in financing the Franco-Prussian war in 1870 and was casting around for a fresh project. He considered that a New Guinea colony would enrich Germany's prestige throughout the world. Perhaps he also had a premonition of the fine cigars that would be produced there from the excellent local tobacco. The Deutsch Neu Guinea Compagnie was formed in May 1884 and the directors chose Dr Otto Finsch to pioneer the settlement.

Otto was born in Silesia, qualified as a doctor of medicine, and was a largely self-educated naturalist, ornithologist and ethnographer. His many drawings and collections of Melanesian art (long before it was known as such) now enrich museums from New York to Vienna. He had met Miklouho-Maclay in Berlin and knew aspects of his studies on the Maclay Coast. They met again in Sydney in 1884 at a meeting of the Linnean Society, where they managed to agree on certain aspects of the controversial Garagassi bandicoot. However, an element of duplicity crept into their relationship. Finsch mentioned to Maclay that he was setting off soon for New Guinea on a scientific expedition but failed to mention it was also a reconnaissance mission for

German annexation to be set up under the auspices of the Neu Guinea Compagnie. Maclay in his generous, unsuspecting fashion briefed Finsch on what he could expect to find on the Maclay Coast. He taught him some of the indigenous language from the village of Bongu and some Russian words he had introduced into it. If Finsch would but use the expression 'Aba Maclay'[1] he would be guaranteed a warm welcome as his 'brother' by the local people.

The steamship Samoa commanded by Captain Dallman and carrying Dr Finsch was disguised as a trading vessel. She sailed from Sydney to Mioko in the Duke of York Islands, a small group lying between New Ireland and New Britain. Finsch later headed for Port Constantine on the Maclay Coast, his true motives now becoming clearer. The villagers cheerfully built him a house near Garagassi as Maclay's 'brother'. In October the German corvette Elizabeth joined the party. The German flag was unofficially raised over this hut for the first time on the New Guinea coast on 17 October 1884. Finsch wrote of the local people: '. . . in every face there was joy.' They clearly had no idea what was in store. The Reichsflagge was officially raised at Mioko claiming the Bismarck-Archipel for Germany by Kapitan Schering of the corvette Elizabeth on 4 November the same year.

Maclay was always fearful of the annexation of land. In October 1883, apprehensive that the second largest island in the world was about to be taken over by the Queensland Police Magistrate, he sent a telegram to Lord Derby, Secretary of State for the Colonies: 'Maclay Coast natives claim political autonomy under European Protection' (presumably Russian). In January 1885 he cabled Bismarck: 'Maclay Coast natives reject German annexation.' But the British and the Germans were already coming to an agreement as to how New Guinea might be divided between them. Anglo-Russian relations were disintegrating and spies were already playing the 'Great Game' in Central Asia.

---

[1] Bongu language for 'brother or friend of Maclay'.

Maclay's own raising of the Russian flag over Port Grand Duke Alexei fifteen years earlier was discounted as it had been by the merchant marine and not an official naval operation. On his return to Russia, the 'Baron' was invited by Tsar Alexander III to the palace at Gatchina in St Petersburg to meet the Minister for the Navy to discuss the possible setting-up of a Pacific naval station. He petitioned for 'Russian protection, with recognition of their [the indigenous people's] independence'. This heroic rear-guard action on behalf of the 'Archipelago of Contented Men' is the reason his memory is held in such reverence on the Rai Coast to this day. His deep, particularly Russian humanism, makes him a venerated figure in the country of his birth.

The Neu Guinea Compagnie began commercial exploitation of the north coast of New Guinea, properly known as Kaiser Wilhelmsland, and the nearby Bismarck Archipelago in 1886. It was protected by Imperial Charter but was made responsible for all administrative and judicial decisions, thus neatly removing the onus from the German government and planting the seeds of its future difficulties. Despite his pompous pronouncement, 'I will have no colonies,' a mountain range, archipelago and sea were named after Chancellor Bismarck with almost indecent haste. From the capital Finschhafen on the Huon Peninsula, the Compagnie administered the islands of Neu Mecklenburg (New Ireland), Neu Pommern (New Britain), Neu Lauenburg (Duke of York Islands), later Buka and Bougainville, in addition to Kaiser Wilhelmsland, their territory on the north coast of the mainland. Unlike the British, whose interest in New Guinea was mainly strategic and missionary, the Neu Guinea Compagnie hoped to exploit the cheap local labour force on their plantations to further their economic interests and establish profitable trading companies. Neither nation had much interest in the feelings of the indigenous population towards this invasion of their land and culture, although later German colonial policy went a significant way towards recognising Melanesian cultural differences in the administration of justice.

For the first couple of years matters proceeded well under the administration of Jan Kubary, 'the Lord God of Astrolabe Bay', yet another Polish national. Coconut plantations were laid out, tobacco, rubber, cotton and kapok were assiduously cultivated. The sea surrendered trepang (bêche-de-mer), tortoiseshell and pearlshell. But disease and death demoralised the employees and violence was an ever-present reality. A nephew of Bismarck who worked in the colony, Stephan von Kotze, wrote in his grimly entertaining book *Aus Papuas Kulturmorgen* (*Cultural Mornings in Papua*) that the unmarked cemetery mounds and the hotel were the most visited places in the capital. The customary approach to work was immediate postponement of any given task followed by a period of seriously hitting the bottle. The company flag in the German colours of black, white and red featured a black lion clutching a red torch, its appearance cynically referred to by the employees as the *blutige knochen* or 'bloody bone'. It became a macabre symbol of their doom-laden existence. The term ultimately came to refer to the *Kaiser* beer in which they drowned their *tropenkoller* or tropical madness. Visitors to Finschhafen commented on the employees' desire to simply 'sit and soak'.

Von Kotze reports the following vignette on being asked a question by his great leader.

'Where is secretary Muller today? I imagine at the hotel!'

'Herr Muller died this morning,' Kotze replied sadly.

'*Ach so*! So it is not as bad as I thought! But today is mail day, so I forbid the funeral beer!' And so it was that poor Herr Muller was carried to his grave by local tribesmen without a Compagnie escort.

Problems over the acquisition of land could have been anticipated from the first report in 1886 by the Neu Guinea Compagnie which arrogantly comments:

It may therefore be confidently assumed that there are large areas of uninhabited and unclaimed land which falls within the domain of the Company.

The local people entirely misunderstood that enormous tracts of land were being 'purchased' *permanently* for a few trinkets. This confusion finally led to violent confrontation. Scenes took place reminiscent of Conrad's *Heart of Darkness* where a warship fires incomprehensibly into a continent as a gesture of French imperial frustration. In the summer of 1893, as a result of severe disturbances, the SM Cruiser *Sperber* fired a number of shells 'at the enemy position' – into the jungle. 'However the shells, fired from a distance of 5–7 km, had the effect of killing only one native and paralysing another with fear,' reads the po-faced report for that year. Maclay had already warned the villagers about the arrival of the destructive 'white ants' from across the sea.

The climate was lethal and the capital was moved a number of times further up the coast. The adoption of the Mexican silver dollar current in Singapore, as well as the use of British currency and Dutch guilders, led to commercial confusion. The value of financial transactions fluctuated markedly, leading to wide abuse. By 1894 distinctive copper and silver coins featuring the bird of paradise had been issued, gold coins being introduced by the Compagnie the following year.

But gold coins could not save it. Mindless bureaucracy from the Landesbureau headquarters in Berlin, such as voluminous ordinances concerning the correct erection of flagstaffs and the exact constituents and utilisation of artificial manure, contributed greatly to the disease-ridden, white-suited employees' hatred of the Compagnie. It became fatally inefficient and the staff chronically demoralised by the catalogue of deaths from malaria, dysentery and smallpox. Its ships were wrecked on the treacherous reefs and the local villagers refused to work. The employees particularly feared the treachery of those whom they termed *Spitzkopfe* – people whose heads were bound at birth resulting in elongated skulls. Indentured labourers from China and Malaya committed suicide in droves. *Strafexpeditions* did little to endear the Compagnie to local people – one savage reprisal resulted in over forty deaths to revenge the killing of a solitary German bird of paradise collector. When the Kommissarischer Landeshauptmann (Administrator) of the company, Curt von Hagen, was killed in 1897, the two ex-police murderers from Buka, found drowned in the Gogol river, were decapitated as a reprisal and their heads vengefully displayed in public. *In fine*, this early colonial experiment in feverish exploitation and iron-handed punishment was an unmitigated disaster.

In spite of this there were some lasting achievements. The great river systems of the Sepik and the Ramu together with the mountainous interior were mapped by fearless German explorers. But administrative problems persisted, and by 1899 the Imperial Government had relieved the besieged Compagnie of its onerous political responsibilities. The administrative capital of Deutsch Neu Guinea was transferred to Herbertshöhe (the present Kokopo) on the Gazelle Peninsula in the north of the island of New Britain. The Compagnie continued to trade but under redrafted Articles of Association as a commercial colonial company.

New Guinea continued to attract the cream of the world's eccentrics, many now German. The autocratic 'Baron' Heinrich Rudolph Wahlen ('Rudi'), a politically powerful planter, kept a harem of indigenous young girls who had been gifted to him by custom at his improbable stone-built Rhine castle, Schloss Wahlenburg, erected on inaccessible Maron Island in the Hermit Group just below the equator. Here he imported fine horses, deer, kept a domesticated shark and a species of fish that were fed on tropical butterflies. His nubile lovers drank vintage French champagne and played croquet and tennis on the manicured lawns. This remarkable coral atoll boasted early telephones, electricity and air-conditioning. It has been asserted that he imported golden Polynesian girls and fierce warriors from the Solomon Islands to further his experiments in eugenics. This formidable character finally married a Swedish marquis's daughter and was appointed Swedish Consul to New Guinea, remaining in office up to the beginning of the Great War.

The Imperial Governor himself, Dr Albert Hahl, was not

averse to dressing in a grass skirt and having his naked torso rubbed down with sweet coconut oil. Travelling by pony carriage, he was a charming social companion, fond of picnics and reportedly a brilliant conversationalist. More constructively, he supported a high degree of autonomy for the local people against the planters, studied their many languages, and was sympathetic to their claims for justice. His marriage to the aristocratic Baroness von Seckendorf did not appear to hamper this lively Bavarian.

These Pacific possessions had no strategic significance for Germany and posed none of the profound difficulties encountered in German Africa. According to the last government doctor in Yap, Ludwig Kohl-Larsen, they were considered by Berlin more as a sign of national maturity, a coming of age, 'beautiful jewellery, which is of value only to its collector and does not yield a profit, but which gave us pleasure'. Consequently, following the rapacious period of the Compagnie, the Pacific empire flowered under a more relaxed regime, allowing Germany to take its rightful place beside the 'enlightened' British, the French and the Dutch as the available world was carved up and distributed. But as the German anthropologist Dr Richard Thurnwald bitterly observed, 'All the best parts of the world were taken long ago.' The Pacific administrators were not drawn from the aristocracy as in Africa, but from the professional and university-educated middle classes. This resulted in a far greater tolerance towards the indigenous population and a maturing interest in the ethnography. Allied to this pursuit, there grew a desire to preserve the culture and collect exotic works of art for the great Berlin museums.

Many European men fled the sackcloth of Protestant Europe for these exotic islands. The German authorities did not normally grant residence to European women, so the white men married local women or enthusiastically embraced the joys of polygamy with virgins purchased two or three at a time, or if short of cash, exchanging them for a rifle or axe. They became passionate *amateurs* of Pacific sexuality. In 1903 a nudist colony of sun-worshippers and avant-garde 'proto-hippies' led by a Berlin musician named Engelhardt lived a pagan lifestyle on the remote island of Kabakon in the Duke of York Group. Sexual jealousies flared, and the limited diet of bananas and coconuts devastated their health (Engelhardt was nicknamed Mr Kulau or Mr Coconut by the locals). He was murdered in mysterious circumstances and the colony soon disintegrated.

As might be expected in this riot of licentiousness, a state of undeclared war existed between the libidinous German settlers and the austere Lutheran missionaries, their moral conscience. The religious were intent on saving souls and curbing forbidden passions, yet often fell victim to the temptations of the flesh themselves. A sexually permissive hothouse of some magnitude evolved in the Bismarck Archipelago which profoundly affected the lives of all who chose to settle there. The British plantation managers were often abandoned misfits who lived miserable and lonely lives, all but disowned by their companies. The German managers, in contrast, were well paid and lived a civilised lifestyle of some distinction and considerable hedonism. The level of German integration with the local female population was significantly greater than the more puritanical settlers from other colonial powers permitted themselves.

On my last evening in Madang, a large number of health workers descended on the hotel for a conference, which sent me scuttling for my room. Most of the delegates were obese and smoking heavily. Squadrons of flying foxes were squeaking and frantically seeking their roosts as I packed my bags for the flight to Kavieng in New Ireland the next morning. Later, I sat out on the terrace overlooking Dallman Passage and fell into conversation with a visiting magistrate from Mount Hagen in the Highlands. He was short and powerfully built, a dark forbidding face, stern with justice, but he had an excellent command of English. Among the remarkable collection of anecdotes that drifted through the tropical night, a case of assault and sorcery caught my attention. It illustrated that belief in the occult is still widespread, despite the influence of Christian missionaries.

The drama involved some villagers on the Sepik river. Whenever an eagle flew over a particular house on the river, a villager's wife would vomit blood. This was a terrible omen for the family. She had been engaged to a man in Madang, but had jilted him to marry another who lived in this Sepik village. The jilted lover arranged a beating, then an evil spell to be cast on his former paramour. She suffered this torture until fear of madness approached. The family desperately wanted to rid the woman of this terrible *onim*, so an old sorcerer in Madang was engaged to expel the evil spirit. The sorcerer warned that the process was very dangerous and that either the wife or he could die. The family decided to go ahead.

The exorcism was to take place at midnight when everyone was asleep. The sorcerer placed his head on the pillow and fell into a deep slumber, but some in the house were awake and spied on him. His face became transfigured with emotion. Footsteps were heard descending the steps of the house, indicating that the spirit was departing the body. The spirit flew over the Sepik river and destroyed the evil spell in a fierce battle high above the crocodile-infested waters. Meanwhile, the body of the sorcerer in Madang strained with tension, his face grimacing in pain and exertion. Footsteps were then heard ascending the stairs, clearly the spirit returning to the body. When the sorcerer awoke totally exhausted, he was able to describe in detail the river and the position of the huts in a village he had never seen. The wife never vomited blood again and a ginger plant, unknown in the area of Madang, miraculously sprouted at the base of the steps.

The magistrate drifted away after this story, swallowed up in the velvet gloom. I sat there for some time pondering this tale, watching the moon dancing on the placid water. The flocks of several thousand flying foxes had given way to an isolated few flapping gothically overhead through the silver light. Nowhere I knew in the world was remotely like this astonishing country.

10. Minotaurs on Gilded Couches

The long flight to Kavieng gave me time to read, and I delved into my notes taken from Richard Parkinson's account of the Bismarck Archipelago in his monumental study *Dreissig Jahre in der Südsee*[1] published in 1907. He was a pioneer planter, collector of artefacts and chronicler of traditional life. So few Europeans had visited the area at this period, the record is an invaluable portrait of societies before they were overwhelmed by European influence. We would be flying via Manus Island, the largest in the scattered group known as the Admiralty Islands. Parkinson considered the inhabitants among the most interesting of the archipelago and the Moanus tribe by far the most intelligent. He commented on the delicacy of the women and their fondness for ornate and finely-worked ornamentation. The men were not to be neglected, although their approach to dancing decoration was, on certain occasions, minimalist to say the least.

Finally, the curious penis shell must also be reckoned among the decorative items; covering the glans of the male member in battle or in dancing. This shell is always a medium-sized *Ovulum ovum*, with cross-hatched patterns incised on the outer white surface. The internal spiral is partially excavated, and the glans and the foreskin are squeezed into the fissure thereby formed. The shell

---

[1] *Thirty Years in the South Seas.*

is always carried in a small woven pouch on a cord around the
neck or under the arm . . . so that it is always in readiness.

I had read of the penis-swinging dance performed at large
exchange ceremonies and in past battles. The men swung their
penises with the shell attached violently to and fro, catching the
shell between their thighs and then catapulting the penis upwards
in a wild arc. They taunted the receivers of the exchange or the
enemy, encouraging them to fight or compete in the power of giving. In other parts of the islands, the warriors cultivated 'macro-
dontism' (literally, 'the cultivation of large-scale teeth'), a practice
that allowed tartar to build up massively on the front of their teeth
from chewing betel and lime. These growths even pushed out the
lips, and were carved into decorative designs with obsidian tools.

As we flew along the coast of the largest and most mountainous
island of the group, we passed low over many tropical atolls cast
in a sea of pale jade. Huts on stilts hugged the shore. As I was
admiring this idyllic scene, the plane was suddenly engulfed by
dense cloud and torrential rain. The wheels threw up huge sheets
of spray as we touched down.

Before the Air Niugini jet could come to a halt, a group of
local people streamed across the tarmac and ran under the
moving wings, almost entangling themselves in the undercarriage.
The plane shuddered to a stop, packages flew across the cabin
and a stewardess sprawled in the aisle. Two women became
detached from the crowd and threw themselves onto the bitumen,
crawling and writhing at the base of the gangway. They were
beating their heads with pieces of wood. Blood and rain streamed
down their tattooed faces. It was only then I saw the coffin being
unloaded, draped in a white sheet. One of the women sprang up
and with a piercing wail launched herself at the casket, waving
a white handkerchief. The bearers waited as she clawed hysteric-
ally at the wood, the sheet sliding to the ground. They then
moved off towards the bleached-blue Department of Health van
parked nearby and loaded their grim consignment. Wind tore

through the palms, the fronds thrashing themselves in the storm. She climbed into the cramped interior, and through the steel mesh covering the windows, I could see she had collapsed over the coffin. The van was sardined with other mourners, then the door was slammed shut. It lurched down the muddy road through a plantation and disappeared in the murk.

Chopin's 'Raindrop Prelude' was playing as a bizarre accompaniment to the drama over the cabin speakers in the plane. Such extraordinary moments evolve totally without warning in this unpredictable country. People boarded as if nothing had happened; a smiling Melanesian child in a white shirt and red bow tie sat next to me and chatted merrily away. I had briefly landed at Los Negros Island in Manus Province en route to Kavieng, the provincial capital of New Ireland Province.

On a map, New Ireland has the appearance of a flintlock pistol recently recovered from centuries on the seabed. The Dutch navigators Jacob le Maire and Willem Corneliszoon Schouten in the ships *De Eendracht* and *Hoorn* navigated the east coast for the first time in 1616. The butt lies along the St George's Channel, so named by the Englishman Lieutenant Carteret when he sailed through this passage separating New Britain and New Ireland in the storm-battered *Swallow* in 1767. This old 'sloop of war' had fourteen cannon and twenty-two deck officers, a vessel entirely unsuited to the purposes of exploration. Carteret named his discovery Nova Hibernia. The encrusted barrel forms the spine of the island, some 290 miles in length, leading to the muzzle and the capital Kavieng. This slender, rugged land is intersected by valleys, streams and torrents that tumble down from the rainforested mountains in the wet season. The East Islands spill from the muzzle like discarded shot. New Ireland has not been studied ethnologically in any exhaustive academic sense, which makes the island a rare and relatively-unknown case. Complex rites known as *malagan* are carried out on the occasion of death, which are essential to the social, aesthetic and economic fabric of the society.

From the air, the reefs look like lace ruffs encircling islands of green velvet. Oases of milky green shelter lakes of turquoise giving way to the stippled brown of the underlying coral shelves. Around the airstrip, vast plantations of coconut palms replace the oil-palm plantations seen in Alotau. The terminal is a tidy building with carved posts supporting the roof – robustly-worked birds, skulls, fish and snakes writhe up and down the poles. A large crowd of local people had gathered to meet the plane and stared intently at the rare visitor. I had scarcely unloaded my bags when I heard the German of the original colony spoken. Clearly, the foundations of the outpost remained. An attractive blonde German woman was gathering together a diving group who were heading out to Lissenung or Paradise Island. The road into town was full of potholes, but it was a pleasant enough drive through plantations of formally-spaced coconut trees and picturesque hamlets.

I had arranged to stay in Kavieng at The Malagan Beach, located in an idyllic position on the bay. My room was built of traditional materials and opened directly onto a secluded beach with pandanus palms and old kerosene trees gnarled into fantastic shapes. Canoes with faded blue hulls were drawn up under the spreading branches. Across the water, the low island of Nusa with a long strip of white sand and dense stands of coconut palms hinted at mysterious activity behind the walls of tropical vegetation. Distant drumming floated over from the shore. The bar was deserted save for a young villager mowing a swathe of lawn, a blue cocktail umbrella stuck nonchalantly in his hair. I sat in a wicker armchair in the shade, out of the gruelling heat and humidity, taking refuge in that solace of all men in the tropics, cold lager beer.

My panama had slipped down over my face and I awoke with a start. A group of local children were standing around me on the beach and giggling wildly. Their striking blonde hair against their dark skin was beautiful. The day was fading, the harbour had become a mirror of gold, an outrigger thrust like an abstract sculpture towards the sky, clouds of folded silk slashed by the sun were slipping below the line of palms. The vision only lasted a moment, for as soon as I adjusted my hat above my eyes the children ran away, screaming with excitement, down the beach towards their village by the shore.

'You're quite an attraction in that hat.'

Two men were having a beer under the spreading kerosene trees. A European was smartly dressed in a tropical silk shirt while the other looked like a young local man. I wandered over and introduced myself.

'The sunsets from this beach are the most beautiful in the islands,' the European murmured dreamily.

Grey clouds reclined like minotaurs on gilded couches above the horizon. The silhouette of a woman silently paddled a canoe across to her island. Fishermen in antiquated goggles stood waist deep in the shallow water and threw their nets in wide embroidered arcs, drawing them in to catch their evening meal. The silent equanimity of the scene served only to magnify my tropical lethargy as the sun languished.

Justin Tkatchenko, my newest acquaintance, is a formidable young Australian betraying all the persuasive charm and grace of the successful Eastern European entrepreneur, an outcome perhaps of his Polish and Ukrainian parentage. A rather cynical outlook was in keeping with his fastidious appearance and trimmed nails. He wore a heavy gold watch and other gold accoutrements with surprising discrimination and taste. I had met such sophisticated Slavs before in the *faux-luxe* hotels of Białystock on my travels in the northeast of Poland, and in Kiev in the Ukraine. He had never visited Eastern Europe, but the genetic influence was inescapable.

'I saw this fellow in an extravagant hat and thought, now what in God's name is he doing here? I must talk to him,' he smiled and shook my hand.

'Travelling the islands and writing.' I had a sudden instinct to be open about my intentions, although this is usually not advisable when travelling.

'Ah, well, you must have a drink and dinner with us. We're having barbequed fish and crocodile on the beach tonight.

Charles 'Cannibal' Miller was born in Java and became an air ace during the Great War. His wildly sensational accounts and photographs of cannibal ceremonial in his book *Cannibal Caravan* (Museum Press, London 1950) are a fine example of the 'machismo genre' of male writing about New Guinea. Characteristically, he preferred oriental slippers in the jungle to boots.

Men performing the penis-swinging dance on Manus Island, one of the Admiralty Islands in the Bismarck Sea. This sensual dance is offered as a challenge during large-scale wealth exchanges. The men attach an *Ovulum ovum* shell to the glans of the penis and swing it violently to and fro, catching the shell between their thighs and then catapulting the penis upward in a wild arc.

Left Charles Bonaventure du Breil, Marquis de Rays (1832–93), self-proclaimed King Charles I of *La Nouvelle France* and Emperor of Oceania. In the 1880s he perpetrated an audacious fraud selling useless land to hopeful colonists in the malaria-infested fever-hole of Port Breton on the southern tip of New Ireland. The ensuing debacle cost hundreds of lives, evolving on a Napoleonic scale of absurdity and madness.
Right 'Queen' Emma (1850–1913) was born Emma Eliza Coe of a Samoan princess and an American father. This formidable woman was proud, arrogant, sexually alluring and possessed brilliant business acumen and political judgement. Her adventures in German New Guinea among cannibals, malaria and rough traders, while building a huge copra empire, were as astounding as they were improbable.

Gunantambu, Queen Emma's famous mansion at Ralum near Herbertshöhe (Kokopo) in Neu Pommern (New Britain), was famous throughout the Pacific for its luxurious furnishing and appointments. She entertained on a lavish scale with enough champagne to bathe in, feasts of twenty-eight courses for the German naval officers, theatricals, dance, and music. The house was destroyed during the Japanese occupation, 1942–45.

Impressive men from the island of New Hanover photographed by the Hungarian scientist Lajos Biró in the late 1890s. Of particular note are their hairstyles, pierced ear-lobes, colonial belted *laplaps* and elegant necklaces with *kapkap* discs. The raised scars of body decoration were formed by cutting the skin with obsidian tools and introducing sand into the wound.

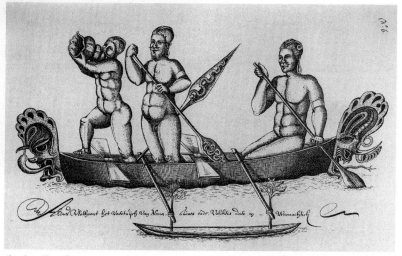

Shark callers from New Ireland wearing elaborate hairstyles and paddling an ornately-decorated canoe that contains the distinctive 'propeller' and conch shell used for shark hunting. This superb engraving is from the Journal of the Dutch navigator, Abel Janszoon Tasman, who explored the islands of New Guinea in 1642 for Anthony van Diemen, Governor-General of Batavia.

The Melanesians of German New Guinea were fascinated by German patterns of behaviour, and some dressed in European style. The Germans in their turn became passionate *amateurs* of Pacific sexuality and embraced polygamy with enthusiasm.

The author and his guide perched on the wreckage of a Japanese 'Betty' bomber submerged in ash, Rabaul, East New Britain Province. In what was one of the foremost theatres of the Pacific conflict, aircraft wreckage and other detritus of war continue to be found in the jungles and on the seabed surrounding Papua New Guinea.

Oh, yes, this is Ariel, a friend of mine from New Ireland.'

A gently-spoken young islander took my hand in a soft grasp. Intense eyes gazed at me from a curiously vulnerable yet dark face that pressed forward in its need to communicate on a level somewhere beyond the merely articulate.

'You must come on Friday to Tsoi Island. We are opening a supermarket.'

I must have appeared diffident at this less-than-exciting prospect and glanced over at the fishermen.

'It'll be like nothing you've ever seen before. This'll be the first shop *ever* for the people of New Hanover. No need for them to come all the way into Kavieng. We've invited dance groups from all over PNG, choirs from distant islands and sorcerers who will perform magic to protect the store from evil.'

Justin's enthusiasm for the project was infectious and so I agreed to go. The distressing subject of banana boats arose, as it was a two-hour voyage from Kavieng. The problem was tactfully shelved until nearer the time.

'Do you believe in *puripuri*?' Ariel asked, his eyes drifting mysteriously above my head out of reach over the bay.

'Sorry? I don't know the word.'

'Sorcery. Everyone fears them, the sorcerers. I wear a charm to prevent bad spells cast on me. If you believe you will die, you *will* die.' His voice was silky as he held the charm forward for me to inspect.

'Well, I do believe in evil.'

'We have a sorcerer to make *puripuri* against evil coming into that store. He eats glass bottles. You will see!' His mood was suspended somewhere between elation and trepidation.

'While we were building the supermarket we uncovered the grave of five sorcerers buried head to head in a star pattern. A fire had been made in the central area between the skulls. Let me show you.' Justin began to draw skulls and bodies in the sand and explain the exact position of these macabre discoveries.

'The skeletons lay along the points of the star like this. The people don't know the significance. It's all been forgotten, unfortunately.' He appeared disheartened. Ariel became agitated at the mere mention of these things, looked uncomfortable and his eyes clouded over.

'Sorcerers are trained to it, not like the flying witches who inherit their skills. Of course, sorcery is illegal these days.'

The whole idea of sorcery being 'illegal' was still a novel one for me. We sat in silence for some time listening to the waves lapping on the shore, preoccupied for a moment with the claims of the occult.

Justin had begun his early career not among skulls and bones, although, ironically, he now ran a funeral parlour and embalming service in Port Moresby as part of his business empire. A talented son of a wealthy Eastern European family, he had begun his colourful life in Papua New Guinea in 1993 as the Curator of the National Capital Botanic Gardens in Port Moresby, specialising in breeding rare orchids. He presents a gardening programme on television, runs a landscape garden business, owns the upmarket artefact shop in the luxurious Park Royal Hotel in Moresby, and has set up an excellent restaurant near the Royal Papua Yacht Club. He told me that in recognition of his services to the community he was awarded the British Empire Medal – Civil Division, but had chosen to receive it in Port Moresby rather than London. More controversially, he had been a senior policy advisor during the brief interregnum in 1997 of the former Prime Minister Bill Skate, who had been forced to resign under accusations of corruption.[1] For someone

---

[1] Bill Skate, the controversial leader of the People's National Congress Party, was elected Prime Minister of Papua New Guinea in July 1997. He was forced to resign from office under threat of a vote of no confidence in July 1999 concerning an alleged cash for recognition deal with Taiwan. After the violent and chaotic election campaign of 2002, he was re-elected as the MP for the National Capital District. His finest achievement was to secure peace on Bougainville Island.

not yet thirty this was an impressive catalogue of achievement.

The fire had been burning for some time and all types of fish were cooking shrouded in perfumed smoke – sweetlip, squid, swordfish, red emperor and a marooned piece of sirloin steak. The crocodile was disappointing, rather rubbery in texture and bland in flavour. I sat listlessly in the balmy air and cool breeze that came off the water. Erratic drumming drifted across the bay from Nusa Island. Tiny lights glinted within the darkness of the plantations.

When the first copra trading settlement on Nusa was established by the German trader Eduard Hernsheim in 1878, New Ireland was notorious for the ferocity of its cannibals. The tribes lived in an atmosphere of constant hostility. Missionaries kept detailed records of their gruesome experiences. The Reverend James Chalmers, who was eventually eaten by the Goaribari, describes in his book *Work and Adventure in New Guinea* practices following death that must have seemed horrifying to the Victorians. He observed that in some parts of the country fires were lit at the head and foot of the corpse, the skin was stripped away and the grieving wife, husband or son would smear the 'juices' over themselves before the smoked body had fully dried out. As death was almost always considered to be the result of a malevolent spell, revenge was exacted on their neighbours. In search of an occult 'sign', blades of grass representing nearby villages were spread out and watched closely for when a fly might land. When an insect inevitably alighted on a particular blade of grass, the inhabitants of that village would be massacred.

The formidable Wesleyan missionary, the Reverend George Brown, addressed the Royal Geographical Society in London in 1887 and reported that on one visit to a New Ireland village he had counted:

> thirty-five lower jawbones suspended from the rafters, most of
> which were blackened with smoke; but some of them were quite

clean and had not long been there. A human hand, smoke-dried, was hanging in the same house; and outside I counted seventy-six notches in a coconut tree, each notch of which, the natives told us, represented a human body, which had been cooked and eaten there. The name of the chief was Sagina, which means 'Strong Smell' and was given him because the smell of cooked pork or human flesh was said to be always perceived in his village.

Yet the Reverend George Brown underwent a remarkable 'conversion' and by 1910 had decided that, 'Many cannibals, indeed, are very nice people . . .'

Throughout my travels in the islands I noticed a tolerant integration of indigenous beliefs with Christian principles – all except one activity, considered to be the most horrific manifestation of the 'fallen state' of man, the practice of cannibalism. The European history of Papua New Guinea is littered with nauseating cannibal stories.

During my journey I thought a great deal about the anthropophagi as they shook their bloody bones throughout the historical account. The cannibals of the South Seas, Melanesia in particular, became legendary for their singular ferocity. Isolated tropical islands were transformed into symbolic hothouses of the European imagination, all supporting 'monstrous rites'. The practice is surrounded by a profound fear of the dark side of our human psyche, yet simultaneously evokes a prurient fascination.

The word 'cannibal' has a particularly complex semantic origin, deriving from the Arawak *caniba*, and probably originating in what were believed to be the anthropophagous horrors of the Canibales or Caribes, a fierce nation in the pearl islands of the Lesser Antilles situated off the coast of Brazil. Columbus commented of the inhabitants, 'Human flesh to them is a command viand.' It has recently been discovered through the discovery of gouged bones that cannibals roamed Scotland in prehistoric times. Some scholars see all societies as symbolic cannibals, while others judge it as the troubled response of civilised man to deeply buried desires he instinctively recognises in the 'savage' races. In an imperial context, colonialism *is* cannibalism. For the Marquis de Sade, a consummate breaker of tabus, the cannibal is the truth stripped bare, the erotic foundation of the European finally revealed. Voltaire makes the cannibal a woman in his *Dictionnaire Philosophique*: 'I asked if she had eaten men; she answered very frankly that she had done so.' In over forty years in New Guinea, the pragmatic Lieutenant-Governor, Sir Hubert Murray, wrote that he could come up with no rational objection to it. Scholarly written analyses are of recent date and expanding rapidly. The subject is fascinating.

On the afternoon of 23 November 1773, the astronomer and meteorologist William Wales stood on the deck of Captain Cook's ship *Resolution* watching a ghastly experiment. The scientific party were attempting to determine the 'ocular proof' of cannibalism by the Maoris of New Zealand. He wrote of the experience in his *Journal* that the sailors 'shuddered' and were 'frozen into statues of Horror' at the sight and further:

> I have this day been convinced beyond the possibility of a doubt that the New Zeelanders are Cannibals ... another Steake was cut off from the lower part of the head, behind, which I saw carried forward, broiled and eaten by one of them with an avidity which amazed me, licking his lips and fingers after it as if afraid to lose the least part, either grease or gravy, of so delicious a morsel.

A fascination and deadly fear of cannibalism preoccupied all those who sailed these southern latitudes in the days of the 'savage native'.

European writers were not immune to the grotesque allure of it. Gustave Flaubert wrote excitedly to the French businessman Jules Duplan concerning progress on his novel *Salammbo* in September 1861, 'I'm just finishing the siege of Carthage, and I'm coming to the grilled kiddies.' Here cannibalism achieved legitimacy through art.

Charles Darwin records in his *Journal of Researches into the Natural History and Geology of Various Countries* that, on his visit to Tierra del Fuego, an Englishman asked a local boy how they dealt with the food deprivations of winter. He answered that they kill and eat the old women, pursuing them even into the mountains, long before they ate the village dogs. 'This boy described the manner in which they are killed by being held over smoke and thus choked; he imitated their screams as a joke . . .' On being pressed as to why they do this, the boy is alleged to have replied, 'Doggies catch otters, old women no.'

In *The Shipwreck and Adventures of Monsieur Pierre Viaud*, the narrator, Monsieur Viaud, is lost and starving in the jungles of Florida early in 1766. Despair and hunger cause his 'roving eyes' to gaze longingly at his negro slave. Viaud battles the all-too-human 'voice of pity' but shamelessly reports, 'I became a wolf again, a crocodile, an hyena!' His reason is impaired, explaining that 'hunger had gripped me in its talons, my bowels were at civil war within'. Encouraged by the supplicating, impatient and eager gestures of his female companion, Madame la Couture, together they murder the negro for his flesh, Viaud 'roaring out, at the same time, to increase my frenzy and smother his cries'. After a prayer, Madame la Couture assists him to light a fire where they roast the head of the unfortunate victim, but the intensity of their hunger dictates they 'devour it when it was but little more than warmed through'.

George von Langsdorff in his book *Voyages and Travels in Various Parts of the World during the Years 1803–1807* asserts that many cannibals 'feed upon human flesh merely on account of its delicacy, and as the height of *gourmandise*'. He claims that especially delicious is 'the flesh of young girls and women, particularly of new-born children'.

A 'nutritional argument' is advanced by many anthropologists. The Trobriand Islands of Eastern Papua New Guinea are extremely fertile, producing an excess of yams for consumption, whereas the Dobuans, situated on their volcanic rock in the nearby D'Entrecasteaux Archipelago, were deprived of adequate areas to cultivate and were thought to cannibalise their neighbours to supplement their meagre diet. In other parts of Papua and New Guinea, the protein-deficient diet of the staple sago crop was similarly regarded by Sir Hubert Murray as an 'understandable' reason for cannibal practices.[1]

One of the most likely causes of cannibalism is thought by some anthropologists to be the desire for vengeance exacted in battle. This was a popular explanation for the cannibalism of the warlike Dobuans. One Dobuan victim, while being ritually burnt alive, loudly enumerated all the enemies he had despatched in battles against the Trobrianders. In other parts of the country, trees would be hung with the body parts of defeated enemies and licentious dancing would take place before the ghastly feast.

Almost all the male writers I have encountered feign a curious macho romance associated with describing New Guinean cannibals. Such writers assume a complicit 'masculinity' as their readers shudder in gentlemen's clubs and private libraries

---

[1] A family link has been established between Creutzfeldt-Jacob disease (variant CJD – the human form of BSE or 'mad cow disease') and a fatal spongy-brain disease called *kuru* ('shivering'), contracted by cannibals in the Okapa region of the Highlands of Papua New Guinea. Women and children of the Fore tribe began to ritualistically eat the brains of relatives at mortuary feasts, partly because their diet was lacking in protein. Symptoms begin with joint pain and headaches, progressing to tremor, loss of coordination, uncontrollable laughing and dementia. The face finally freezes into a mask-like smile. Painful death follows within two years. Few men contracted the disease as there were *tabus* against men eating human flesh. More than 2500 villagers have died since cannibalism became illegal in 1957. That elderly members of the Fore are still dying of the disease indicates a long incubation period. Research on *kuru* is crucial as scientists are worried that vCJD may also have a long incubation period (perhaps thirty years), particularly if transmitted across species. The future death toll from vCJD may be far greater than originally thought.

throughout the land. They describe courageously 'being there' to witness the horrors. The books of the forgotten American writer Charles 'Cannibal' Miller, writing in the 1950s of his adventures in Dutch New Guinea, are a perfect example of the genre. We find him standing on a 'savage' New Guinea shore in a white pith helmet, loose white shirt, cigarette dangling from his nether lip, hands in pockets in an attitude of studied nonchalance, surrounded by his fearsome 'cannibal crew'. He affected bedroom slippers in the jungle as superior footwear to boots.

'Cannibal' Miller is in his element when describing a vengeance raid by the Digoel tribe of head-hunters. After accidentally splashing a bowl of blood from a freshly-murdered child on his white shirt in a stinking hut amid drums and shrieks, the heavily-armed warriors choose to fall upon an unsuspecting village of the Marind-Anims. Miller tags along in mainly a journalistic capacity. Both cannibalistic imperatives of vengeance and lust are covered here.

> In cases where the victim was already dead, the ceremony was brief. A quick slash across the throat with a bamboo knife, and then a deep cut to the spinal column as close to the shoulders as possible. When all the neck muscles were cut the head was seized in a hammer-lock and given a sharp wrench that snapped the vertebrae with a report like a pistol shot. Clawing in the crowd surrounding each decapitation would be the favourite woman of the knife wielder, and to her would be tossed the head. She would clasp it to her bosom with both arms, and nothing short of death would get her to release her hold . . . The work of harvesting the heads lasted little more than half an hour. By that time the odour of blood hung heavy on the air . . .

Can this account be believed? Certainly such descriptions, not rare in the literature, confirmed an image of New Guinea difficult to dislodge, even in modern times. It is as if the European *needed* the man-eating myth as he does the myths surrounding the death of Mozart. Cannibalism depicted as vengeance and lust became acceptable as an almost noble form of peculiarly masculine reprisal, a violent fantasy of absolute freedom from the fear of creeping feminisation.

Superstition and ignorant degradation were not fashionable interpretations until missionary activity began in the nineteenth century. Cannibals believed that the consumption of flesh allowed them to absorb the potent spirit of a revered chief or destroy the malign influence of an enemy. In one region, after killing a warrior, you were permitted to drink the water in which his heart was boiled, as long as you balanced on a coconut placed under each heel. The Reverend Samuel Marsden drew some uncomfortable parallels with divine revelation. He noted in this context the unsettling words of Our Saviour. 'He that eateth my flesh and drinketh my blood dwelleth in me and I in him.' In the Catholic mystery of transubstantiation, the bread and wine do not remain symbols, but change their essence at the consecration of the Mass into the true body and blood of Christ. In his argument, Marsden presents cannibalism as a deformed expression of Christian ritual, not as true religious observance.

The anthropophagi were probably most successful in their use of the practice as a terrorist weapon. I encountered traces of this theatrical use with my friends on Samarai and New Ireland. Naturally, they used it in a humorous sense, but how they enjoyed my discomfiture at their mock gnashing of teeth. There are numerous accounts in exploration literature of horror, disgust and shuddering aversion at pantomime cannibals feeling prospective arms, admiring white chests, licking their lips and stroking their bellies with a gourmet's anticipation of a feast. Genuine and quite justifiable horror was felt by pioneering missionaries such as the Reverend Samuel MacFarlane. In his book *Among the Cannibals of New Guinea*, he writes of tribes that severed the feet of their victims to prevent their escape, so as to keep the captured flesh fresh on the bone. He describes the use of terror by the 'renowned cannibal chief' Thakumbau:

On one occasion he cut out the tongue of a captive chief who had used it to beg for a speedy death, and jocosely ate it before his face.

The final answer to the charge of cannibalism was of course that it was 'too monstrous to be believed'. Some anthropologists deny sufficient 'ocular proof' for its existence at all. In his final summation, the Reverend MacFarlane offers us consolation, the heartening intelligence of 'the many noble qualities' of cannibals, of their 'courage, manliness and even humanity', and further as being 'distinguished for their hospitality. Indeed they are as a rule a good-tempered, liberal people ... exhibiting great skill and taste in carving.'

I woke in the early hours of the morning to absolute silence apart from the waves lapping on the shore. A power cut had rendered the island as silent as the tomb except for the ringing birdsong. Stillness prevailed as dawn broke, and I went for a swim in the lusciously warm water. Masses of dark cloud against the palest blue reared above the palms. The heat of the day began swiftly working into the island. I sat on the sand and had a breakfast of piercingly sweet pineapple, lemon-perfumed paw-paw and lady finger bananas. More beautiful children with mahogany skin and blonde hair ran down from the village and dived among the tropical fish. A delicately-built canoe glided past.

Kavieng is a slow town with a Somerset Maugham feel of somnolent colonialism. The closest meaning in English of the word 'Kavieng' I obtained was 'Gardens over there'. The colourful market on Harbour Drive under the ancient casuarinas begins early in the day. Many local people were wearing bright, woollen Highland hats despite the heat. Lips stained red gave a disturbingly louche appearance to the more decorative males. Men dress in a manner reminiscent of pirates straight off a schooner – muscular and bare-chested, stained ivory trousers rolled to the mid-calf, blue-and-white bandanas. A few had tied vines around their foreheads, with aromatic leaves tucked into the centre. Their skin has more gold in it than other parts of the country. Women from nearby villages sat on mats with meticulously-arranged bunches of peanuts or areca nuts. A few smoked pipes, others picked their teeth with slivers of bamboo. Yet others had stalls in the covered section displaying raw meat, fish and vegetables that attracted myriad flies which they mechanically brushed away with bunches of fragrant leaves. The atmosphere was more sleepy-social than actively commercial.

Further down the harbour drive, the ruins of a once-grand staircase leads up to the site of the former residence of Franz Boluminski, a Prussian who built one of the finest, straightest highways in New Guinea. This still magnificent road runs along the New Ireland coast through the oil-palm plantations. He arrived in Kavieng around 1900 with his wife, ducks, chickens and eight police boys to take up the position of *Bezirksamtmann* (District Commissioner). There were no other volunteers owing to the fearsome reputation New Ireland had for cannibalism. Born in the medieval town of Grudziądz in Poland, then Graudenz in East Prussia, Boluminski was greatly respected by the Melanesians who thought he was a 'good, fierce, just, and inexorable' administrator. When he died in 1913 a spiritual cult grew up around the massive concrete cross that had been raised over his grave by public subscription.

A modern Catholic cathedral lies on a hill above the harbour. I had pressed my face to the grill at the entrance and was imbibing the cool fragrance of incense when I heard a disturbance behind me. A graceful woman was sitting on the ground leaning against a pillar with two toddlers hiding behind her dress. She was wearing a loose turquoise top and a magenta sarong printed with yellow starfish. Brown hair escaped from beneath a pale blue shawl cascading to the floor, almost Moorish in effect. She had elegant fingers and ebony bangles rippled along her slender arms. Her lips were perfect, melancholic brown eyes, a small indigo tattoo on her forehead and another at an angle across one cheek. She had poise despite her poverty.

'Help me, please,' she whispered.

'Why, what's the matter?'

'My husband beats me.' How could anyone beat this exotic creature? I sat down on the step.

'Tell me what happened.'

'The men are terrible to women these days. They are drinking beer all the time. I'm from Madang and want to go home.'

It must be around six hundred kilometres across the Bismarck Sea to Madang. I realised I was going to have to give her money, but wanted to hear the story.

'What's your name?'

'Ludi.'

'Tell me about it, Ludi.'

'My husband is from Kavieng and he came to Madang on a ship to work. You cannot find work here in New Ireland. We met and he wanted me, he took me, so I went with him.'

'But why did you come back to Kavieng?' Her children kept tugging her sarong open, revealing her beautiful, sculpted thighs. She covered them with a gesture of sensual, tired lethargy.

'He was lonely and had a fight with his boss. He beat me when he was drunk. He brought me here to his family. They don't like me because I'm from Madang.'

'Are you still with him?'

'No. He kicked me out and now is with another woman. The nuns help me with the children. I'm saving money to go back to Madang but it is so expensive by boat and impossible by plane for me.' She gazed sadly over the tremendous blue of the sea. She asked for nothing, simply presented me with the situation.

'Men are terrible to women in our country today, terrible.' She stroked the curly blonde hair of her children. I gave her some money. She smiled with exhausted emotion and thanked me. Another casualty of modern change – a familiar story of unemployment, alcohol and violence. The new cultural mobility had tempted her from her own district and now she was stranded in his land trying to return to her own province with two small children. She could only have been twenty-five.

On a remote tropical island you do not expect to come across a golf course, but as I rounded a bend in the road, one unfolded before me. It appeared in surprisingly well-groomed condition. The public use it as a park when competitions are not being held. A number of local men were playing, dressed in golfing costume that attempted to look sporting and fashionable with scant success. The Kavieng Club, established in 1927, had been popular during colonial days and although quietly decaying in the harsh climate, it is still one of the social magnets of the town.

Errol Flynn must surely have drunk at the club, perhaps even fought there. Born in Tasmania, he originally came to Rabaul in New Britain in 1926 at the age of seventeen as a 'Sanitation Inspector' for the Australian Administration, to teach the local people about personal hygiene. He dressed ostentatiously in a white suit, white jodhpurs, wore a white helmet and carried a white walking stick. This theatrically antiseptic costume and its inescapable 'hygienic' associations greatly upset the conservatives at the Rabaul Club. While working on a tobacco plantation, he 'bought' the beautiful fourteen-year-old native girl Tuperselai for two pigs, a 'fuse' of English shillings (coins wrapped in tubes resembling dynamite fuses) and a few strings of shell money. He would read Maeterlinck as this sulky 'slave' washed his back with fragrant herbs. The fatalism and mysticism of this poet and playwright attracted the unconventional Flynn. When he lived in New Ireland as a young man he was accompanied by another of his native lovers, his 'puppy' as he called her, the young Melanesian girl Maihiati. He described her in his autobiography, *My Wicked, Wicked Ways*.

> She had a perfect figure, and the most glorious pair of breasts you ever saw – the classic ski-jump type, a lovely little hollow and then the line goes way up into the air and doesn't dip. She was bushy-haired of course and lightly tattooed.

Flynn later purchased a schooner called the *Maski* which means in Papuan, 'I don't give a damn,' or more aggressively, 'Screw yourself!' He began to make a living dynamiting fish as he was too poor a navigator to trade between the islands. He left Kavieng owing money to everyone and was rumoured to have abandoned a natural daughter he conceived with a Selapiu girl.[1] Susy Flynn disappeared during the war after being spotted on New Hanover, an island north of New Ireland.

Wrestling irate crocodiles on the Sepik river for a film documentary finally got him a part playing Fletcher Christian in an Australian film called *In the Wake of the Bounty*. He was paid six pounds for three weeks' work. It comes as no great surprise to learn that he was a direct descendant of Fletcher Christian on his mother's side. The lad finally fled New Guinea with some diamonds stolen from an ageing *inamorata* concealed in the hollow handle of his shaving brush. Ting Ling O'Connor, his casino lover in Hong Kong, welcomed him with open arms.

Compared with Rabaul, Kavieng has always been relaxed and casual. I rattled the gate and was allowed into the club, although not a member. Warning signs dominate the entrance in an attempt to make it respectable.

Shirts or T-shirts with long sleeves must be worn after 6.00 p.m.
No hats in the Bar

I was sweating profusely in the humidity and ordered a cold beer. A weather-beaten face with half-closed eyes engaged me in desultory conversation.

'G'day, mate.'

The bar was almost empty. An important golf tournament was proceeding outside.

'G'day.'

The bearded figure was glued to a wall, flowing over a barstool.

---

[1] Selapiu is a mangrove island fourteen miles south of Kavieng Harbour.

'What do you do around here?'

'Fuckin' very little.'

He had the thin legs and the detached lack of interest of the alcoholic. A burst of cackling laughter. Clearly he was very drunk.

'Yes, but where do you work?'

'I used to work in the fuckin' mine on fuckin' Lihir, but since the fuckin' gold price is now shot to shit . . . fuckin' nothing.'

He struggled up and lurched towards the gate, violently waving to me which immediately put him off balance and he stumbled down the stairs. I saw him get into a utility truck and weave down the potholed road like an intoxicated snake.

In the clubhouse there was great activity around the blackboard as scores were being chalked up amid intense discussion after competitors' rounds. This was obviously a crucial tournament with competitors from around the country. Club Rule 3 read, 'Walk briskly between shots' – an amusing imperative in this oppressive climate.

A number of expatriates were competing and calculating the scores, together with the new entrepreneurial class of Papuan businessmen dressed in distant approximations of a golfing theme. Their precocious young sons watched with boredom, drank pop, punched each other, and tripped over the golf buggies. Everyone seemed excruciatingly self-conscious except the expats, scattered loudly and confidently around the tables in the beer garden. There was not a woman in sight. I tried to make conversation but was greeted with monosyllabic responses or drunken astonishment. Well, I was a stranger in a panama, not a member, and it was late in the day.

The rest of Kavieng was rather run down apart from this extraordinary statement of colonialism surviving at its heart. The humidity and alcohol had made me drowsy and I wandered without much sense of mission along Coronation Drive. The slightest effort brought me out in a profuse sweat. Drifts of perfumed frangipani had collected in holes by the road. A stuffed crocodile stared out at me from a dusty glass case at the entrance to a run-down hotel. Rumbles of distant thunder. Sunday afternoon in Kavieng.

A couple were bowling along the harbour drive on a bicycle, she sitting on the handlebars and screaming, he pedalling furiously as they descended the curve past a memorial. This was the first courting couple I had seen in Papua New Guinea since my arrival. Of course they may have been brother and sister. It is rare to see a romantic couple in this country, especially in daylight. Even married couples are seldom seen out together. Associating with women in public seems to be a sign of weakness to Melanesian men. Assignations take place at night and in secret. The general atmosphere of defensive masculinity that hovers in the air becomes oppressive after a time. I found this secretiveness among young people an astonishing contrast to the wild exhibitionism of the London clubs where public intimacies are taken for granted.

As dusk settled over the harbour and solitary canoes made their way across the burnished water like a scene from a Japanese watercolour, a great peace descended. The famous Kavieng crayfish were grilling on the fire, smoke rising deliciously in the still air. Drumming drifted over as usual from the distant shore of Nusa Island. The world about me slowed and time dissolved as the sun settled into the sea. I had begun to see and think in images rather than the rational progression of analytical thought. The islands were demanding acceptance rather than analysis. I had begun to experience the pantheism that is so seductive in the tropical Pacific, my mind gently becoming one with nature. Tomorrow I would be wrestling with the art of the *malagan*, a profound and complex expression of the soul after death through carvings and mortuary festivals, both manifestations unique to New Ireland. My visit would be to the village of Libba and one of the few master carvers remaining on the north-east coast. One of his friezes already protected my room.

11. Feverish Nightmares

'Culture is people.'

Demas Kavavu, Paramount Chief and a former Premier of New Ireland, made this pronouncement as we sped down the arrow-straight Boluminski Highway.

Early morning. The weather was superb – brilliant sun and tremendous heat. We had already passed through the outskirts of Kavieng and were heading down the coast to the village of Libba on a smooth, metalled road discussing the complex problems of Papua New Guinea. We passed through plantations of oil-palm and stopped for water at a tropical beach – a crescent of sand, driftwood and coconuts, a sapphire sea fringed with palms. We sat on a fallen trunk and gazed out to the horizon.

'There is a threefold aspect to all problems in Papua New Guinea,' he said firmly. 'Culture, government and religion. When considering any difficulty you must take account of all three.'

'How do you mean?'

English was not his language and although it was accurate, he was not given to expansive sentences. His work in government ministries had made him more succinct than most educated people. He spoke the Nalik dialect, one of the twenty or so languages spoken on New Ireland, but had a good command of English.

'The culture must be preserved but young people are not so interested.'

His face registered disappointment. He kicked a coconut husk with his Nike trainers.

'Yes, they do seem to be rather lazy about continuing the traditions. It's a problem all over the world.'

'But people always turn to traditional culture and not to religion when they become sick or approach death.' He laughed.

It was the continuity of the culture that was most important to him. Culture should not be dissected but considered as homogeneous. He respected missionaries, but I could sense he was sceptical of the religion they professed. A battle of loyalties raged within him.

'I must be sure of my son before I can give him my position.'

'How can you make sure of him? Do you give him some type of test?'

'Yes, of course. In one test he must survive for twenty-four hours being completely ignored!'

'Twenty-four hours is not too difficult! Being ignored for most of your life is normal where I come from in London!'

'Well, in a village this is very difficult, yes, and painful. We are very close in the families.'

He became serious and did not appreciate or perhaps understand my English levity. I began to comprehend that this man was saturated in his ancient culture and constantly felt the distress of being forced to compromise with modern times. An educated warrior from another era, another world, demanding to be taken seriously in all the conflicting aspects of his divided nature.

He had retired from public life a few years ago. Despite being dressed in European clothes – a salmon-pink cotton shirt, jeans, Ray-Bans and trainers – he commanded all the presence one would expect of a chief. He wore his grey hair shaved ultra short which accentuated his noble profile and fine head. A mysterious necklace of coloured shells sat rather tightly around his neck. His expressive, enigmatic Melanesian face became fierce when he was emotional. Subtle shadows of aristocratic Polynesia also flickered there; his skin was the colour of old polished walnut, quite different to those ominous night faces encountered in the Highlands. It was a striking head that gave him great charisma.

As a Paramount Chief he carries a grave responsibility as a custodian of the spiritual content of the traditional *malagan* culture, an ancestor cult. The term is a Pidgin word that refers to the ornate carvings produced on the north coast of New Ireland and on the islands of Tabar and New Hanover. It also refers to the complex of ceremonials and feasting utilising these carvings and associated with the death of relatives. Social life and religion in this province are dominated by the whole idea of *malagan*.

New Ireland society is broadly divided into two groups designated as 'Sea Hawk' or 'Sea Eagle'. These are further subdivided into clans ruled by a totemic animal symbolising a cosmic influence. Severe penalties result if members of the same totem have sexual relations. Originating on the island of Tabar, *malagan* sculptures are associated with particular clans and represent the idea of the life force of the animal. The totem could be a bird, fish or snake and they are represented most often in wooden sculptures. Tall poles with figures stacked vertically are carved in openwork designs. Elaborate horizontal friezes may terminate at each end with the heads of fish. All show an obsession with metamorphosis, transformation and ambiguity, which endeared them to European surrealist artists.

These symbolic representations act in a similar way to a family coat of arms. Their value lies in the history of ownership and the way the design has been passed down through the generations. The *malagan* carries with it ideas of honouring the dead, love, initiations, the discharge of social and financial obligations, and the cementing of family ties created through marriage. The ceremonies record the trace of a human being through this world. In addition to honouring the dead, the ceremony is also a celebration of the capabilities of the living and the power of the men of the sponsoring clan to organise the *malagan*. A mortuary feast is ruinously expensive to mount with the requirement of carved masks, dancing groups, the raising of taro in the gardens, the slaughtering of many pigs and the contributions of money to the families of the various deceased. Years of effort and work go into these rites and this is what is valued and remembered by the old folk.

Demas was supposed to accompany a theatre troupe of dancers that was performing at the Festival of Pacific Arts in New Caledonia but something happened at the last moment and he did not go with them.

'Don't come and say goodbye, Demas, it will be too sad,' they had said when they left.

Demas had restructured the troupe's masculine *tatanua* dance, which forms an important part of many mortuary celebrations. The colonial period under Germany and Australia, followed by the horrors of the Second World War, severely disrupted cultural traditions on the mainland of New Ireland. Much has been lost owing to the sensitive and complex nature of these mutable ceremonies. The powerful *tatanua* dance is a typical casualty, rarely seen now outside of cultural festivals or feasts for important chiefs.

The *tatanua* masks (from the word *tanua*, meaning soul or spirit) are spectacular and each one is unique. They are believed to carry great spiritual energy when in use. The wooden face is meant to display the exaggerated characteristics of male beauty: an open mouth with, according to the ethnologist Richard Parkinson, 'a healthy bite', prominent teeth, a broad, flattened nose, greatly elongated ear lobes and the characteristic shaved hairstyle of bereaved warriors. The unnervingly real eyes of the mask are made from the valve of the sea snail *Turbo petholatus*, and are a distinguishing characteristic of much *malagan* sculpture. Alberto Giacometti believed that in this 'extraordinarily lively gaze, almost intrusive' lay the secret power of the Oceanic mask. The face is surmounted by a yellow crest of 'hair' which gives the mask the form and appearance of an ancient Greek helmet. Sometimes a dance ornament in the shape of a bird is placed in the mouth.

The contrast in gender which distinguishes so many cultural practices in Papua New Guinea means that the preparations and the making of the masks take place in secret out of the sight of women and their baleful influence. The male dancers are forbidden concourse with women for at least a week before the ceremony. At dawn on the day of the dance, the *tatanua* wade into the sea where they are purified with incantations and drops of seawater as the first rays of the sun strike the masks. The dancer has now given up his individuality to express the masculine life force at its purest. Demas concentrated and focused this power.

He had had a difficult childhood during the Second World War. He told me of being captured as a six-year-old boy by the Japanese and wetting his pants in terror. Someone had falsely accused his parents of 'collaboration' with the Australians. The Japanese threw the family into prison with their hands tied behind their backs. They thought they would be shot. Many males in certain areas of Kavieng had already been rounded up and executed in a large pit. The women were then taken away by the Japanese 'to use'. This terrifying experience has haunted him all his life.

'How have the missions affected the culture?' I asked. Sunlight glared off the glassy sea.

'The influence of the missions is a sensitive subject, Michael. I'm torn by it. I went to a Methodist Mission school but I also try to defend the old culture. They can live together.'

'I went to see Kwato last week. What a tragedy it's finished! Charles Abel seemed such an enlightened man.'

'Abel tried to preserve traditional values as well as give us Christianity. Do you know the meaning of the letters in Kwato?'

'I thought it was simply the name of the island.'

'For me they represent Knowledge, Wisdom, Action, Truthfulness and Obedience. A good formula.'

We climbed back into the four-wheel drive and spun along the Boluminski Highway once more. What a remarkable injection of disciplined Prussian spirit this road was, slicing through the island communities like the blade of a sabre. For years it was the finest highway in Papua New Guinea. Occasionally, as we passed a truck loaded with passengers and produce, Demas would say, 'That is the vehicle of (such and such) village.' He never mentioned individual ownership, only the possessions of the community or clan. Melanesian society is passionately egalitarian.

'Tell me, Demas, what is required of the Paramount Chief?'

We were passing through immense plantations of oil-palm. Some of these huge trees beside the road were a sickly yellow and piles of orange husks were smoking with a fetid odour that made me gag.

'The Paramount Chief must understand all the cultures of the island, and speak many languages. I speak about ten because I was a teacher in West New Britain for many years. The chief is initiated only for certain clans and must be initiated four times. The *masokala* clan initiated me, but I also belong to the *moxokirin* clan. Turn left here!'

The halt was his own village of Fatmilak – 'Fat' meaning stone and 'milak' meaning red. Villages are often hidden from the road, but huts and swept paths open out like a fan as one passes beyond the jungle barrier. He showed me his own house and, placed almost at the front door, the graves of his wife and mother. They were covered by a small, hut-like, fenced enclosure of palms with a roof over the plots. A black Madonna wrapped in white linen was attached to a corner of the fence, symbolic cordyline plants grew at the foot of each grave. This enclosure seemed to be another hybrid product of Christian and pagan practice. When I mentioned my mother's recent death from cancer during my journey through the islands, he appeared not to have heard me and stood silent, looking down at his own mother. His lack of what I would have thought was conventional sentiment manifested itself on more than one occasion. We set off again.

'Turn left again here!'

We drove down a winding path through cleared jungle to an idyllic spot by a beach. There were a number of huts of different sizes nestling among hibiscus gardens, kerosene trees, coconut palms and pandanus. The open areas had been raked with perfectly parallel striations in the sand. Tidiness, so any passing sorcerer would be prevented from fastening on a human nail or hair and casting a spell of death. This was his small business, a 'guesthouse'. An ornate *malagan*-carved frieze protected the entrance to a large two-storey building with walls of diapered and woven palm. It was comfortable with screened windows. He dreamed of the imminent arrival of tourist ships, tourist 'cargo', like Wallace Andrew of Samarai.

Curious, I drew aside a cloth from the door of a small hut, and was shaken to see in the gloom a group of the forbidding *tatanua* dancing masks placed on posts in a circle. A dank odour of mould engulfed me, as if from the grave. The grinning wooden masks faced each other in anticipation of some future *malagan* death ritual. Their masculine aura was powerful. These were the masks of *Panatalis Dodor*.

'Ah, yes, the masks,' he said with a tone hinting at volumes left unsaid. I always felt that he was disinclined to share any deeper-level meanings with me, the 'secrets'.

'This is all well known. But there is a particular dance not often seen called *suksuk*. The men must fast and abstain from sex with their wives. The backs of the dancers are pierced with sharpened bamboo sticks and thin rope or twine is pulled through the piercings. Masks are attached to the ropes and the dance takes place. Bleeding only happens if the preparations have been compromised.'

'What does it mean, this dance?'

'Mean? Mean? We have always done this, and our fathers before us. This is the way it is done, that is all.' His penetrating eyes flashed at me. My seeking after rational explanations had struck another brick wall.

'Turn left again!'

We had finally arrived at the village of Libba, a centre of master carvers on the mainland. *Malagan* at its purest originated on the offshore islands of the Tabar Group, where death rituals are still practised in their traditional form. The great artist and carver Edward Sale lives on Tatau Island in this group and his carvings are scarce and passionately collected by connoisseurs. *Malagan* practices in New Ireland differ according to the location. On the north coast and Nusa Island, the body is cremated. On Tabar and New Hanover there is cremation, interment or burial at sea depending on the totem of the clan. Mortuary festivities consist of many phases and may continue for years. During the periods of rest, the soul is given time to accustom itself to its new state. Death is the defining reality of life in New Ireland.

Libba village gave me an intense feeling of remoteness – expanses of firm, manicured sand glaring in the savage sun, sea drained of blue bleeding to green, a gnarled tree with pendulous fruit. The grave of a Paramount Chief with a plain concrete headstone lay close by the hut of the master carver. These people experience a fervent intimacy with their dead. Ben Sisia emerged to greet us, a spare figure reminiscent of the slender elegance of a Giacometti sculpture, dressed in a white shirt and lime-green *laplap*.[1]

Demas and the old master carver talked quietly in the shade for some time, seated on a low bench. It was understood I should stand apart. Blonde children swam in the balmy water, shrieking with delight. Dogs slept in the sun. Demas and Ben eventually wandered over to me through the shimmering haze as if in slow motion. The distinguished old man had a thin and fine-boned face. His skin was the colour of ebony and wood shavings had caught up in the dark hair on his arms. This aesthetic being could only have been an artist, but here also was a man who emanated strong spiritual qualities of goodness and love. The inner demons of cultural conflict seemed to have been vanquished. He had

---

[1] An ankle-length piece of cloth wrapped round the waist. It is worn throughout the island provinces, also by more traditional men. It was originally introduced by Europeans offended by traditional dress.

become a devout Christian, yet those eyes still contained the wisdom of the ancients. I clasped a dry, soft hand.

Ben did not speak but gestured that I should enter his carving workshop. It was built on a bank of sand immediately above the water and had a pandanus-leaf window opening onto a tiny coral atoll. Demas came in and we sat in the semi-darkness and began to discuss the intricacies of the clan and *malagan* carving. Ben Sisia is considered one of the few carvers left who make fine *tatanua* masks. Early European interpretations of these carvings described them as the products of the 'feverish nightmares' of schizophrenia residing in the infernal side of the unconscious mind.

There are numerous types of *malagan* masks. The totemic animal of each clan is depicted in various stages of transformation together with other perplexing symbols. The clan alone possess the rights to any 'design'. A mask may also represent the soul of the dead, such as the fearsome *ges* mask with its tusks and slanting eyes. In this Notsi-speaking area, the *ges* represents the sometimes invisible, sometimes destructive, spiritual double of the deceased person that perishes with him. The past tradition was to burn the masks immediately after ceremonies. Today these precious carvings are kept and rented out for other *malagans*. German ship's doctors before the Great War were among the first to collect and preserve carved pieces, placing them in the great museums of Vienna and Berlin. But such carvings cannot be separated from the society which utilises them. Museum artefacts, collected solely for their aesthetic beauty, have been struck dumb.

Full *malagan* ceremonies, so gloriously described and photographed by anthropologists in the 1930s, are still mounted on Tabar, but on increasingly rare occasions. European art connoisseurs now commission large carvings and even subsidise traditional ceremonies by purchasing pigs. Dances and carvings are taken to Pacific Island cultural festivals as New Ireland's contribution to Oceanic culture, but my observations on the village scale were rather different to the picturesque image portrayed in academic texts.

Financing redemption within two religious beliefs, one pagan and one Christian, is a burden few villagers can accomplish when living at a subsistence level. Taking a couple of weeks off work to attend a mortuary festival is not a priority in the modern cash economy. Speedy banana boats make travel between islands rapid which reduces the traditional ceremonial time-scale of months to days. Commissioning masks places an excessive financial burden on the poorer clans. The resulting hybrid cultural evolution is not peculiar to New Ireland, but is to be found in all the island groups.

Carvers are finding it increasingly difficult to survive solely as *malagan* artists in this farrago of conflicting religious beliefs. They are now forced to sell their work for its value as artwork or souvenir. This is not a particularly modern phenomenon and has been done throughout the period of European contact, but it is increasingly becoming their sole source of income. They are no longer an integral part of the traditional community as it begins seriously to fragment. Christian additions to traditional spiritual beliefs demand a Christian iconography. Traditional sculpture has largely disappeared from regions evangelised by Protestant missionaries, but the Catholic faith is more tolerant of indigenous imagery. The number of traditional carvers is dwindling and the subjects of their carvings changing. Yet the true spirit of *malagan* lives on strongly in the souls of the local people.

'Ben, how do you know the design a clan wants? Do they describe it for you?'

I was interested to see if he carved masks according to a programme of features.

'It begins with the felling of the tree and the bringing of it to the secret place. The clan pay in shell money for this. The wood is dried and there follows a small feast when the first cut of the carving is made. The clan tell him what they want in general terms. He gets an instant image of it in his mind. Of course, the masks do not remain exactly the same down through the generations, but it is the same *malagan*.' Demas translated from the Notsi dialect Ben spoke.

'It is a very, very serious matter if a man fakes a *malagan*. He can go to the prison or become a victim of sorcery!'

The ownership of *malagans* and the knowledge of the connected ceremonial constitutes a type of 'copyright' that can be bought or sold. Each owner receives the rights to his *malagan* in public and special musical chants bring it to 'life'. The symbolic meaning of these designs was lost long ago as the *malagans* are handed down through the generations. Stories are associated with the carvings and in this way New Ireland people pass on their history.

'How do you know if they are dishonest?' Demas smiled at this.

'Easy. Ben gets a "block" and cannot continue carving. The solution is to take a mouthful of ginger and spit upon the carving, thus expelling the evil spirit which is blocking you. The carver can then continue with his work.' Ben gave an amusing pantomime display of spitting on the carving. He deferred to Demas as a Paramount Chief on many occasions of explanation.

'When are they painted?' They looked nervously at each other as if revealing trade secrets.

'First the wood is polished, the eyes of the sea snail go in and then the colours go on. The *malagans* then go into their special enclosure.'

'What are you working on now?' I could see unfinished sections scattered around the hut.

'He's working on a copy of an historic canoe for a famous museum in Berlin. They might pay him six thousand *kina* for this!' Demas was proud of his friend's work.

'I see. A copy.'

I could not help reflecting on the artistic bargain the German curators were acquiring for fifteen hundred pounds.

'He also wants to show you some posts his son is carving for the new church.' In fact Ben seemed more enthusiastic about this than anything else. We wandered through the beautiful village along a line of huts. His son's hut had unusual windows shaped like huge almond-shaped eyes looking out to sea. We went behind it to where there was a large workshop. Two enormous carved poles lay on trestles.

'These are usually carved in a secret place far from here, but it is more convenient to work on them in the village,' Demas informed me.

A cloth was removed to reveal relief carvings of sacred subjects from both the Old and the New Testament. Biblical quotations such as 'The Crown of Righteousness' hovered above carved globes, flaming torches and doves. Under the theatrical head of an Old Testament prophet with prominent mouth, powerful hand and pointing finger was inscribed '*I am giving you the word. You must speak.*' *Jeremiah 1–9*. They had the rich and exuberant style of a naïve but gifted artist. I photographed Demas and Ben seated together on these poles. Ben looks questioningly at the Paramount Chief as if asking for his approval at this new phase of Christian carving. Demas looks down at the pole rather diffidently, with an expression that indicates inner conflict, but rueful acceptance of this departure from tradition. He seems reluctant to confide his true thoughts.

Was it that I had been brought up a Christian, was it the familiarity of this imagery that caused me to seriously question the artistic value of this work? The 'feverish nightmares' seemed infinitely preferable and artistically valuable, but that may be simply the European fascination with a more primitive aesthetic. Ben Sisia left us, a solitary figure among the palms, carrying a plaited bag containing his betel-chewing equipment.

We took to the Boluminski Highway once more and returned over a fearfully potholed section to Demas's guesthouse where he promised to show me his personal *malagan*. He disappeared into a hut and emerged dressed in the complete regalia of an island Paramount Chief, carrying a large *marua* mask called a *vaneriu*.[1] He wore a *laplap* of pale blue, pink and yellow cotton printed with various crustaceans and sea ferns. A beautiful *kapkap* symbolising his chiefly status and his powers as an orator hung around his neck on a thin string of shell money. The *kapkap* is a solid white disc of polished Giant Clam shell not unlike ivory, with an intricately-worked fretwork design in tortoise-shell applied to it. The quadrant design is known as the symbolic 'eye of fire' and the lacework is in superb contrast against the pale shell. They are very rare. White lime powder was dusted across his chest and he wore a crown of cockerel feathers fastened to his forehead with a broad vermillion bandana. Another streak of lime powder decorated the right side of his face. We sat by the timeless tropical shore and ate soft, sweet paw-paw. The heavy mask was propped between us.

'Demas, can you explain the symbolism of the mask to me?'

'I can, but remember, only I can do this because I own the rights. We normally do not do this. Everyone understands the images except the anthropologists! Ha! Ha! They explain to us what we believed! They explain our culture to us!'

'Well, can I ask you when it was carved?'

'In 1980. It was to commemorate my mother and one of my babies who had died some years earlier. My ancestors got the rights in a secret place behind Fatmilak. On the "ears" of the mask are two figures of the dead, both of them women wearing the traditional, pointed pandanus cap of the islands. The stars above the figures are the morning star. The two green lobsters you can see on the "ears" belonged to another *malagan* which came to the island from Simberi Island. I acquired the rights to this part and made them a part of my own *vaneriu*. The face in the centre of the mask with the fierce teeth and long tongue pointing straight up between his eyes is frightening, isn't it? This warns other clans not to steal my *malagan* design!'

---

[1] *Vaneriu* are large *malagan* masks carved in various styles, each type possessing complex symbolism.

I could not help feeling puzzled at how pagan animism can live so comfortably alongside humanist New Testament phrases such as 'Thou art my beloved son. With thee I am well pleased'. Villagers' minds must surely be crammed with the various 'gods' they seek to placate. Again, it is a testament to their great adaptability that they have not gone mad under the pressure of this acceleration from Stone-Age to modern beliefs.

Demas showed me a photograph of himself in fully-costumed splendour, seated atop a heap of forty pigs during his second initiation ceremony. He spoke again of hordes of phantom tourists whose arrival was imminent, showed me the VIP toilet (the AusAID mnemonic for Village Improvement Project) which consists of a concrete pedestal and plastic seat placed over a hole in the ground. I learned of the many ways of opening coconuts depending on the custom of the local area – the top can be simply sliced off or the nut can be sliced laterally, the halves forced apart allowing a stream of juice to pour directly down the throat. I drank mine from a glass like a good European after drenching my face and clothes with the juice attempting the local method. A young chief was introduced to me, blew his nose on the tail of his T-shirt, spat a jet of blood-red betel juice in the dust and shook my hand.

'Pleased to meet you. Lunch is ready.'

The meal was a surprisingly-delicious mixture of baked taro and spaghetti cooked in chicken stock laced with the ubiquitous chunks of Spam.

'Do you know the meaning of this mark on my face?' Demas asked.

'No. Tell me.'

'It means that we are chatting and eating together in a friendly way but at the end of the meal I will kill you. It is a secret sign. Most guests are proud to die in this way, accept it and have excellent conversations before death.'

Our last stop on the Prussian highway was at the home of Cathy, the Eel Lady. She is a celebrity around the Kavieng area.

We drove into another neat and well-pressed village and made for a house at the far end. It was adjacent to a pure freshwater stream that cascaded down to the sea from the New Ireland mountains. Cathy had been an Air Niugini stewardess for many years, had travelled the world, but had decided to return to her remote village on New Ireland to live. A glamorous and sophisticated figure dressed in a French silk shift resplendent with scarlet, green and blue tropical flowers, emerged from the gloom of the primitive dwelling through a cloud of wood smoke. A dazzling smile lit up a rich, dark complexion, framed by a halo of sculpted frizzy hair. She was like a tropical bird emerging from the jungle.

'I had given you up, Demas!'

Cats and kittens were asleep or playing on the earthen floor; a chow gambolled about in a frenzy at our arrival. This dog was the healthiest and happiest I had seen during my entire stay in Papua New Guinea. An elderly lady could be dimly seen in a corner of the large hut. An old man in a mauve *laplap* sat like a gnarled piece of antique ebony on his tattered sleeping mat, leaning against a post. He took my hand in a dry, claw-like grasp. The whites of his eyes were suffused with yellow, the irises had simply faded away.

Cathy began to attend to the fire in the 'cookhouse' and put a kettle on the bright flames.

'Would you like a cup of tea?'

Smoke filled the entire area, but this had the benefit of keeping away the myriad flies attracted to the region by the rotting husks on the oil-palm plantations. Making tea in a smoky bush hut can hardly be compared with serving caviar on ice and pouring champagne on a jet to New York. She was in her village without electricity or conveniences of any kind and seemed supremely happy. An atmosphere of friendliness and warmth prevailed overall.

'Did you bring anything to feed the eels, Demas?'

'No, I forgot.'

'Well, we can probably get them to come out, but the biggest won't venture out unless something is up for grabs.' Her command of idiomatic English was impressive.

At the rear of the cooking area was a clear stream snaking towards the sea. Dappled sunlight fell through the palms onto the gently swirling water, so clear it was almost invisible. New Ireland has many of these freshwater rivers that contain delicious small crayfish. She waded ankle-deep into the water and immediately an enormous eel – I would estimate it at about five feet long and six inches broad across the head – lethargically undulated from beneath a bank and brushed against her legs like a cat, insinuating itself into her graces. Other smaller eels appeared until she was surrounded. She bent over and gently stroked their heads, looking up at me and smiling.

'They're affectionate in their way, but they're really after a good feed!' Her young daughter was begging her mother to let her plunge among the eels, jumping up and down on the bank in extreme excitement.

'What a pity Demas did not bring some meat. They like meat. You're a naughty boy, Demas!'

By this time he was chewing betel and had drifted into a mood of repose quite beyond the barbs of criticism. She emerged from the stream and wiped her hands.

'Don't you find it difficult living in a village after the jet-set life?'

I remained amazed at this ability Melanesians have to calmly embrace extremes that would drive a European into a frenzy of frustration. Even the High Commissioner in London, Sir Kina Bona, had become nostalgic for Kwato Island and the lapping waves when we spoke of his eventual return to Papua New Guinea from London. Cathy was anxious to explain.

'It's wonderful to be with my clan. We all return to our villages after the career has run its course or even during it. We are drawn back by the natural magic. I know a pilot who lives two separate lives – one flying between continents and the other chewing betel and singing in his village where he is the chief of his clan. No time to get bored!'

She did not seem inclined to offer any intellectual justification as I might have done, rationalising the superiority of the 'simple life' of the village over the metropolitan cascade of consumerism.

The tiny kitten I was stroking mewed and strained, sinking its razor-sharp claws into the back of my hand. Flies settled and rose. Demas appeared to be asleep in a wicker chair.

'You've just missed a huge *malagan* on Tabar Island. It was for an important chief and quite incredible. Masks and dancing for days. We all stayed in the guesthouse. What a pity you missed it! Demas, Demas, is there a *malagan* you can take him to?'

'Yes, yes. There will be one in Panafiluva on Friday.' He lurched awake.

'When we next have a *malagan* here, I am going to include my mother. She is getting on a bit now.'

This was surprising as her mother was sitting near the fire and looked happy and remarkably healthy.

'Is she ill?' I kept my voice to a whisper in case she understood some English.

'No, no. But we can include her with the dead when we next have a *malagan*, all for the same price. It's so expensive having *malagans* for every individual who dies. All those pigs!'

'But doesn't this mean she attends her own funeral?'

'Well, she attends her memorial celebrations actually. My mother is loved in this village. The feast will be a celebration of her life and of those who have already passed on. When she actually does die there will just be a simple ceremony of interment.'

This seemed to be another eminently practical, surprisingly pragmatic, modification of the traditional *malagan* rite in light of the financial pressures of modern life. Like a living organism, the *malagan* ceremony inexorably continues to evolve under modern constraints.

Demas remained silent on the return to Kavieng. On the last

part of the journey I noticed that lumps of mud seemed to have dropped from a truck onto the tarmac. The sun was setting and night rapidly enfolded the plantations in its dark wings. Small kerosene lamps and fitful fires burned in the hamlets. Local people loomed up alarmingly out of the roadside gloom like figures in a shadow theatre. I noticed that whenever I ran over a lump of mud there was a strange hollow sound.

'What is that on the road, Demas?'

'Toads. Hundreds of them sit on the warm surface. Just keep on driving.'

And so I returned to Kavieng over a carpet of exploding toads. Dazzled by the headlights, their heads lifted erect, they made no attempt to avoid having the life squashed out of them by the wide tyres of my jeep. Demas would never have understood my squeamishness at their violent death beneath my wheels.

12. Grand Opening – Tsoi Island General Store

The banana boat had promised to collect me early in the morning but it was late. According to my agitated boatman, the price of fuel had doubled overnight. Instead of being annoyed, I thought the whole notion quite an amusing way of relieving me of *kina*. I put my arm around his shoulder as we stood on the beach.

'Look, Elisha, I simply cannot go to Tsoi for that price. The whole thing is off. Sorry. Just can't afford it.'

We soon returned to the original figure, but I knew there was not much profit in it for him. Petrol was expensive, sometimes difficult to obtain, and it was a long trip. It is difficult to balance a sense of justice with avoiding being taken for a soft touch as you haggle for transport in the islands. Demas had decided to accompany me to the grand opening of the general store at Mansava on Tsoi Island, situated just three degrees below the equator. The ceremony would be unique as this was the first time in history that a store had opened there. Word had spread quickly around the nearby island groups that Justin Tkatchenko had decided to open a joint venture with Bosston Tusikai, the leader of a local clan and owner of a small guesthouse. Many speculated it would fail miserably and mistrusted his motives. Others saw it as a sign of rare genuine concern for the local community. Boats had been ferrying supplies to the island all week and excitement was running high.

On this trip I at least had the luxury of a plank to sit on. The outboard motor opened up and we skimmed across the clear water between the small islands that litter the approach to Byron Straits. The surface changed from mirror smooth to choppy about an hour into the trip. The wind rose and then rain squalls hit us hard. The dinghy began to buck and slam into the waves. I covered myself in a plastic tarpaulin but Demas continued to stand bolt upright at the prow, riding the swell with all the presence of a chief being delivered to his people.

As we arrived at Mansava, the sun broke across the beach. A hundred canoes rested lightly on warm crystal water the colour of pale viridian. Bleached driftwood littered the sand. The cool verandas of huts, built within groves of coconut and pandanus palms, were festooned with swathes of magenta bougainvillea. Miniature white orchids cascaded from the eaves while ladders led up to sleeping rooms opening off airy atriums.

Local people ambled listlessly along the carefully-raked paths. Many were wearing the classic fashion statement in the islands, scarlet cloth against ebony skin. Faces were subdued from pure Melanesian to island mixtures that reminded one unexpectedly of Tunisia or Morocco. These were sculpted, strangely Semitic faces with aquiline noses of great nobility. Raised scars cut by obsidian splinters, the wound filled with sand, worried across men's chests. Foreheads and cheeks were decorated with simple tattoos like crusader blazons from the ancient deserts of Jerusalem. Skins of midnight black from the Northern Solomons vied with honey-coloured shadows from Polynesia. Beautiful girls crowned their brows with miniature cream orchids, strikingly blue eyes and blonde hair betraying perhaps the echo of a trader's love child. Children came up to inspect my skin, touching it with wonder and cries of delight. Was it possible that some had never seen a white person? The aroma of roast pork drifted by on the sound of drums and laughter. The sweltering heat and humidity were desperately enervating.

The discovery I was to be an honoured guest came as a complete surprise. The VIP pavilion was furnished with bamboo arm-chairs placed behind a fringe of split palm fronds threaded with orchids and fragrant pink frangipani. It overlooked the dancing area and a speaker's platform draped in the black, gold and red flag of Papua New Guinea. The store was a long, narrow building made of pale green prefabricated metal sheets which faced the beach.

TSOI ISLAND GENERAL STORE
C & J TKATCHENKO

I lay back in the bamboo armchair, adjusted my panama, and inevitably felt like an erstwhile planter from the turn of the century. A programme of events on yellow paper was handed out as the Master of Ceremonies began a long address in Pidgin. This, after all, was the first general store in the history of New Hanover.

'*Nau Tsoi bai kamap wanpela ples gen insait long hap blong yumi.*[1] No longer will you have to make the long trip to Kavieng for your goods.'

The first piece of theatre was Pastor Haludal. Dressed in a blue shirt, spotted tie, long, brown trousers and jacket but wearing no shoes, he climbed the podium and lifted his black Bible and brass cross high into the air. His evangelical fervour was awesome. He waved his arms and raised his voice to a prophetic pitch, leaning forward and shaking the sacred book. Local people, standing in colourful groups three or four deep, sitting on the ground, squat-ting on their haunches, hanging from the trees, clustered around the store, were transfixed to this demented display. Children lying on their fronts with sand still clinging to their cheeks gaped in terror.

'*Yu mas redi nau! Yu redi ha? Jisas emi stronpela ston!*'[2]

---

[1] 'Now Tsoi will be sewn back into the local scene.'
[2] 'You must be ready now! Are you ready? Jesus is the rock!'

I had never thought seriously about the persuasive power of theatrical Christianity, but here it was in full force, reminding the faithful of what should be their true priorities in life.

'A wise man builds his house upon a rock. The winds blow and the rain beats upon that house on that rock.'

He glanced behind him at the store.

'It fell not! Oh no, my brothers! *Nogat!* The store will not fall! Jesus is that rock! He is that rock on which we build! Are you ready now? *Yupela mus redi* . . . Are you my brothers . . . ?' I allowed myself the wicked reflection during this harangue that everything on the island was in fact built upon sand.

The pastor left us in a suitable state of heightened animation and he was followed by some fine close harmony singing by choirs from neighbouring islands. The performers were mainly young women dressed in cotton shifts with delicate voices hard to hear. Pink and cream frangipani leis were placed over the heads of the honoured guests. I had not anticipated becoming the centre of attention and felt once again the pressure of a thousand Melanesian eyes. As I leaned back in my bamboo recliner, I was served pineapple, paw-paw and the familiar luminous green cordial, greener than the green sea. The scent of frangipani released by the heat was like a drug preparing me for the dance groups that would follow.

Demas, dressed in a rainbow-coloured, tie-dyed *laplap*, white T-shirt and Ray-Ban sunglasses, was the next speaker to mount the podium. His political charisma was powerful before the crowd, the sentiments he proclaimed in Pidgin suitably high-flown.

'*Dispela nupela stoa we iop nau em wanpela bikpela samting we ikamap long New Ireland.*[1] It shows the qualities of trust and self-reliance. Trust because Mr Justin has been taken into the clan as *wanpela* equal-partner. He wants to help the people of New Hanover and it shows the possibilities of human develop-

---

[1] 'The opening of this store is a great achievement for New Ireland.'

ment. He has done what I could not. It is a defining moment. The basis of equality in life.'

He stepped down and a group playing plangent guitars began to sing a sad melody for Justin that was quite moving in its simple appreciation and charming lack of self-confidence.

'Thank you, *tru*, Mr Justeen ... Mr Justeen ... Thank you, *tru* ...'

There followed an emotional speech by his island partner, the former schoolteacher, Bosston Tusikai. Justin had been granted use of the land to build the store by the Ngakuskusing clan. Given the fiercely-defended rights of land tenure, this agreement was a remarkable achievement and an optimistic sign of future developments throughout the country. The complex claims of land tenure lie at the heart of so much of the violence that has erupted in the Highlands. Here was an example of intelligent cooperation rather than a hostile demand for 'compensation'.

'I am confident all the way *tru* and believe in this dream *blong* me. Nothing bad will happen here.'

Justin himself spoke in English.

'In opening this store I hope it will serve as an example of what is possible given faith. The community now comes together as one.'

Catherine, his wife, distributed certificates to all those involved in the project. She came from Manus in the Admiralty Islands and was one of the most beautiful young women I saw during my travels. She had a fine-featured, exquisite face with almond eyes and straight dark hair in a long ponytail. She was refined and elegant, gold jewellery glowing against mahogany skin, sparkling blue nail polish, fragments of the sea clinging to her fingers and toes.

'For Eramis the builder, this certificate ... a small man with a large heart ...'

Tables were then brought and papers were signed and countersigned and signed again and countersigned until one was dizzy with it all, the terms agreed by the many members of the clans involved. After what seemed like hours, a pig tethered to a pole, struggling and squealing, together with baskets of taro, were dumped unceremoniously on the ground. Justin dramatically appeared from behind some palms dressed as a chief. He had been admitted to the clan. He was dressed in a crimson *laplap* tied with a palm frond, his bare chest and face dusted with lime powder and a crown of red croton leaves encircled his head. His white skin and slightly generous proportions made him appear faintly ridiculous but everyone smiled and took the gesture of solidarity in good part. There was a movement towards the entrance to the store and the MC grabbed the microphone.

'Now *cuttim* ribbon!'

And so the Tsoi Island General Store was officially open. Women were not permitted inside before they placed some money on a tray in front of the entrance. This neutralised the sexual taboo. The real celebrations and feasting could now begin.

Lunch of roast pork and taro was unpacked from a *mumu*,[1] the steaming meat and tubers removed from their leaf packets. Portions were cut up and distributed according to complex clan rules I was unable to fathom. Pork cooked slowly in the ground overnight, wrapped in banana leaves and buried among hot limestone rocks, is wonderfully soft and moist. The taro was bland but went very well with the meat. Accustomed to drinking wine with food, it suddenly struck me that the island was alcohol-free and only cordials and coconut juice were being drunk. Yet high excitement was in the air, everyone drunk on the adrenalin of the event, of 'something happening' in their restricted lives.

The scene by the turquoise sea was colourful and extravagant. Old women wandered through groves of tropical flowers puffing pipes, heads enveloped in clouds of harsh smoke. Men carried plaited bags containing their betel-chewing equipment tied to their waists by rose-pink sashes; a man wearing a huge sun-hat

---

[1] A type of oven excavated in the ground which uses hot limestone rocks to cook meat and vegetables slowly to tender succulence. It is used throughout Melanesia and Polynesia.

tied under his chin with a magenta ribbon, gathered at the crown into a festive puff-ball, gazed at the new store in wonder; blonde children with amber skin and huge eyes wore clusters of white shells; young bucks in European shirts smoked cones of palm leaf filled with rough tobacco, fastened with a sliver of bamboo. Each reveller proclaimed his individual sense of the decorative with a flower, leaf or piece of old plastic – a joyful occasion pregnant with anticipation.

The dances began. A group of male drummers laid hollow bamboo posts on the sand, sat in a line and began to strike them rhythmically with sticks. The sound was penetrating and insistent. The performers were dressed in yellow bark skirts with rosettes of the same in their hair; bunches of bleached seaweed were fastened to their shoulders. Excitement overcame the onlookers as the rhythms increased in intensity. The main characters appeared on the sandy stage. A story, it could have been of marital infidelity or courtship, was enacted with squeals from the children and shouts of encouragement from the adults. Lovers discovered by husbands or rivals, beatings and rushings about, pleadings for mercy – all this stylised erotic panache created vast amusement.

A group of female dancers in all-concealing smocks was the next item. There was no attempt at costume, only pink-and-white nylon fringes placed around their hair. Any physical display of sexuality or suggestion of eroticism had been totally smothered. They moved in a desultory fashion to some recorded, vaguely Oceanic music. Occasionally, one of the younger dancers would make an infinitesimal 'sexual' movement which evoked a wild response from the crowd. It was deeply dispiriting. Christian missionaries seem to have leached all the erotic energy from these women. How ironical that the so-called 'civilised' races are enthusiastically clubbing in Europe with all the raw sexual energy of pagan dance and here at the nexus of inspiration, the flame has been trimmed to a pathetic flicker.

A ripple of excitement ran through the crowd, a tremor that was quite different to anything that had gone before. Children screamed and craned for a better view. An emaciated old man surrounded by a keening group of dancers made his way towards the dancing place. The dancers were dressed in cloaks of split palm and wore conical headdresses of ferns with curious vertical attachments in red and yellow carved wood. They looked like radio antennas for collecting metaphysical emanations. The dancers were working the old man into a trance, turning him, taunting him, circling him, firing him up. He staggered backward raising his outstretched arms towards them in the posture a seated corpse might have been displayed in the past. The palm-leaf capes became sodden with sweat, bright green giving way to dark, evil strands dripping down their backs, lime powder running in streams. They chanted while he moved as if possessed by spirits. A sense of danger filled the air.

This was the sorcerer. He was swallowing oil and an assistant was smashing beer bottles and passing them to him to consume. His mouth became bloody, oil and blood dribbling onto his soiled *laplap*, his hollowed stomach caving in to meet his backbone, the dancers whirling about him urging him on. He bent double and straightened his withered frame rhythmically, his grey eyes seeking another world to this, his mouth gaped. The shards of glass slid down his gullet and he was close to me now. A terrible power and sickly perfume emanated from his heated body. The local villagers were petrified. An assistant offered him a woven grass basket and he spat some of the bloody shards into it, vomiting them up, his face glistening, his rake-thin body at the mercy of the force, invoking who knows what spirits to cleanse and protect the store. He collapsed into the arms of his sweating acolytes. The crowd were alight by this time, and I felt the authentic feelings of a pagan people like a punch in the abdomen; harsh, inexplicable, and far from Christ, demanding their original souls. The sorcerer from New Hanover was helped away, dragging his spindly legs over the blood-spattered sand.

Another dance for which New Ireland is famous concerns the visit of a bush spirit to a hamlet. The local people, birds and animals are frightened by the arrival of a *ges*, the spiritual double of a deceased person. Dancers decorated with coloured feathers moved with quick vibrations in mimicry of birds. They had attached painted herons, representing the totem of their clan, to their headbands. A dancer of great muscular strength and agility was wearing the fearsome *ges* mask, decorated with tusks and upward slanting eyes. His body was painted in chalk to resemble a skeleton. A long battle of complex choreography between good and evil began. The *ges* finally danced victoriously over the dead bodies of the clan, their animals and spiritual defenders. The whole performance was as carefully choreographed as a classical ballet. I found it impossible to discover anyone to explain it in detail to me, even the main dancer himself, who simply said it had always been danced like that and he did not know why.

As a gentle chaos took over, I decided to have a look inside the store. The shelves seemed somewhat bare apart from row upon row of neatly arranged tins of meat and tuna fish, tinned milk and biscuits. A brisk trade was already in progress, which seemed to augur well for the future. The effect on the local population must have been similar when the first Europeans set up their trading posts and introduced 'cargo'.

The opening had been a magnificent spectacle but the day was drawing in. I waded into the warm sea, the sand as soft as talc under my feet and clambered into the banana boat. About nine island boys crowded into the bow of the craft. 'Treasure Island' and its exotic visitors faded into the distance as we sliced through the placid, cobalt-green network of channels and straits, treacherous reefs and open ocean that separate the numerous islands. The sun was setting behind the craggy mountains and dense vegetation of New Hanover. Demas's austere profile was etched into this ostentatious backdrop as we sped across the glassy surface. It was gloriously cool on the move.

As darkness fell, I noticed one of the boys staring at my feet. He seemed fascinated by them, would look at mine then inspect his own, silently and thoughtfully comparing them. Mine, the feet of Western urban man, white, protected since birth, well-formed, tender, undamaged and cosmetically trimmed. His, bare from birth and tough, ravaged by coral cuts, dropped stones and hammer toes, machete wounds and crushed nails. Both of us looked down occupied with significantly different conclusions concerning this odd moment of communication. The comparison of feet seemed to speak volumes about the significance of colonialism on these distant shores.

We had been travelling through the gloom for about an hour without lights. Dim white fringes materialised around silhouettes of rocky reefs. The boys were a dark, humming huddle in the bow, the pilot stood firm and aloof in the stern, his hand on the tiller of the outboard, perfect night vision guiding us home. The lights of Kavieng were coming up distantly on the left. It was then we were told that we were running out of petrol.

As we steered away from the welcoming lights of the town into total darkness, the consternation could be felt rather than heard. In the distance a storm was approaching, lightning splitting the horizon, thunder distantly trembling. The pilot occasionally shone a powerful torch into the sea to check the depth. Without life jackets we could simply disappear in the approaching storm and be devoured by sharks.

Hours seemed to pass as our New Ireland Charon guided us slowly across the tropical Styx. A glimmer of light suddenly appeared ahead and we headed towards it.

'A man there has petrol I think. *Sopos nogat bai yumi bagarap olgeta yah!*'[1]

We approached the deep gloom of an island. The pilot shone the torch towards the shore, briefly lighting up forbidding mangroves and a ghostly beach. A naked lamp burned outside a hut. The place was completely deserted. Mosquitoes were beginning to bite ravenously.

Demas and the boatman, carrying a jerry-can between them,

---

[1] 'If not, we are completely buggered up!'

stumbled across the sand and disappeared in the direction of the light. As we waited for them to return, one of the boys began to strum his guitar softly and sing a sad tune. I rifled in my bag to find some insect repellent, convinced that this was my moment to contract malaria or dengue fever. They were away a long time and clouds of bloodthirsty marauders had by then descended upon us rapaciously. Occasionally, a solitary canoe would glide past in a disembodied fashion as if levitated above the water, and was then swallowed up by the darkness of the mangroves.

At length they returned and we sped off once again towards Kavieng after a bit of spluttering from the engine. The storm seemed about to break upon us. The silhouettes of islands were lit by jagged bolts of lightning. Yet it did not rain, despite the deafening thunderclaps that left my ears ringing. The boys sang with more enthusiasm as the lights of the town grew brighter. We arrived back at The Malagan Beach very late and it was clear many had feared for our safety. I immediately ordered a beer and began to reveal the glorious day to those who had decided not to attend the grand opening.

A group of three divers had arrived from the Discovery channel[1] a few days before, and I fell into conversation with them. Diving is the only significant form of tourism that exists in these islands. Groups arrive in a diving 'cocoon' with huge amounts of equipment, go out each day to see the incredible displays of large pelagic fish, barracuda and sharks on the reefs, and then after a week or so return to their own country. Although photographing rare black corals and barrel sponges, they rarely encounter the indigenous culture of the islands in any real sense. The previous evening I had noticed the group assembled in *sotto voce* discussions with a guide, poring over maps. They had come to film the famous shark-callers of Kontu.

---

[1] Discovery Channel is a US television channel which explores science and technology, geography and human cultures, history, the natural world, exploration, adventure travel and health.

The people of New Ireland are great shark-hunters or 'callers'. The men ('calling' is strictly *tabu* for women) prepare by living in the *hausboi* (men's house) of their village, fasting and abstaining from sexual relations and family contact for some days. They then paddle their canoes to the reef and with magic chants select a fishing site. The sharks are attracted to the sound of bunches of coconut shells being rattled in the water. When the predator swims near, the callers slip a noose attached to a type of wooden propeller device over the head. The shark exhausts itself struggling to escape and is finally clubbed to death. The haunting wail of a blown conch shell signals a capture. A certain amount of myth-making attends 'calling', but a more worrying problem is the dwindling number of sharks. Taiwanese trawlers are engaging in the brutal practice of 'shark-finning'. The sharks are caught and their fins are violently hacked off for use in soup and other culinary delights. The mutilated animal is then released back into the sea with a gaping wound on its back to die a desperate death, unable to control its swimming, sometimes eaten by other sharks who scent its blood. The formerly large groups of silver-tipped sharks which abounded in the Bismarck Sea have now decreased in numbers. After many attempts and promises by their guide, our American divers from the Discovery channel had so far failed to see a shark. But I was telling them about Tsoi Island.

'I knew we should've gone, goddammit!' one of them exclaimed.

The previous evening I had suggested they come and bring their cameras, but they had rejected the opening of a trade store as a boring idea.

'Do you need to go diving every day?'

'That's what we came here for! And to interview some shark victims.'

I reflected that shark victims rarely have much to say for themselves.

'Yes, but the island culture is so fascinating, particularly if you have good camera equipment to record the dances and music.'

'Yeah. I suppose you're right. We need a few seconds' rush of culture. Look, can you tell us where we might see some of the stuff you saw. A flash of culture in the film, you know, lighten up the shark victims!'

I turned the phrase over in my mind as I lay back exhausted with the adventures of the day. The frangipani lei had wilted around my neck and my shirt was damp with sweat. Gorged on pork, mind blown apart by the sorcerer and the excitement on Tsoi, I was still dealing with the stressful return in the frail boat. 'A flash of culture.' The opening of the general store represented in microcosm so many of the complex forces operating in Papua New Guinea at present – foreign investment and local land claims, compromises with resident European traders, Christian evangelical belief in conflict with paganism, the raising of the Slavic ghosts of Pacific exploration, idealism and the realities of the fledgling tourist industry, the joy of beautiful children mesmerised by the theatre of their own folk mythology – I perversely thought that the opportunism behind the reductionist phrase, '*a flash of culture*', could remain beneath the water with the tropical fish. I suddenly became very angry.

'No, I don't know about any other cultural events. Sorry, I can't help you,' I said, coldly.

But I had already planned to go to a *malagan* mortuary festival in Panafiluva with Demas early the following morning.

13. An Account of the Criminal Excesses of Charles Bonaventure du Breil,

Marquis de Rays, Grand Cord of the Negro Republic of Liberia,
Consul for Bolivia at Brest & etc, King Charles I of *La Nouvelle
France*, Emperor of Oceania, as recounted to me on the island
of New Ireland in the Bismarck Archipelago by the
Honourable Henry de Little of Dublin

Many strange individuals remaindered from former colonial
times inhabit these shores. None more so than the Honourable
Henry de Little, nicknamed 'Aristo' by the local Europeans. He
is an impermanent dweller who moves through a panorama of
constantly-changing scenes. In this way he avoids commitment of
all kinds, a noticeable characteristic among those social casualties
who have fled to Papua New Guinea in search of adventure,
escape or relief from suffering a 'civilised' existence.

Some years ago he described himself in the local two-page
newspaper as 'slight, vivacious, *soigné* in dress and uniformly
courteous . . . an elusive recluse'. He claimed to be the son of
John de Little of Dublin and related to the de Pettits, an old
emigrant Huguenot family. It is rumoured that an ancestor was
transported to Australia for stealing waistcoats in Vauxhall Gardens. De Little read Law at Trinity College, Dublin, before a
spot of ravishing the local female gentry put the constabulary on
his tail. The antipodes beckoned and he took passage to Sydney
but found life there as a solicitor insufferably prosaic, so this

adventurous dreamer set out for Papua New Guinea to try his luck as a planter.

A thoroughbred and entirely noble figure, Aristo's reserve and culture disintegrate markedly under the influence of cheap cognac. I remember his elegant figure gliding through the palms towards the lagoon; a face marked by solitude and fatality under an ivory *Montecristi* panama, boots of Russian calf salvaged from an eighteenth-century sloop wrecked on the reef, a cigar between his lips. His plantation at Kavieng had been sold long ago and is now under oil-palm. Suspected of smuggling bird of paradise feathers in the Highlands, De Little was subject to visions of the most violent and apocalyptic kind. Now in his late fifties, he was an authority on the notorious Marquis de Rays expedition to Port Breton on the south coast of New Ireland. Any visitor possessed of the faintest culture would set him alight recounting this notorious tale as the sun settled behind the palms and the *garamut* drumming drifted over from the shores of Nusa Island. Often drunk as a sponge, he would appear at the bar of The Malagan Beach in the early evening, dressed in black tie despite the tropical heat. This is the scarcely credible story he recounted to me one balmy evening as the village fishermen waded across the lagoon beside their canoes, gathering in their nets. The full account in all its outrageous detail has seldom been related in modern times.

It is a story of fraud and paranoia on a Napoleonic scale. The seat of the noble Marquis de Rays was the feudal castle of Quimerc'h in Brittany. During the upheavals of the French Revolution, the family, who were descended from a chivalrous Norman line, fled to Holland and later to London. The chateau was plundered of its Gobelin tapestries and harpsichords and turned into a barracks. After the Restoration, their fortunes revived and a rather architecturally dull mansion replaced the magnificent ruined fortress.

The Marquis de Rays, Charles Bonaventure du Breil, was born in 1832. Country life did not agree with him and he was subject to severe bouts of ennui. These dismal moods finally drove him abroad in search of adventure. The young Marquis combined a love of the exotic with a rare talent for failed enterprises. In America and Senegal, in Madagascar and Indo-China, his poor judgement was breathtaking. A fortune teller once told him that he would 'reign over an empire' which fed his paranoia if not his bank balance. Idly musing in the library of his manor one fatal afternoon, he read in a volume by the French Captain Louis Duperry of 'idyllic' Port Praslin in New Ireland, part of New Guinea. His unerring instinct for yet another doomed scheme was instantly aroused. Like a moth to the flame, de Rays was drawn ineluctably to disaster. He set about the outrageous task of establishing an empire on this tropical island, situated four degrees below the equator. His realm was to be called La Nouvelle France and was to include the vast area occupied by all the adjacent islands. The very idea of New Guinea has always attracted deluded dreamers, but the Marquis de Rays surpassed them all with the audacity of his vision. In the spirit of Napoleon III, he saw himself as the future Emperor of Oceania.

France had endured the terrible consequences of the Franco-Prussian War in 1870 and the even bloodier days of the Commune when Paris was put to the torch. It was a propitious moment to launch an overseas scheme. In July 1877 an advertisement appeared in various newspapers which played upon the French propensity to take *voyages imaginaires*. It read, 'Land in the Free Colony of Port Breton *terme à cinque francs l'hectare, fortune rapide et assuré sans quitter son pays.*'[1]

The Marquis addressed a meeting in Marseilles in messianic language, and the infant colony soon assumed, as he loftily termed it, 'the sacred character of its religious baptism'. Port Breton was painted as a veritable Garden of Eden that would earn investors a fabulous fortune and assist the Trappists and

---

[1] '. . . five francs per hectare, a rapid and assured fortune without leaving the country'.

the Chartreux in their conversion of the cannibalistic heathens. He proclaimed himself King Charles I of La Nouvelle France and bequeathed titles: investors who purchased twelve square miles were made aristocrats of the first class, a duke or a marquis. Those who purchased half that amount were deemed second class, an earl, viscount or baron. If the property was purchased inland, these third-class aristocrats were elevated to mere baronets or knights. A newspaper called *La Nouvelle France* was launched illustrated with engravings of the phantom settlement – a great harbour with ships riding at anchor, schools, churches, boulevards watched over by birds of paradise and pious nuns. Engraved portraits of the Marquis, maps and the music of 'The March of Port Breton' composed by the Liberian Consul to Paris, could readily be purchased by subscription. The paper gravely warned prospective investors that 'Ballet dancers are lacking, also night clubs'. By 1879 almost half a million pounds had been subscribed to the venture.

The first ship, a three-masted sailing vessel called the *Chandernagore*, was purchased and a crew recruited. The French Minister for Foreign Affairs immediately forbade it to sail. The crew were forced to sign off and the Marquis sent the vessel to Antwerp. A second crew were forced to abandon it by the French Consul there. De Rays now engaged any tawdry crew he could muster. They came under the command of the cruel and pretentious Monsieur le Baron Paul Titeu de la Croix de Villebranche, first Governor of the colony, and an alcoholic Belgian, Captain Seykens.

The poorly-provisioned ship with some eighty colonists aboard surreptitiously slipped out of the harbour at Flushing in Holland in the early hours of 14 September 1879, bound for Hell. At Madeira it was quarantined and forced to sneak away again. The heat was insupportable, the putrid food by now inedible and Captain Seykens navigated in a drunken state tied to the mast to prevent him from reeling into the sea. The 'gentlemen' supplied themselves generously with champagne and delicacies on the voyage to the chagrin of the starving colonists who had paid for it all. Discipline began to break down but was maintained by the use of thumbscrews and other 'medieval' tortures.

In the middle of January 1880, after a horrifying four months at sea, Port Breton was sighted. What a forbidding scene met the eyes of the benighted colonists! Instead of the Garden of Eden with crystal streams purling through lush valleys, in reality Port Breton was a terrible fever-hole. This region has the highest rainfall in Melanesia, carnivorous mosquitoes, blood-sucking leeches, scorpions, spiders and huge cockroaches. Only a tiny amount of arable land lay between the glowering rocky coast and rearing jungle-clad mountains, continually wreathed in low cloud. Even endemic species have difficulty thriving in the trapped air of that . miasma. With unerring instinct, the Marquis had chosen the worst place for a settlement on the whole extensive coastline of New Ireland. *En plus*, the native people who inhabited the area were enthusiastic cannibals and considered to be the most savage in New Guinea. Hot pebbles were known to be introduced into the rib cage of their enemies through incisions under the collarbone.

The plucky colonists made a creditable beginning. However, one night during a violent storm, the *Chandernagore* slipped away and sailed some miles up the coast to a mudflat. The colonists panicked when they awoke to find the bay empty and slashed their way overland through the jungle, bleeding and injured, desperate to find the ship. They found it riding calmly at anchor at a place called Likiliki. The area seemed superior to Port Breton, so they voted to stay there and marked out a large square which they named Place de la République.

The water turned out to be brackish and the malarial mosquitoes just as deadly. As they unloaded the ship, the surrealism of their situation became apparent. The Marquis had sent thousands of bricks and a marble altar to build a cathedral in the jungle but no mortar, there was machinery for sugar-refining but no sugar-cane could thrive, incubators yearned for non-existent chickens, the sawmill could not cut the hundred-year-old trees, a huge millstone had no grain to grind, cases of knife handles were without blades, the axe-heads were without handles, the wheelbarrows without wheels. There was no quinine but there was a statue of the Madonna. The colonists lacked even the basic necessities and were in desperate need of food. When they tried to board the vessel to escape, Captain Seykens met them at the companion[1] brandishing a revolver. The *Chandernagore* treacherously stole away under cover of darkness with the Governor Baron de Villebranche on board, abandoning the colonists to their ghastly fate. Upon reaching Sydney he promptly 'abdicated' never to return. He sent the Marquis a cynical telegram: 'Likiliki occupied. Friendly relations entered into with natives. Send money and orders.'

Their plight came to the attention of the formidable Reverend George Brown who had established a Wesleyan mission station in 1875 at Port Hunter in the Duke of York Islands, not so distant from Hell. He came to their rescue.

> About twenty men came staggering down to the beach, and stood there awaiting us; some with bandaged legs, all emaciated, sunken-eyed, and mere skeletons, a ghastly sight . . . They could not take their eyes off the ship.

After a period of recuperation during which many died, the colonists were returned to Likiliki where to their fury, a brig despatched by the Marquis arrived from Sydney loaded with thousands more bricks.

Many now tried to escape including a young Italian named Boero who stole a canoe and together with five companions set sail at once for the Duke of York Group. Unfortunately the north-west monsoon drove them towards Bougainville where the

---

[1] An old nautical term for the type of wooden hood placed over the entrance or staircase to the master's cabin in small ships.

leaking canoe fetched up on Buka Island. His companions were promptly eaten by cannibals at which point Boero burst into tears of despair. The warriors had never seen such a human phenomenon and indicated that if he continued to weep and tear his hair for their entertainment they would spare his life. The absurdity of his situation suddenly made him burst into laughter which equally fascinated the cannibals, as they were much accustomed to glowering. Thus was Boero forced to embark on an endless cycle of weeping, singing, laughing and dancing to preserve his life. He was thrown rotten fruit and some unpleasant boiled parts of the human anatomy to maintain his strength. Boero was finally released from this circus a few months later by the second ship of the expedition, but never recovered his sanity.

This ship, the small steamer *Genil*, sailed from Barcelona with a Spanish crew under the Liberian flag in March 1880. Captain Gustave Rabardy had a villainous reputation and imposed discipline with some medieval instruments of the Inquisition he loaded in Spain. He had a rack, thumbscrews and a few steel-studded whips. His favourite punishment was to string men up by their thumbs. Faced with this demon, in Singapore the colonists and crew deserted as a body, apart from the cook. Rabardy managed with difficulty to recruit a Malay crew with sundry Europeans, but like many cruel men developed a morbid fear of assassination. He began to take his meals with a loaded revolver on one side of his plate and a slave girl on the other. A bodyguard of Buka warriors, men with the blackest skins in the world, padded about the deck. His 'slave' was a pretty native girl from Bougainville called Tani whom he had purchased from her father. She followed him about and lay at his feet like a dog. He arrived at Port Breton in August 1880. The wretched remnants of the colony, often referred to by the Marquis as the 'Light of the World', came under Rabardy's 'protection'. The *Genil* rocked at anchor in the amphitheatre of death.

De Rays had also sent some representatives to market his

venture at Udine in the north of Italy. Italian farmers were suffering from political strife, so some three hundred and fifty optimistic colonists, more than half of them women and children, were lured to the specious paradise. This would be the third ship of hope. They paid their fees and were duly transported to Barcelona to embark. The *India* was a large steamer of three thousand tonnes, converted from an old sailing ship. She set sail in July 1880 under Colonel J. A. Le Prevost, the next Governor of the colony.

All three ships had departed so close together there was little time for reports to filter back to Europe that the colonists were in the clutches of a madman beset by millenarian visions, although some warnings did appear in the Geneva newspapers. The engine often broke down, water ran low and the food was filthy (the very dogs refused it). Salt horse provided the delicate consommé in which the ship's biscuit was lightly poached together with the fat weevils that infested it. Rumours of a poisoning plot flew through the ship. When Port Breton hove into view four months later, some one hundred colonists were dead. The survivors had sailed optimistically down the east coast with its delightful tropical bays and flat terrain. Their spirits were high until they rounded Cape St George and their pitiable colony came into view. The *Genil* lay simmering menacingly in the heat. Their horror can only be imagined when instead of the broad avenues, churches and mansions of the prospectus, a few pathetic huts and a blockhouse graced the shores of a sinister bay. The pharmacist of the expedition set off to explore the hinterland. Only his scalp was returned to the settlement.

The colonists worked in terrific heat and humidity, surviving on three ounces of rotten meat or putrid fish per day. Half of them went down with fever and other illnesses. By November the colony was desperate for food. The *Genil* sailed to Sydney for supplies, promising to be back within two months. She finally arrived at Port Breton four months later, to find the settlement deserted apart from some lunatic dogs dashing along the beach.

Three thousand unused jewelled dog collars lay in a heap beside a number of Louis Quinze lamp-posts and a fountain.

De Rays' insidious little journal continued to report 'A nation is born' at Port Breton. The provisioning of the *Nouvelle Bretagne*, the fourth ship of the series, completed the cast of the operetta. The Marquis was seen posturing on the wharves at Barcelona in a coat heavy with gold frogging. Wine flowed, dinners were held and generals wept with the emotion of it. A number of Punch and Judy shows complete with puppets and stage were loaded together with countless boxes of pink and white satin slippers for the Court of Oceania. Twenty-two cases of official stationery, stamped with the coat of arms of the Marquis bearing the motto of his House, 'I spare the weak, I humble the mighty', were lowered into the hold. Gilded lions rampant adorned a collection of uniforms for the Court.

By chance, there were two excellent officers on board this ship to care for the one hundred and fifty colonists, Captain Jules Henry and the surgeon, Dr Baudouin. The passengers were mainly elderly folk, women and children, together with the normal complement of fierce dreamers. A prosperous barge-owner and his family from Lorraine had sold everything to invest in the scheme. He named his future property 'Nancy' and the stream that he supposed ran through it, 'Moselle'. Another polished his shotguns in anticipation of the hunt. Yet another planned to erect a luxury hotel to which visitors would flock from all over the globe.

At Singapore Captain Henry purchased a 'pontoon', an old sailing vessel infested with vermin, 'especially cockroaches flying all over the ship', as noted by the colonist Octave Mouton. It was renamed the *Marquis de Rays* which misled the French authorities that the Marquis himself was indeed on his way to New Ireland. This horror was towed behind the *Nouvelle Bretagne* and moored in the port to function as a hospital ship.

The nefarious Captain Rabardy aboard the *Genil* had meanwhile been diverting himself by mowing down native warriors in their canoes with a Gatling gun fastened to the deck, watching plumes of blood slowly spread over the pristine reefs. As 'provisional governor' he greeted the arrival of the horrified colonists aboard the *Nouvelle Bretagne* in sepulchral tones: 'Unfortunate beings, what brings you here?' In dismay they viewed the lines of black crosses on the beach. The promised peacocks were not strutting on the lawns of the Grande Maison, nor was the fountain splashing. The only harsh cries came from their own throats as giant centipedes nipped their toes.

The madness crossed cultural barriers. A renegade native chief called Maragano styled himself 'King Kanaka' and 'sold' the whole of southern New Ireland to Rabardy in a ludicrous ceremony for some tobacco, clay pipes and colourful scraps of cloth. Rabardy began to suffer from *tropenkoller* and would not stir from the *Genil* for fear of assassination. He crouched behind the Gatling gun, refused to communicate with the shore and he set signals that forbade any visiting ship to call.

On 25 August 1881 the Feast of St Louis, a High Mass, was celebrated on the deck of the 'pontoon' *Marquis de Rays*. There was a salute of guns and hymns were sung. One could almost have believed in the successful establishment of the colony. At the same time at Lyon in France, a banquet was being held. An engraved menu maliciously described the fare:

BANQUET OFFERT À MM. ROUBAUD ET SUMMIEN,
MEMBRES DU CONSEIL COLONIAL
DE LA NOUVELLE FRANCE LE 25 AOÛT 1881
(*Fête de Saint-Louis, Roi de France*)

MENU

Potage	*Nouvelle France*
Saumon	*Chandernagore*
Volaille	*Genil*
Filet	*India*
Champignons	*Liki-Liki*

Dr Baudouin did his duty and elected to remain behind with the remaining forty Port Bretonians who seemed to have fallen under the spell of a *fata Morgana*[1]. They wanted to rise to their mad challenge at whatever price. Rabardy had received secret orders from the Marquis that the colonists were to be kept there at all costs, even execution being permissible. 'By *divine right* I command you to do so in my name,' he had written. Rabardy took to having his food and drink tasted as his dementia increased. His faithful Tani, the native girl, slept on a mat outside the door of his cabin and attended to his every insane need.

The settlers now began their final downward spiral as thirty-five of the forty fell ill with fever. Too ill to move, soldier ants began to attack their eyes and ears as they lay on their palettes. In the Duke of York Islands, Dr Baudouin and Captain Rabardy enlisted the assistance of the neighbouring trader Thomas Farrell, lover of the notorious planter 'Queen' Emma. The remaining colonists were loaded with their pathetic goods aboard the *Genil* and the hospital ship *Marquis de Rays*, and the sad convoy steamed towards Mioko. The first signs of civilisation they encountered as they entered the pretty harbour were the strains of the 'Blue Danube' waltz floating through the open windows of a mansion, played on the piano by the 'Queen'.

Captain Rabardy died at the plantation in mysterious circumstances. He may have been poisoned. Farrell refused Dr Baudouin permission for a postmortem, pleading the speed of decay of corpses in the tropics. Little Tani was the only person who showed sadness at his passing, crying inconsolably for days then suddenly brightening up like a child.

'Where is your Capi?' asked the good doctor.

---

[1] From the Italian *fata* meaning 'fairy'; *morgana*, sister of the British legendary hero Arthur, apparently located in Calabria by the Norman settlers. Originally the 'spell' was the influence of a kind of mirage frequently seen in the Strait of Messina attributed in early times to a fairy agency. Often used figuratively in modern times to indicate the presence of a hypnotising power that paralyses the will.

'Gone, gone,' she chirruped and ran off to play.

'Queen' Emma later furnished her mansion 'Gunantambu' on the Gazelle Peninsula with the tapestries, grandfather clocks, Louis Quinze tables and other precious objects 'purchased' from the desperate colonists. She turned the marble altar for the cathedral into a bar for fancy Yankee drinks.

Henry de Little paused and gazed at me in the unfocused, heavy-lidded way of the inebriate, dwelling at length on this amusing thought.

'The Marquis never did see the domain that lured so many into the clutches of the grim reaper.'

'How did he end up?'

'Tried for fraud and spent six years in prison. Then he attempted some lunatic scheme of making gunpowder out of Breton granite. His last extravagance was a financially disastrous luxury cruise on the liner *Tyburnia* restricted to titled ladies from the English and French aristocracy. He believed them to possess superior manners. His final laughable delusion.'

The night was warm and soft as I began to gather myself to turn in.

'The story is like an incredible opera without the romance.'

'Well, there was always Tani, the native girl. Hope you enjoyed the tale, *mon vieux*! I'm going to toddle off now. May see you in the morning if the ghost of de Rays doesn't get me first!'

Henry de Little finished speaking with a grand gesture across the lagoon, lost his balance and narrowly avoided throwing himself into it. He lurched elegantly away through the coconut palms towards his lodgings in Kavieng, settling the *Montecristi* panama at a rakish angle and chomping on his dead cigar.

14. 'In Loveing Memory'

Inscription on a grave marker
Panafiluva Village, New Ireland
Bismarck Archipelago

The day of the mortuary festival in Panafiluva began in the usual way with no assurances about any aspect of the event. Melanesians prefer vague arrangements and fuzzy boundaries to timetables and exactitude. Time is a process without scale rather than a quantifiable commodity. 'Jesus knew my grandfather,' I was once assured by a local man. I collected the absurdly expensive four-wheel drive from an exotic mixed-race island woman at her home which doubled as the hire-car office. Demas was waiting for me beside the stuffed crocodile mouldering in a glass case at The Kavieng Hotel.

We hit the rapier-straight Boluminski Highway and were soon on the outskirts of Kavieng. I asked him about yesterday's ceremonies.

'I had tears in my eyes, Michael.' His reply surprised me as I saw him as rather a macho figure quite beyond sentiment.

'Why was that?'

'I had tears in my eyes because Justin had achieved so quickly something I could never have achieved as Premier.'

'How do you mean?'

'The money from national government never gets there. Now

a white *bikman*[1] can come along and give the people something in six months. I was never able to do that. In the end I resigned my office.' Something heavy seemed to hang suspended in the air of the cabin and I remained silent.

First stop of the morning was at the Catholic church, an old, cream, weatherboard building with blue windows and a simple blue cross above the tin-roofed porch. A perfect South Seas mission church. It stands at the end of a lawn among hibiscus and coconut palms with striped cordylines planted on either side of the broken concrete steps. Two of the most beautiful children I have ever seen ran across the grass; a brother and sister about six years old, he in orange shorts, she in a short, orange *laplap*. The girl had a huge shock of blonde curly hair, as unruly as a Renaissance cherub. The contrast against their golden skin was breathtaking.

'Tell me about marriage and divorce in New Ireland, Demas. How do young couples go about love affairs?'

Accounts of 'primitive' customs by the indomitable Reverend George Brown writing in the nineteenth century hovered in my memory. He had described one New Ireland custom of sewing young girls of six to eight years of age into individual conical houses made of pandanus palms, built a few feet above the ground, shaped like candle-snuffers. Only when their breasts had fully developed, which could involve years of this tortuous imprisonment, would they be finally released for their marriage feast.

Charging down the Boluminski Highway caused Demas to open up about many issues that would have been difficult to discuss face to face, although his answers were often cryptic and difficult to unravel. He had been a politician, after all.

'Well, there are two ways – the traditional and the modern. The traditional is dying out with the influence of television in towns. But in my village, if a boy likes a girl, the mother of the

---

[1] Pidgin for 'bigman', 'entrepreneur' or wealthy man worthy of respect.

boy makes up a parcel of taro and fish and the boy gives this to the girl's mother. This is the sign that there is love interest. Also at the dance in the village that sometimes precedes the *malagan* feast, a boy is able to take the hand of a girl he likes. 'Ah . . . so now we know, Ah . . . so now we see!' everyone says. It's dark at night around the villages, only a kerosene lamp or flickering fire, secret meetings can easily take place.' He seemed quite amused by my interest in these things.

'And what happens next?'

'Sometimes we give the boy a "trial". Stand him up against the wall and thrust at him with a spear. If he does not stand firm, he is not man enough for her!' He wheezed gaily with laughter. I was never sure whether he was playing me in the spirit of Margaret Mead's more imaginative informants. Perhaps he was simply saying, like so many Melanesians, what he thought I expected to hear. Boys with T-shirts wrapped around their heads like turbans walked the highway in the solid heat. Girls leaned on the gates to the missions.

'You never see couples, even married couples going around together Demas. Why is that?'

He tried to evade this question. I found that he resisted many of my 'why' questions. Strong clan bonds and densely-wrought kinships prevent an easy understanding of relationships in this sexually-polarised society. Certainly, women and dogs seem to have a hard time of it in the Black Islands.

'We don't have the same idea of romantic love and tender affections as Europeans. On Manus Island the couple are quite ashamed of each other. A man is closer to his sister than his wife there. Men feel they don't look impressive sleeping with women. Their skin dries up.' There was little I could say in reply.

'If people get divorced, what happens?'

His brow furrowed with slight distaste as if this was a familiar question from an amoral European that meant precious little in his culture. He gave a 'democratic wave' to clusters of villagers beside the road who clearly recognised him.

'Before marriage, as part of the bridewealth negotiations in shell money and pigs, land titles are agreed between the clans. Divorce isn't encouraged, Michael. We don't own land individually, so there's nothing to argue over if a couple separate. Your question is . . . well, irrelevant.'

'All right then. And the children, if the worst happens?' I kept swerving around the largest holes, the jeep lurching wildly.

'They are just taken into the extended family of the clan.' The way he put it seemed the most natural thing in the world.

'We're very close in the families. If a child has an argument with father or mother, he just runs to an uncle or aunt somewhere in the village. The parents aren't worried and know the child is safe. We could teach you a few things about morals!'

His dark eyes flashed at me and his voice was suddenly vehement. The oil-palm plantations, an endless monoculture, had begun to unroll like stage scenery. From deep within the groves, workers looked up from their steaming heaps of husks with expressionless faces.

'Yes, I know. Our children argue with their parents, get into drugs, even disappear. Our culture is changing fast too, Demas.'

'So we all have problems.'

'Of course! But you have the "Melanesian Way" of bonding the family. Yours are the happiest children I've ever seen in my life. Secure values, secure families.'

'The greatest social danger we have is young people bringing back new ideas from the city to the village. Computer games. Marijuana. Alcohol. Free sex and stealing the women. It is destroying everything! We must keep *kastom* alive as well as the new life.'

It is impossible to stop the juggernaut of cultural change even on remote islands, but the closeness of the clan remains intact. The lack of electricity in villages has saved them for now. We passed a group of replica 'warriors' dressed in pandanus leaves and carrying spears, standing together on a hillock. They shook their spears theatrically and laughed as we passed. Many tra-

ditional ceremonies have accommodated to modern work schedules and now take place over the weekends.

'Something is happening here, but drive on.'

His unaccountable logic left me nonplussed. The fetid smell
of oil-palm husks steaming in orange heaps beside the road made
me choke.

The village of Panafiluva was spread over a large area opening
out onto a classic tropical shore. There was activity everywhere,
large numbers of women and children sitting around in open
shelters waiting for the ceremony to begin. The arrival of our
four-wheel drive outside the chief's cookhouse was greeted with
great excitement, particularly the children who screeched around
it in droves. Demas introduced me to all his relatives and we
went inside. Everyone was friendly and we drank tea.

After we had chatted for a while an elderly lady produced
some shell money which she sternly laid out in various lengths
on a table. There was intense discussion about the purchase of
it. Shell money differs all over the Bismarck Archipelago. The
type in northern New Ireland is fashioned near New Hanover.
It is made up of hundreds of tiny, rose-coloured discs about 7mm
in diameter and 1mm thick. A minute hole is drilled through the
centre of each disc and groups of about twenty are threaded onto
a thin black string, interrupted at intervals with a smaller number
of white discs. The value is derived from the rarity of this pink
Patella shell taken from the coral reef. My string, costing 10 *kina*,
was almost a metre long and terminated in tiny red and yellow
beads. Shell money is often used when bartering for pigs and in
bridewealth negotiations. Demas tied it around my neck and I
felt suitably festive.

He began speaking to the young chief of the village in low
conspiratorial tones.

'I must educate him into the correct way of some parts of the
ceremony. Excuse me.'

His expression became grave as we walked towards the large
open area in the centre of the village. This was where the feast

was to take place. The young man wore his hair in dreadlocks, had surprisingly white teeth and was dressed in the usual nondescript shorts and T-shirt. He listened carefully to the advice as he chewed betel nut.

The *malagan* mortuary festival is held some time after death, sometimes many years after. The occasion is determined by the financial means of the sponsoring clan and does not hold to a fixed schedule. The spirit of the deceased must be given time to become accustomed to the world of the departed. The living in their turn are to be freed from the claims of the dead. For subsistence villagers, the expenses are great, since a *malagan* 'runs on pigs'. A single feast is usually held to honour a number of the deceased, rather than an individual. As we have seen, even old folk who are not yet ill, but approaching death, may be included to save money. I was indeed fortunate to be able to attend a grass-roots *malagan* as opposed to a large ceremony for a great chief. I would see how the *malagan* of the average person was evolving under the pressure of modern constraints and Christian belief.

We reached the *malagan* enclosure, a rectangular area erected around the graves, surrounded by a tall bamboo fence. A dried palm-frond screen concealed the entrance. Women and children are not permitted beyond this screen, but men can enter through a small opening marked by a forked branch. The *hausboi* or men's house was on the beach behind this enclosure. A stand had been erected to one side for the slaughtered pigs. I counted about thirty of different sizes stacked in rows, their skin mottled brown where the bristles had been burnt away. Blood congealed around the occasional puncture, weeping through the skin or dripping from the stump of a leg where a trotter had been severed. Powdered lime had been brushed in a long streak down each snout indicating that the pig had changed status when butchered at the sacred place of the *malagan*. Pigs are symbolic of the prosperity and standing of a clan. Scrawny dogs were licking up the blood from under the stand and were brutally driven away with rocks, yelping away in pain and cringing with practised intensity.

Beyond the pig stand towards the beach a large earth oven had been excavated. Huge amounts of meat and taro had been cooking overnight. Adjacent to it, live pigs were still being slaughtered, or rather suffocated. I went over to watch this unpleasant procedure. A young man in a flowered scarf, jeans and a T-shirt was sitting astride a small pig that had been tied to a bamboo pole. He was binding black insulating tape tightly around its snout. A bloody contusion near the ear indicated it may have been stunned beforehand. The pig struggled to breathe and then gradually ceased moving. Another boy violently stuck his finger in its eye and another jammed his finger into its anus to see if it was dead. Eventually the prodding produced no reaction and the animal was quickly carried to a bonfire frame on the beach. Dry palm fronds were heaped around it and lit. The animal was engulfed in flames and as soon as the bristles had been burnt off, it was taken to the beach, cut open and the entrails removed. The guts were thrown into the sea where sets from other pigs bobbed about. Small sharks thrashed the surface and pulled them down. The carcass was then carried to the stand and piled on top of the others.

A shout went up as the first screen around the *malagan* enclosure was torn aside. Wild cries accompanied the removal of the palm-frond barrier. The ceremony was directed by a man dressed in a scarlet bandana and *laplap*, carrying a woven basket. He constantly encouraged the crowd with harsh exclamations and tried to whip them into a passion through fierce expressions of disapproval. Five gaudily-painted concrete grave-markers were revealed in the shape of Christian crosses. Two had swollen Sacred Hearts pierced by swords dripping blood, another a lighted candle beneath the letters RIP. These people had died at least four years ago, one as much as seven. I was disappointed that there were no 'feverish nightmare' masks generally associated with the *malagan*, but this was clearly a ceremonial that .

had been modified for reasons of expense and perhaps Christian belief. The use of painted markers with Christian iconography as 'feverish' *malagans* seemed a good solution to the cultural dilemma of coexisting beliefs.

Cloths were spread on the ground before the crosses. Contributions to the clans of the deceased were invited with shouts of encouragement in the Kuat language. A tropical storm suddenly broke over us with powerful thunder and bolt lightning. Demas and I retired to the cookhouse. Torrential rain came through the roof as I handed out dried fruit and cigarettes. We drank black tea. The calling for contributions went on for a good two hours, the amounts of money given reflecting the status of the clan. The crowd around the entrance to the enclosure became quite unruly at times.

'It should not be done like this. This is not very good!' Demas was saying to me in an agitated voice. Other 'big men' were shouting criticisms of the protocol; such abuse is considered quite normal.

'You see, they only throw one or two *kina* or just a few *toea*.[1] This is a small *malagan*.' Demas was quite dismissive. There is a strong competitive element among clans in these ceremonies.

A young girl of about twelve attracted my attention as she was so intensely involved in the proceedings. She had curly, blonde hair, a pearl-shell rosary and crucifix lay around her neck against her cinnamon skin. She was craning around the legs of the adults standing before the *malagan* enclosure with the beads in her mouth, her eyes absolutely alight with excitement. The uninhibited emotions of this beautiful child avidly absorbing her own ancient culture, but without any apparent conflict with her Christian belief, made me optimistic for her future.

Without warning a huge cloud of white, acrid smoke engulfed the village. The earth oven had been opened! A tangible thrill of anticipation rippled through the crowd. I wandered over to watch

---

[1] The PNG currency of the *kina* is divided into one hundred *toea*.

the limestone rocks being thrown out of the pit, still hot and smoking. Huge parcels of pig and taro, brown with fat and white with lime, had been wrapped in banana leaves fastened together with vines. They had been baking overnight. With whoops of delight, the roast meat was carried triumphantly into the 'village square' on sheets of corrugated iron. It was laid out on palm fronds covering the ground. The women dealt with the taro and excitedly decided on the portions for men or women with banana leaves placed in the bottom of plaited baskets.

A *garamut* slit drum sounded distantly across the village and the first dance group appeared. They were a mixed group of men, women and children, less decorated than the groups on Tsoi. They were dressed in pandanus leaf skirts and collars with yellow 'halos' circling their foreheads and a feather ornament in their hair. They danced with fiery energy and commitment, particularly the children, as if their very lives depended on it. They began in a stately and slow rhythm, with endless repetitions of the same sung phrase and beat, until they fell into a hypnotic trance, the emotional abandon driven to a frenzy by the bamboo drum. A senior villager with a blonde moustache berated them without mercy.

'Singup! Singup! Singup!'

The group resumed with even greater abandon, the young girls watching their leader with furious intensity. Suddenly, a man from the audience rushed among the dancers and with a resounding smack, slapped a handful of lime onto the bare back of a performer. A cloud of dust rose and drifted off. Many others joined in until the air was full of powder and the dancers were in a melee. Demas told me later that it was to acknowledge relatives in the dance and disperse evil spirits.

When the performance was complete, the villagers began to mill about the area laid out with food until the next group arrived. Captain Moresby's 'pariah dog' was much in evidence at this time. Driven mad by hunger and the smell of roast meat, dogs were whipped, kicked or stoned if they dared approach the pieces of pig. Puppies yelped. Bitches on heat trailed past, howling when approached by dogs beside themselves with desire, their famished hides stretched over their backbones. I recalled an incident concerning dogs described by Captain Moresby when he visited Milne Bay in the 1870s. Some warriors with faces painted black and white returned to the *Basilisk* after a misunderstanding, bringing 'one of the lean wolfish curs that infest their villages . . . the leader carrying the dog in his arms, dashed out its brains on the quarterdeck.' The gesture was intended as one of peace. Dog's teeth were valued by local warriors as ornaments for necklaces, and would have been extracted from the carcass and offered as a gift. The naval officers, not surprisingly, were upset that their almost sacred quarterdeck had been so bloodily 'defiled'. At the *malagan*, snarling dog fights broke out among the pig bones, taro and oily palm fronds. A man fought a dog for the jawbone of a pig and hung it around his neck on a strip of leaf. A land of beautiful children and wretched dogs, indeed.

'*Dispela em kaiks blong Maiks!*'[1] announced the Master of Ceremonies as he handed me a packet of food wrapped in a banana leaf.

As an honoured friend of 'the boss', I had been given some choice portions of roast pig, taro and banana and retired to the chief's cookhouse like a medieval prince. The slab of pork looked alarming but was delicious and went well with the bland taro, sweet baked bananas and black tea. I showed some villagers a European art book about *malagan* sculptures and ceremonies. They crowded round in great excitement, the old people recognising the places of past feasts, the young amazed at illustrations of the old masks.

'*Wanpela man i save rait lon mipela!*'[2]

They had clearly never been educated in their own culture,

---

[1] 'This is the *kaikai* of Michael!' *Kaikai* is Pidgin for food; '*kaiks blong Maiks*' is rhyming contemporary Port Moresby Pidgin.

[2] 'Someone is writing about us!'

through the written word or photographs. These were words and images cemented in documented time, our tradition. The mission schools offer education, but a Christian education, not based in the rich indigenous traditions of fables, stories and the occult. Oral memories are fragile and the assaults upon them by colonisation, war, missionaries and the market economy have been almost terminal. The current appetite for fantasy, myth and magic in the West would find a deep resource in Papua New Guinea.

'Some people make *planti moni* from us, make a lot of money from these books!'

I tried to explain that this was not usually the case with books but they were having none of it. There is a deep resentment of white people who can 'afford to travel' and 'write books about us'.

A murmur in the crowd and then it was time for the butchering and dividing of the uncooked pigs. Men with huge machetes leaped over carcasses thrown off the stand by a villager dressed in a spectacular scarlet *laplap*, decorated with the occult 'eye of fire'. The flying knives divided up the sloppy, warm meat with superb skill. Some animals were so big the butchers needed to climb inside to dismember them. The size of the pieces and the particular parts of the pig reflect a complex exchange process in the repayment of obligations and debts in kind. Every man knew beforehand what pieces had been selected for him and there was no discussion. The ceremony wound down to its conclusion and the local people drifted away.

Droves of children ran behind the four-wheel drive as we left. We rolled down the Boluminski Highway once more on the long drive back to Kavieng. Demas had fallen silent and was quietly chewing betel.

'Turn left here! I want to show you something.' He suddenly erupted into life.

We lurched down a long, grassy drive, past the mission of St Francis and some half-finished houses, to a clear natural pool

in a jungle clearing. Children were swinging from ropes and
leaping into the water with extreme excitement. Round and
round and round they went in this simple cycle of pleasure, the
squeals of delight and energy expended with never a pause.

'You see what freedom and joyfulness our children can find!'

He gazed at them with real pride and paternal devotion. He
had simply wanted to show me this joy. The Reverend James
Chalmers had noticed this quality too, and written about it a
hundred years ago: 'What happy children! Far happier than most
of our British children.'

The next stop was at his guesthouse at Fatmilak to collect
something. I was waiting in the car when a terrible wailing rent
the air. It was as if all the harpies in heaven had descended.

'What ever is going on, Demas?'

'A *pikinini*,[1] the son of the pastor, has just died,' he said, in
an alarming matter-of-fact voice.

'How old was he?'

'About ten years.'

My emotions were brutally wrenched away from the Arcadian
pool we had so recently left.

'How terrible. How did he die?'

'He got sick with malaria. Fortunately, he died at home and
not while at the river or in the sea. If he dropped down in the
sea they would never have found him again.'

The women were keening wildly as we left Fatmilak.

Demas resumed his comatose state as we rattled towards Kavieng. The astonishing differences in cultural response to death
had affected me. There was such a startling contrast between my
mother's funeral – Requiem Mass in a city cathedral – and the
malagan memorial service in this small New Ireland village. In
the city, mourners had arrived in the sombre luxury of gleaming
black Mercedes-Benz cars. The catafalque lay on the Victorian
high altar, incense rose, organ adagios by Bach and Couperin

---

[1] Pidgin for 'child'.

echoed in the Gothic vault. The priest wore vestments of black and gold and the bells rang half-muffled in the tower.

On this remote island in the Bismarck Archipelago, the ceremony was accompanied by a passionate slaughtering of pigs, wild dancing and explosions of lime powder, the laughter of children and the casting of money, the howling of dogs over graves. Their 'cathedral' was a hot tropical shore of aching beauty and pristine clarity, the spirits of the dead gliding about the jungle. Under an austere Christian vault, I was forced to wrestle a solitary fear of death and the judgement of the great God upon my individual little soul. In these islands there was a liberating collective celebration of death, an entire community bolstering itself against the inevitable, the living drawing strength from within the clan.

On the return to Kavieng, toads again exploded under my wheels in the dusk, heads foolishly alert and erect, facing their own dark hammer.

15. 'The Sick Man Goes Down
with the Plane'

PNG pilot on a flight to Lihir Island

The pale blue hotbox that served as the police van lurched under an optimistic banner stretched across the road which read 'Inspirational Messages to Come'. The truck that was to take me to Kavieng airport had failed to materialise.

'He's collecting chemicals from the wharf,' was the unlikely explanation offered for the delay. The check-in-desk at the terminal was unattended. A luggage tag had come off one of my bags and a blank one lacking a destination was hurriedly tied on. The *Islander* light plane was bound first for Lihir Island and then Rabaul in East New Britain. I had been unable to arrange the necessary security clearances to visit the famous gold mine on Lihir, so I arranged to fly over the volcanic island at low altitude. Unlike the intrepid early miners forced to negotiate treacherous, uncharted reefs, this brief trip allowed me effortlessly to cross the short geographical bridge of the Duke of York Islands that link the cultures of New Ireland and New Britain.

'Mind that pool of oil, sir, as you get in,' remarked the pilot cheerfully as I clambered into the seat beside him. The twin Lycoming engines chattered into life and we trundled down the runway at what seemed a snail's pace before labouring into the air. The water below was a startling ultramarine with reefs appearing as dark stains on the ocean floor. I could see Paradise Island (Lissenung) where I had spent my last day diving on the scattered remains of a Catalina flying boat. The diving centre is run by a German couple – there were rows of gleaming air cylinders in ordered, numbered racks. Dietmar, the owner, was stressed. He was organising a long journey back to Port Moresby to see the dentist. There are only a few good ones in the country. Until recently, the nationals and expats all swore by one who operated a treadle drill and was both deaf and dumb, advantageous attributes it would seem for a dentist. He was finally forced to retire when blindness overtook him.

A tropical storm appeared from nowhere shortly after take-off and we were unable to climb above it sufficiently quickly. Lightning shot across the wings giving our faces a lurid pallor. We were thrown about like revellers on a roller-coaster. The pilot smiled grimly as he fought the controls in zero visibility, and I thought the wings were about to be wrenched off the fuselage. They flapped like a bird. The engines staunchly maintained their 2200 rpm despite straining at their mountings. My own flying licence had lapsed years ago, but I could read the erratic instruments with informed alarm. They had long lost their original paint and had been hurriedly touched up with a brush. The windscreen was an impenetrable river. After what seemed an eternity of bucking on the tattered fringe of death, we just as suddenly emerged into an azure sky. There was audible relief within the cabin. The plane glided down like a blithe tropical bird on the approach to rugged Lihir.

It had been known since the sixteenth century that there was gold in New Guinea. There had been a number of failed gold rushes but it was not until 1888 that significant quantities of gold were found in the Louisiade Archipelago in the far east of the country. Suddenly four hundred men crowded onto the tiny island of Sudest, followed by similar gold rushes to Misima Island. Samarai in East Papua became the base station and soon filled up with the tents of alcoholic and pugnacious prospectors, both government and private. Melanesians were treated as virtual slaves. The next discovery was on Woodlark Island but this resulted in a terrible death toll from dysentery and malaria in the damp climate. The macho miners stubbornly refused to prepare adequately for the tropics. Gold strikes on the mainland were hampered by defensive, murderous attacks by the local people. The British New Guinea Annual Report of 1899–1900 offers this graphic description by the magistrate:

> The boneless leg was wrapped carefully round a three-foot stick, and the foot secured to the stick by a piece of vine. In this manner the flesh could be carried comfortably on one's back. Two of the bundles contained only the buttock and thigh of a man . . . the girl's (severed) left side and arm and hand presented a heart-rending spectacle as it lay on the ground; the arm extended, the pretty, shapely little brown hand resting on the ground . . . The smaller intestines were trailing down the creek with the oscillatory motion imparted by the running water.

Such reports created revulsion at the very idea of New Guinea.

The exploration of the Mandated Territory was predominantly driven by this desire for gold. The legendary figure of Mick Leahy started out in a number of dead-end jobs in Queensland, working as a clerk, cutting cane and hewing railway sleepers. While at work one day in 1926, he heard of the gold strike at Edie Creek (in the present Morobe Province of Papua New Guinea). Mick immediately dumped his decrepit Model T Ford truck on the wharf at Townsville and jumped aboard a steamer. His two brothers followed soon after. The Mandated Territory of New Guinea was an extremely violent place compared with Papua. The miners were unregulated and certainly unsympathetic to local people. There was often trouble. In 1931 on a trip into the country of the greatly feared Kukukuku tribe, Mick's camp was attacked and he was struck by a stone club. Like many of these adventurers, he wrote graphically of the events in true 'Indiana Jones' style.

> I dragged myself upright just as Paddy burst through the front of the tent, hauling an arrow out of his arm and calling for cartridges. I shoved them into his hands and heard Paddy exclaim: 'My God! They've brained you!' Then he was outside again, blazing away.

The first large gold strike in the Mandated Territory came in 1922 in the Bulolo Valley (Morobe Province). This strike also launched the most heroic chapter in the history of civil aviation in New Guinea. Canoes packed with excitable warriors circled the first 'supernatural' seaplane in wonder, inspecting it closely to find the genitals and determine its sex. Alluvial gold was too difficult for the individual miner to extract, so dredges were needed to exploit the rich field. One mining engineer considered Bulolo 'the world's most important placer deposit . . . since Klondike'. As there were no roads, the Bulolo Gold Dredging Company made the formidable decision to transport the entire operation by air – dredges, horses, food, trucks, nails, wood and wire. In 1931 it was the world's largest airlift of cargo, some eight times as much freight as shifted by all the airlines in the United States. Pilots would begin at dawn and step out of one Junkers G31 into another. These aircraft made thirty flights a week to transport eight dredges each weighing four thousand tons. The aircraft had open cockpits, trimotors of nine cylinders each with twenty-seven bellowing exhausts. The dredges were assembled by a team of riveters who had previously worked on the Sydney Harbour Bridge.

The mutilation of the land of the Anga-speaking people erupted in the usual carousel of warrior attacks and punitive reprisals. Many indentured labourers died from exhaustion and malnourishment. Ela Gofton, daughter of the notorious Flora 'Ma' Stewart who owned the Samarai Hotel and then later the Wau Hotel in the main town of the Morobe fields, remembers:

I used to work in the hotel, in the bar with my mother, and I would work until 11 o'clock, we would work to 2 o'clock in the morning. The miners used to drink their rum, and when they'd had enough, they'd call out 'Buick!' 'Chevrolet!' and their boys would come along and they'd climb on their backs and be piggy-backed along the steep tracks home.

Bulolo Gold Dredging was far and away the most profitable New Guinea company before the war, but set the pattern of exploitation and violence for the huge mining operations that were to follow.

We stayed at Lihir long enough to load an injured Papua New Guinean mine worker. He was on a drip and catheter with his nurse en route to the hospital in Rabaul and seemed to be in some pain.

'There are lifejackets under the seats. If we ditch, you can push out the windows after pulling the red handle. You should all be able to squeeze out. The sick man will have to go down with the plane.'

This remark extracted a wan smile from the patient. The pilot happily agreed to my request to fly over the gold mine which had been excavated out of the crater of an extinct volcano.

The mine at Lihir is a vast amphitheatre of wealth. It began production in 1997 and is connected to the 'town' by roads that radiate like the tendrils of a huge malevolent sea anemone. The excavation is encircled on three sides by precipitous, almost verti-cal slopes covered in a dense cloak of tropical vegetation. A sinister brown stain bleeds into the sapphire sea on the fourth side. These tailings from the mine, allegedly contaminated with cyanide and other chemicals, are disposed of from a pipe deep within the ocean and should be carried away by the deeper cur-rents. Allegations of respiratory complaints, environmental dam-age, exploitation of the workforce and the funnelling of profits offshore have accompanied its development. In response, Lihir Gold, managed by the Australian interests of Rio Tinto, have implemented an extensive programme of environmental monitoring and education. They have invested significant resources to distinguish between, as they see it, 'real' or scientifically proven injury and the distortions of untutored or superficially 'perceived' environmental damage. Company reports thoughtfully refer to 'Deceased freshwater prawns'. Recent independent impact studies on fish stocks by the Australian CSIRO (Commonwealth Scientific and Industrial Research Organisation) tend to support their denials of significant harm. Despite these efforts, claims for compensation are on the increase by local people. The familiar development cycle of mining in Papua New Guinea is being played out yet again – scientific research, followed by reassurance, culminating in recrimination. As I flew over the wounded earth I could not help but sense a malign Alberich labouring deep within the mine of the Nibelung.

On Lihir, the local people were well informed about their rights and the consequences of the mining operations. This is not always the case in Papua New Guinea. The full implications, scientific complexity and massive scale of modern technology is not well understood by naive and inexperienced villagers. The positive benefits of mining in terms of employment are tangible enough, but not always the benefits of the cash itself to a subsistence farmer who has nothing to buy in his village or hamlet. Stories abound in the past of the glittering treasure of mining royalties being buried and forgotten. The particularly close Melanesian attachment to clan land and its profound metaphysical significance has been largely ignored by modern business practice that assesses land as a commodity in conventional monetary terms. Calamitous results follow such cultural arrogance. Friction between mine workers from different parts of the country erupts into violence; disruption of value systems leads to alcoholism and drug abuse.

Investment, not only in mining, is crucial if the country is to become independent of the massive foreign aid it now 'enjoys'. Loans from the World Bank and the IMF are contingent upon what is effectively a dissolution of Melanesian cultural values. All around me, in the cities in particular, I noticed the apathy and lack of national pride that results from this severe dislocation of *kastom*. Western culture works on the principle of external governance and rights of the individual. Melanesian society is stateless and egalitarian, a society that has effectively solved the problem of authority through consensus community decisions. Aesthetic sentiment seduces us into believing the Pacific is pacific. The recent history of development shows the grotesque character of a Faustian pact with mammon, the indigenous culture being the sacrificial lamb or, perhaps more appropriately, pig.

16. 'Rabaul i blow up!'

Local people during the 1937 eruption
in East New Britain

The large island of New Britain is divided into two provinces,
East and West. The journey from Tokua airport to Rabaul in
East New Britain is over forty-five kilometres of horrendous road,
passing through coconut plantations and the town of Kokopo
('the landslide'). This thriving administrative capital has grown
up since the catastrophic volcanic eruption of 1994 which
destroyed Rabaul, one of the most beautiful cities in the Pacific.
Kokopo is a busy centre of supermarkets, banks and small resort
hotels. Newly-paved roads, street signs, a bustling market and
building work give the impression of a growing community.
There was massive relocation here from Rabaul after the eruptions. At the beginning of the lunar landscape that begins at the
base of the volcano Vulcan, a telling sign reads:

FOR SALE
BEDS & COFFIN

The sides of the towering cone are fissured and cratered like
cracked and weathered skin. New growth struggles like shaving
stubble through the crust. The scars of *lahars* (rivers of mud)
that flowed down from the hills have failed to heal, and resemble

enormous piles of mouldy chocolate cake. Buses career over the impossible surface, avoiding craters by scaling the inclined banks.

Across noble Simpson Harbour, the primeval volcanic caldera ringing Blanche Bay comprises smoking Tavurvur (Matupit), the relatively new Vulcan, Rabalanakaia, the affectionately-named Kabiu ('Mother'), Tovanumbatir ('North Daughter') and Turangunan ('South Daughter'). Recent research indicates the Rabaul Caldera is one of four interconnecting volcanic calderas ranging north-south along the eastern Gazelle Peninsula. Tavurvur is still pumping out thick clouds of ash at intervals of about three minutes. The wind carries the prodigious plume across the town, the ash falling in grey swathes reminiscent of the vortex of elemental violence in a 'sublime' painting by Turner.

Powder as fine as talc hangs suspended in the air. Utilities packed with children dissolve in it. Local people go about their business at the market as if the catastrophe had never happened; a faint smell of sulphur drifts over their ghostly movement. Orange plastic sheeting has been nailed across the windows of the few houses not abandoned, in an attempt to exclude the worst of the fallout. The gaunt skeletons of ruined buildings rear out of drifts of black ash and concrete rubble, while abandoned streets radiate in every direction. The surviving trees are grey, giving a curiously etiolated appearance to the whole district. Yet there is an air of excitement here, the feeling of survival against impossible odds. It is a scene from science fiction, nature at its most destructive and sublime, the most incredible place I have ever seen. I drove down deserted Malaguna Road to Mango Avenue and The Hamamas Hotel, one of three in Rabaul that survived the eruption, largely through the Herculean efforts of their owners.

A 'charming' Prince Andrew once stayed in the same room I occupy, and brought his own water and toilet seat – not much of a defence against a volcano I would have thought. A gigantic First World War Vickers coastal defence gun guards the entrance to the hotel. It has recently fired a coconut in anger. The colonial building and the palms in the garden are covered in dust like an abandoned filmset. Inside it is beautifully appointed, an extra-ordinary contrast to the apocalyptic vision outside.

In the halcyon days before the eruption, The Hamamas was known as The Ascot, and owned by Sir Julius Chan, a former Prime Minister. Because of 'imperialist associations' he changed the name to Hamamas which means Happy Hotel in Pidgin. It was destroyed by fire in 1985 and rebuilt in a vaguely Chinese style by the formidable Gerry McGrade, a Scottish builder who came to Rabaul in the 1950s and constructed many of the major buildings in the town. Sepik carvers were engaged to produce the pagan erotica at the entrance. As Sir Julius was half-Chinese, the hotel was constructed along *feng shui* principles. He included an excel-lent Chinese restaurant called The Phoenix. The symbolism of the name increases significantly with every passing hour spent in the shadow of erupting Tavurvur. The Hamamas is managed by Gerry's son-in-law, the adrenalin addict Bruce Cameron Alex-ander and his wife, a vivacious and attractive Australian named Susie. Bruce is also of Scottish descent and comes from an old grazing family in Western Queensland. This remarkable dynasty has now spent three generations in Papua New Guinea.

With typical Scottish stoicism, the hotel was open for business the day after the eruption while the rest of the town fled for their lives. It was buried three times by flash floods. The levee banks constructed behind the building burst, and water roared through the hotel swamping the entire place in liquid mud. Two months were needed to dig it out of some forty thousand tons of volcanic material. The McGrade family and their staff worked without proper power or water through the blistering heat and falling ash. Volcanic ash is very dense and similar in weight to concrete. The feathery manner of its falling is insidiously deceptive. Ash continues to fall on the building from the present activity of Tavurvur, and Bruce has an army of local people constantly engaged in hosing and sweeping it clear. The infrastructure of the town has been all but demolished. A heartless provincial government appears to have abandoned those expatriates and local people courageous enough to be hanging on in Rabaul by their fingertips.

To live day by day in the shadow of an erupting volcano, under continual siege from the forces of nature, seems madly heroic to the visitor. But contemplating heroism is surely the luxurious feeling of a voyeur. Daily life is grim here, the inconveniences of a 'normal' urban existence reduced to negligible significance. The looting of Rabaul began even during the eruption, despite the setting-up of police checkpoints and the arrival of the army. Properties that had been abandoned and already creaking under the weight of ash were stripped of everything by villagers. They came over the hills like locusts.

Within minutes of arrival I was whisked away in a dusty Toyota Hilux on a 'tour' past lines of gutted mansions, clubs and offices, over roads of deeply-compacted ash, lurching across pillows of bulging black, negotiating cracks and seams gouged by rain, out onto the abandoned airstrip of Lakunai Airport. It lies on the edge of Simpson Harbour, opposite coughing Tavurvur. Our drive over the old material had been curiously soft and silent as if driving on black sponge rubber.

Tavurvur sat before me in all its ill-natured splendour. Every three or four minutes the earth rumbled from a profound depth, a huge plume billowed voluptuously into the sky followed by the clatter of rocks falling onto the sides of the colossal cone. The plume rose higher and higher, forced out by incalculable power, to slowly disperse over the town, the ash falling lightly like discoloured snow. The strong smell of sulphur that accompanied this bad-tempered display was strangely exciting. I stood rooted to the spot by the primeval scene.

Such mild coughing was of course a shadow of those terrible days in September 1994 when the massive eruption began in sinister silence. The sun was soon obscured, thunderstorms raged over the craters and lightning struck trees, the earth shook and power supplies were cut. Tidal waves and rivers of mud destroyed everything in their path, ash and pumice the size of small cars ('ballistic ejecta') rained down on the town and surrounding villages from Tavurvur and Vulcan. The roads were choked with fleeing families.

Papua New Guinea is one of the most seismically active areas in the world. The landscape has been violently fashioned, thrust up into jagged peaks, folded and detonated by the movement of tectonic plates. The island of New Britain lies on the unstable Pacific Rim running through New Zealand and Japan. The ground often trembles with *gurias*[1] but the locals remain amused and indifferent. The 'old' Rabaul, charming colonial capital of the Mandated Territory, was destroyed once before in 1937 by the eruption of Vulcan and Matupit. The mighty cone of Vulcan had previously been a low island in the harbour. In 1937 there were hundreds of violent deaths in the surrounding villages, due mainly to seared lungs and suffocation when fiery avalanches superheated the air. Diehard Australians refused to leave the town, continuing to play billiards on increasingly dusty baize, more concerned about the lack of orange bitters in the bar of The Cosmopolitan Hotel than the chaos outside. *Roro*[2] from movements of the sea bed were also common. Bruce kept up a running commentary of observations on what he felt was the wimpish behaviour of most of the expatriate population in 1994.

'They just left their homes for the looters. Looting is a national sport here in PNG. Most of the roofs were flat so the buildings collapsed under the weight of ash. The stuff's as heavy as concrete. You could hear the buildings groaning just before they fell in.'

'Everyone was terrified, I suppose.'

'Sure were, but you've got to stick it out! The buildings took a long time to collapse – days, weeks even. There was time to save them. If they'd kept on removing the ash and defended their properties, Rabaul could've been saved.'

---

[1] A word in the Tolai language of the Gazelle Peninsula meaning 'trembling' – of a person or the earth.

[2] Tidal waves.

We drove across the old golf course that lay who knows how deep beneath the black sponge.

'The golf course could be open again soon.'

Looking at what appeared to be a nuclear war zone I reflected again on the power of dreams over the minds of those who choose to set up in Papua New Guinea.

'People left their dogs behind to starve. I must've shot about three hundred that had gone mad with hunger and were eating each other! You needed sticks to ward them off, particularly the guard dogs.'

Few of the original expatriate businesses survive. The Rabaul Yacht Club has been rebuilt but there are no yachts moored in the harbour. The provincial government wanted to abandon Rabaul and relocate the entire population to Kokopo. Now it's a war of attrition between the survivors, the provincial government and the volcano. Plantations were reclaimed from the white owners, put up for sale and squatters moved into their houses. Bruce forced them out.

'Six million *kina* had been left in a bank vault which would've been so easy to open! Everything had been abandoned. If only I'd known about that!'

The night was full of bizarre mefloquine dreams and rumbles from the volcano. A wave of heat and sulphurous ash punched me as I emerged from my room for breakfast, but instead of apprehension I was oddly exhilarated to be in a place seemingly on the edge of annihilation. We were to climb Tavurvur in the afternoon, but the morning was devoted to war.

I wandered out into the bleak landscape and sweated along Mango Avenue towards the centre of town. A white Toyota Hilux pulled up. The wife of the Hospital Adviser gestured for me to climb in. The air-conditioned comfort almost made me faint.

'Where are you off to? It's safer with me,' she said in that friendly expatriate way.

'I'm going to catch a PMV to Kokopo to see the Bitapaka War Cemetery.'

'God! That's miles! I'm going into Kokopo. I've got time to take you if you like. First we have to check the road's safe. There have been a few hold-ups recently.'

She negotiated the nightmare into Kokopo, stopped at the Tourist Office and went inside.

'Well, is it safe?' I asked as she emerged.

'It's *a little bit safe*. That's what they told me.' She emphasised the words with amusement.

'Where were you before this?' I asked the most common questions.

'Mount Hagen in the Highlands. Magnificent country.'

'And you were posted to Rabaul?'

'We left the day after my husband received a bullet in the post. He must've upset someone. We decided not to take any chances.' She smiled without a trace of unease.

The cemetery is entered through wide bronze gates and is superbly maintained within an extensive garden of indigenous trees and flowering shrubs. Areas of manicured grass separate row upon row of grave markers from the Second World War. It is a piteous sight. There are more than a thousand named burials of Australian, British and Indian troops and well over a thousand Australians and eight indigenous policemen with no known grave. Severe earthquakes in the region have meant that each is marked not by a headstone but by a bronze plaque on a low concrete plinth. These lives were squandered in the calamitous withdrawal from Rabaul in the face of the Japanese onslaught.

The war cemetery stands on the site of the German radio station at Bitapaka that saw the first Australian casualties in a small engagement in 1914. The German colonies were extremely shocked at the outbreak of war. Their main reaction was one of consternation, as they were almost entirely reliant on Australia and New Zealand for food and supplies. No troops were stationed in Deutsch Neu Guinea, as Berlin naively believed the colonies would be 'left out' of any European hostilities. After all, these were 'show colonies' with no political value. The only cannons they possessed were for firing salutes.

The entire Australian fleet including two submarines were despatched to 'subdue' the enemy. A group of Melanesian plantation labourers was hurriedly rounded up to oppose the Australians, together with a few half-trained policemen and five Germans. They held out for five hours against four hundred poorly-trained Australian troops, after which six Australians including two officers lay dead, together with thirty Melanesians and a German NCO.

The hero of the battle was Captain Hermann Detzner, an army officer who had been posted by the German Government to New Guinea to survey the border between Papua and Kaiser Wilhelmsland. He refused to countenance surrender and hid out for four years playing *kat und maus* with the authorities. True to type, he wrote a gloriously fictional account of his exploits attempting to reach the Dutch New Guinea border for the German Geographical Society. He told of raising the red, white and black German colours over numerous Stone Age villages while singing the patriotic marching song *Ich hab' mich ergeben mit Herz und mit Hand*.[1] Dressed in an immaculate white uniform and tropical helmet, he finally surrendered in 1918, a symbolic and theatrical gesture that marked the end of the German hegemony.

Young Tolai boys adore Bruce's sense of adventure, so after lunch the back of the utility was miraculously filled with them along with Rosie, his young daughter, and Buzo the Rottweiler. We were going to find the wreck of a Japanese bomber near the airport and explore some of the 580 kilometres of tunnels that had been constructed by forced Indian labour during the war; this included fifteen hospitals, one of which was four kilometres in length. After the fall of Singapore, some six thousand Indian

---

[1] 'I have surrendered myself with heart and hand (to you the German Fatherland).'

troops fighting for Britain were brought to the Mandated Territory as POWs by the Japanese. There is no record of any surviving and being repatriated back to India. Some estimates put the number of Indian dead as high as ten thousand.

Beside Sulphur Creek Road, girders carelessly twisted like spaghetti sprang from the skeleton of the notorious shipping firm, Burns Philp & Company ('BP' were affectionately known in the old days as 'Bloody Pirates' for the high prices they were thought to charge). A once-fine house with palatial balustrades struggled out of the ash and regenerating grass. Magenta flowers bloomed outrageously on a carpet of black while delicate pink and cream frangipani gave a surreal appearance to the savage landscape. Bruce drove wildly and drifted the Toyota in the ash of the airstrip to screams of delight. He pointed out the possible site of a bunker used by the commander of the Japanese Combined Fleet, Admiral Yamamoto. We headed down a riverbed overgrown with bananas and palms to the wreckage of a Japanese 'Betty' bomber, nestling among freshly-sprouting coconuts. Large sections of the fuselage, wings and engine nacelles lay half-submerged in the volcanic debris, lying at crazy angles, the metal tattered like leaves chewed by manic caterpillars.

Rabaul before the Second World War had been an elegant town. Life was far more sophisticated than at Kavieng in New Ireland, with telephones, electricity, refrigerators and a movie theatre. The headquarters of German New Guinea had been transferred there from Herbertshöhe (Kokopo) in 1910. The Germans had built broad avenues and roads lined with rain trees. A fine residence for the *Gouverneur* (Imperial Governor), boasting distinguished colonial architecture, was erected on Namanula Hill at the foot of the volcano Kabiu. Wide verandas were designed to catch cooling sea breezes on all sides. The planter Richard Parkinson laid out a celebrated botanic garden on the slopes of Tovanumbatir – as he wrote, 'with all sorts of knick-knacks'. A wharf as well as native and European hospitals were constructed. Leisure was catered for by the Rabaul Club.

Perhaps the elegance of the expropriated mansions after the Great War led the Australians towards a decidedly-refined lifestyle in Rabaul. Most families had at least seven household servants, but ate mainly tinned food. *Meris*,[1] washboys, driverboys and cookboys served the morning and afternoon tea, polished the walnut floors and tidied the house, pressed the men's freshly-starched, white tropical suits and the ladies' light frocks. They changed them without fail twice a day. A sherry and Radio Australia before dinner was *de rigueur*. Visitor's cards were extended, hats and gloves worn to Government House, once the former German Residency on Namanula Hill.

Mrs Rhoda Coote, wife of the General Manager of Burns Philp in Rabaul, lived in the superb House Rakaia overlooking Blanche Bay. This imposing residence had originally been built by the Norddeutscher Lloyd shipping line. On formal occasions, her staff wore liveried *laplaps* in black-and-white check to match the funnels of the BP ships. Phyllis Cilento, the first doctor listed on the register in New Guinea, gives an indication of the general patriarchal and racist colonial tone in her book *My Life*.

> One day I found Meiabo, an older 'monkey'[2] we employed, flashing his penis for the benefit of three-year-old Raffles. I could not allow that sort of thing so I gave Meiabo a good beating with the rubber tyre broken off the baby's pram. The rubber was useful for this purpose as it hurt but did not leave a mark or welt to report: only police were allowed to thrash boys.

Some *sinabadas*[3] were bored with the endless rounds of bridge parties and fell into love affairs or adopted renegade lifestyles. It was a colourful period now totally erased by war and volcanic eruptions.

---

[1] Colonial term for Papua New Guinean women.
[2] Colonial term for house boy.
[3] Colonial term for European women.

Leaving this scene of failed hopes, we drove up Namanula Road which rises over the rim of the caldera. This was more than a little inadvisable, considered by many to be foolhardy. Some attempt has been made to rebuild the road, but the edges have collapsed alarmingly on either side. Deep trenches were cut by rivers of mud and huge trunks of uprooted trees decorate the drive to the Japanese war memorial. The buckled rails of the earliest narrow-gauge railway in Papua New Guinea, built by the Germans before the First World War, thrust into space across a ravine. Only the gateposts remain of the once-imposing German Residency. A lookout was constructed for the last Royal visit of *Misis Kwin* to Rabaul. The view over the town and Simpson Harbour is suitably imperial in splendour.

Suddenly we were 'off road' and driving through high grass and hub-deep ash to an obscure Japanese tunnel. The entrance was a small opening high on a mud wall, which forced you to climb then crawl on your stomach. This was quite different from the tunnels at Karavia where numerous Japanese landing barges lie rusting. The floor was uneven, the ceilings low and the marks of pickaxes still fresh after sixty years. Tiny bats flew in our faces and one of the torches gave up the ghost. Rosie and the Rottweiler (over whose welfare she showed the greatest concern) knew all the side tunnels intimately, even if there had been recent earth falls. Bruce uses the tunnels for pistol practice – he had cunningly shot a sea-snake while fishing with me the previous night and likes to keep his eye in. Japanese characters are still legible on the walls above the crumbling sandstone steps leading to a gun emplacement. I emerged into the tropical sun covered in grime, sand and sweat. The Tolai boys had already climbed a palm[1]

---

[1] Bruce told me that the majority of accidents in New Britain involve boys falling out of coconut trees. They often fracture their back or neck, requiring the skull to be pinned and immobilised. Medical facilities and equipment are limited so that when the next boy arrives, the frame is removed from the first who may then die. There are few highly-specialised rehabilitation facilities for this type of severe injury in New Britain hospitals.

and brought down some coconuts which we opened on a handy Japanese spike set in concrete. The soft, cool meat and drink were marvellously refreshing.

'All you need travelling in PNG is a boy and a bush knife,' Bruce observed.

The treacherous descent to the neglected Japanese war memorial caused more shrieks from the bouncing crew behind the cab. This important memorial to the Japanese Pacific campaign is an abstract concrete structure with a relief map of the theatre of war on the ceiling. An opening containing a piece of blue glass allows a shaft of sunlight to fall on a block of alabaster at the precise time of the anniversary of surrender. An old Tolai stared at us from the road like a mad sorcerer, hair wild, face grimacing. The Rottweiler slipped into a pool of black water and infuriatingly shook the filth all over us.

For the Japanese, the Pacific War was simply part of the larger strategy of the 'Great East Asian War' begun in China in 1937 and fought in the name of the Emperor Hirohito. It was intended to free Asia and the Pacific from the shackles of European colonialism and create a greater East Asian Co-Prosperity Sphere under Japanese rule. Before Pearl Harbor, there had been an American proposal to fortify Rabaul as a naval base. The sudden outbreak of war meant this plan was abandoned. A garrison of only fourteen hundred Australian servicemen were abandoned by their command to defend the idyllic colonial capital in the face of the Japanese juggernaut of twenty thousand troops. Only four hundred would see the coast of Australia again.

The evacuation of civilians was far more stressful in the Bismarck Archipelago than in Port Moresby. The Australian government adopted a vacillating and unclear policy. All the European women and children were to be compulsorily evacuated on what was code-named by the authorities, 'Z Day', 18 December 1941. The already-chaotic situation became emotionally fraught as they boarded the ships, hurriedly kissing their menfolk goodbye and cramming a few meagre possessions and their ballgowns into suitcases. These women were given little financial assistance or help by the government when thrown upon Australia's shores. Chinese and mixed-race women and children were shamelessly abandoned to their fate, as were the civilian white male population. Missionaries courageously elected to stay with their flocks and many were killed or humiliated defending them as a result. The Tolai of the Gazelle Peninsula displayed great bravery in this merciless foreign war.

The Japanese bombing of Rabaul airfields had begun by 4 January 1942. By the morning of 21 January, at the height of the attacks, only three Australian aircraft remained operational. Kavieng had already fallen. The withdrawal of Allied military personnel through the jungles of New Britain is a horror story of malaria, typhus, dengue fever, dysentery, rotting boots and rotting feet. Fighting in the sapping, inescapable heat and humidity was monstrous. Known as Lark Force, the remnants of the Australian army split into small groups, a great number being killed, captured, murdered or drowned. Over one hundred and fifty Australians were executed at Wide Bay. Rabaul was captured on 23 January; four hundred Australian troops escaped into the impenetrable jungle and some eight hundred were taken prisoner.

At the conclusion of the war, Rabaul had all but been destroyed by the Japanese and the intense American bombing. Fifty ships lay at the bottom of the once-magnificent harbour. Then in July, Australia suffered the cruellest single tragedy of the Pacific War. The unescorted and unmarked Japanese naval transport SS *Montevideo Maru* was torpedoed in error by the *Sturgeon*, a United States submarine. Over a thousand military and civilian prisoners of war were lost en route from Rabaul to a forced labour camp on the island of Hainan, off the South China Coast. After interminable years of waiting, the women of Rabaul were elated that the war had finally ended. Their exuberance was brutally crushed when their husbands, who had embarked on the ship 'with a cheery wave', failed to return. Many only discovered the catastrophe in the newspapers, some were sent a cold, brief telegram from the authorities.

A *waga* (ocean-going canoe) drawn up at Sim Sim, one of the most remote outliers of the Lusançay Islands and Reefs, Milne Bay Province. The canoe has a finely carved Trobriand *lagim* or splashboard of complex iconography.

Lone fisherman at dawn. Kaibola Beach leads to the headland of Bomatu, Kiriwina, Trobriand Islands, Milne Bay Province. When a departed spirit heads towards Tuma, the island paradise where Trobriand spirits repine, they must first linger at Bomatu, the gateway to heaven.

Left: Paramount Chief Pulayasi's House, Omarakana village, Kiriwina, Trobriand Islands, Milne Bay Province. The façade is painted with complex symbols associated with the sea and hung with festoons of pure white *Ovulum ovum* shells.
Right: 'Millton', my guide to the thermal springs of Dei Dei and the prehistoric atmosphere of Fergusson Island in the D'Entrecasteaux Group, Milne Bay Province. The scene at Dei Dei with the young boys holding orchids and raising the spirit of the hot springs with ancient magic was the closest I had ever come to the authentically primitive, either in nature or in human culture.

The author musing far too seriously on the macho career of Errol Flynn during the period of the Australian Mandate, Tovorua Plantation near Rabaul, East New Britain Province. He is wearing his favourite shirt for the tropical climate - a copy of the only known example of a Board of Ordinance Australian convict shirt (small arrow version). Made at the Female Factory at Parramatta c. 1840, it was found in a rat's nest between the floorboard and ceiling cavity of the Hyde Park Barracks in Sydney during the restoration of the building in 1980.

Two children from New Ireland Province near the capital Kavieng. The remarkable blonde hair of many children from this province is natural, although sometimes highlighted by the effects of lime, sea and sun. The hair usually darkens over time.

Dancing at the *malagan* ceremony at Panafiluva, New Ireland Province, prior to feasting on roast pig and taro. The dancers are urged on to extreme efforts by a fiercely critical master of ceremonies.

The active volcano Tavurvur showing its bad temper and spreading its ash plume over Rabaul, East New Britain Province. Fine ash seemed to lodge everywhere – on clothes, skin, eyes, hair and even between teeth.

The male cult figures of the *tubuan* (left) and the *dukduk* (right), East New Britain Province. Men's secret societies continue to have a fundamental influence on the behaviour of Tolai society.

'Daisy' from Buka Island, North Solomons Province. Bukans are reputed to have the blackest skins in the world.

The Japanese had developed Rabaul Harbour into a major air, army and naval base by early 1943, from which they directed all their New Guinea and Solomon Island campaigns including plans to invade north-east Australia. Casualties were not replaced in this tropical hell, where food was always in short supply. But tired military men could always turn to imported 'comfort women' to take their minds off the 'Sacred War'. Among the military horses, military dogs, and military homing-pigeons, a special category was created for these women. They were shipped under the shameful label of 'military commodities'. This meant that if a ship were sunk, there would be no record of their existence.

At The Cosmopolitan Hotel in Rabaul, long queues of enlisted men were supplied with Korean women (often virgins) and the officers with Japanese 'geishas'. Curtains were placed over the windows and each woman was given a mat. They were lined up as if in a sleeping-car and only transparent gauze curtains separated the 'couples'. A sign read: 'We patriotic Japanese women offer all heroic soldiers of the Sacred War our minds and our bodies.' In a recent television interview, a former Japanese military surgeon drew attention to a few of the Regulations for Military Brothels.

- Pay 2 Yen in cash at reception and receive a ticket and one condom
- Use only condom of Japanese Emperor's Army stamped with brand name

Attack No: 1

- Give your ticket to the next available comfort woman but time limit 30 mins
- Leave the room immediately after ejaculation or face military punishment

As the end of the war approached, Korean and some Okinawan comfort women were trained as nursing aides by the Japanese nursing staff so they could return home in possession of at least an official identity. Captured Australian nurses and indigenous women felt grateful to be spared some of the more brutal indignities of war by the services of these hundreds of courageous 'military commodities'.

The entire question of 'comfort women' in the various theatres of the Pacific War is a contentious one in Japan today. Some Japanese authorities flatly deny their very existence. Most of the evidence for those sent to Papua and New Guinea comes from the aged ex-military surgeons, gynaecologists and hospital matrons who cared for their health in Rabaul. The ultimate fate of these women is unknown, as most were considered to be phantom beings with no documented existence.

Emperor Hirohito announced the unconditional surrender of the Japanese on 15 August 1945 and some eighty-three thousand Japanese troops, labourers and naval personnel in the area of Rabaul gave up their arms. The physical legacy of this period remains relatively undisturbed. The wreckage of aircraft litters the jungle floor, ships lie on the seabed and landing barges rot in the labyrinth of Japanese tunnels. This 'Forgotten War' is grim, heroic and largely untold. The indigenous population suffered terribly when their land became a theatre of conflict, a battle in which they could have had no possible interest.[1]

Concerning the war in Europe, many in the Mandated Territory could trace German descent from the period of German colonial rule (1884–1914), which resulted in an equivocal attitude to the Nazis. After the Australian military occupation in 1914, it was only those Germans who refused to cooperate who were deported to Australian prison camps. Those who promised not to assist Germany during the war, mainly planters and

---

[1] The Japanese have recently established a significant financial aid programme for Papua New Guinea.

missionaries, were permitted to stay by the military administrators. Some property was even permitted to revert to the original indigenous owners. In 1920 a three-man Expropriation Board was established to supervise the transfer of property in preparation for the creation in 1921 of the Mandated Territory of New Guinea under Australian rule. Many German residents who had 'stayed on' after 1914 now lost everything and were deported with a mere fifty pounds in their pockets. The previous policy was reversed and former German property was once again forcibly removed from the original owners to become Australian.

In the 1930s, as the Nazis gained strength in Europe, descendants of former German residents came to have divided loyalties. Before their internment, Lutheran missionaries dressed in uniforms and raised the swastika above their mission stations. Over glasses of lager at the Deutscher Klub in Rabaul, it was confidently felt that the Mandated Territory would again become German under Hitler. Many believed that Germany would be superior to Australia in its management of the Territory, as they had always believed 'a colony cannot colonise'.

The banana boat was launched from the slipway in front of the yachtless Rabaul Yacht Club. We were heading for Greet Harbour and the now-solidified lava flow from Tavurvur. Bruce tied the boat to a crag while giving amusing directives to his 'boys' in Pidgin. Bastions of rock lay like prehistoric lizards basking by the edge of the water. The massive crystalline outcrops are razor sharp. After some risky climbing, we headed round to the southern face of Tavurvur, where we beached the dinghy on a shelf of fine, pitch-black sand. Clouds of steam were belching from fumaroles below the rim of the crater, sinister yellow pustules pumping foul gas from the seeping sides. Fumes caught in my throat. The ash was spongy underfoot as we climbed past the burnt trunks of coconut palms, violently stripped of their foliage by the volcanic wind.

Every three minutes or so, rolling clouds of pumice and ash

would be forced out of the crater above us and rise in a colossal plume. This was accompanied by a rumble deep within the earth and the clutter of rocks falling nearby. Immense boulders had gouged scars in the pillows of ash as they plummeted to the ground. Strangely, the climb did not seem in the least dangerous or frightening, rather emotionally electrifying. The air was hot and still, the suffocating heat, noise and smell of sulphur indicating the closeness of fundamental forces below. Close proximity to such power was hypnotic. I could readily believe, along with the local people, that malevolent spirits were about. I climbed higher and higher until the heat warmed the soles of my feet through my boots and rocks began to fall too close for comfort.

We returned to the dinghy and sped downwind towards the northern face. Here the glowing rocks cascading from the crater were clearer to see, their clattering sharper and louder. Water was boiling in stained pools by the shore, shooting in jets from submerged vents. Rotting tree roots marooned in black sand were encrusted with rusty, sulphurous tumours. Unconcerned by the primeval scene around them, men fished from outrigger canoes, searching lethargically in their plaited bags for betel to chew.

The megapode or scrub fowl that inhabits this thermal region is small and dark with horn-coloured feet and a red face. We returned a short distance along the shore to where the eggs, much-prized as a delicacy, were being collected. The people of Matupit have the sole right to collect and sell these elongated, beige eggs. They are about twice the size of a hen's egg.

'*Karim mi igo ol mangi! Mi no inap wetim leg bilong mi!*'[1]

The local boys duly made a seat with their hands and carried Bruce the short distance through shallow water to the shore. The *masta* is alive and well and still living in Rabaul, but there was affection in abundance here, not resentment. His strong personal authority removes the anxiety from decision-making.

---

[1] 'Carry me over boys! I'm not getting my feet wet!'

A primordial scene greeted me. The megapode is a burrow nester and lays its eggs in warm, volcanic earth. The collectors at first glance looked to be a fearsome crew, covered in sweat and dirt from head to foot, scraps of cloth clinging to their loins, filthy T-shirts wrapped around their heads like turbans. Eggs were piled up in their canoes. Thousands of burrows had been dug by the birds in heaped-up earth littered with dead trees. An army of diggers carried shovels, although blonde children would disappear down holes without tools. The birds have learned to burrow deeply, some to a depth of more than six feet. The digger disappears from view and rams himself into a dark, confined space. The soil is light and the danger of a cave-in is constant. Many men are smothered before those nearby can be alerted. Luck and skill are often rewarded, but it is a dangerous and exhausting task. Tensions exist because old men, unable to cope with the exertion of a dig or *kalukal*, retain a right to the eggs on land now worked by younger men. They grimly warn that death and collecting *kiau*[1] are intimately connected. They feel it would be better for the young to direct their energy to gardening or planting coconuts. We bought some eggs and pushed the dinghy off the prehistoric shore. A man was rubbing himself down with masticated coconut, spitting it onto his skin and rinsing himself in the sea. Huge smiles shone from the ash-covered humans, these laughing survivors of the apocalypse.

Two rocky pinnacles known as the Beehives (Dawapia Rocks) thrust out from the middle of Simpson Harbour. We flew towards them, the boys hooting with pleasure as we carved up a Chinese cargo ship and its inscrutable crew in arcs of irresponsible spray. At the turn of the century there was a village of some one hundred and fifty souls on the larger pinnacle, but the volcanoes dealt with that. They are now slowly sinking. Crabs scuttled away in their millions as we approached. Boys dived from precipitous

---

[1] Tolai for 'megapode eggs'.

ledges. Rusting hulks of unbelievable decrepitude lined the harbour, but washing on the line indicated the surprising presence of life. My shower that evening removed more grit and sweat from my body than I can ever remember. The eyelets of my Italian boots had already begun to turn a venomous green from the corrosive ash of mighty Tavurvur.

17. Queen Emma

Queen Emma herself, like the Empress Elizabeth of Russia, could accomplish miracles in love making and drinking.

From the *Memoirs* of the
German trader Eduard Hernsheim

Between the rearing mountains of New Ireland and the volcanic calderas of East New Britain lie the low coral formations of the Duke of York Islands. The broad harbour at Mioko is sheltered from all the prevailing winds and is the finest and healthiest in the Bismarck Archipelago, the water as smooth as glass, the sand as fine as talc. Many early traders and missionaries set up stations in this group, only to find that, as was usual in the Bismarck Archipelago, idyllic appearances concealed threats of disease and death. A recent earthquake in Rabaul has tragically submerged much of what I saw on my excursion and many people have been evacuated as the islands sink. The Duke of Yorks are the nexus of the occult practices associated with the male cult of the *tubuan* and *dukduk* and link the cultures of New Ireland and the Tolai people of the Gazelle Peninsula. These islands were also the place of the earliest European settlements.

Lesley and his black-bearded brother negotiated what seemed an inflated price to take me the eight miles across the St George's Channel separating Kokopo and the islands. The waters of the channel can become suddenly treacherous, but the weather remained fair. Two flawless atolls known as The Pigeons came up on the left. A duel had taken place there one soft tropical morning between two of the principals of the doomed Marquis de Rays expedition. At 6.30 a.m. the good Dr Baudouin and Lieutenant Dessus faced each other across the pristine beach, both victims of false rumour. Their revolvers 'barked beautifully' in the balmy air and honour was satisfied. The friends embraced and prepared for the final evacuation from Hell.

Trying to start or stop this banana boat was an operation requiring the skill of a great improviser. The throttle control on the outboard had lost its handle and Lesley used a pair of pliers to regulate our speed which was either dead slow or screaming across the deep. At Mioko we coasted across the emerald-green water to the beach. It is an exceptionally beautiful island with outrigger canoes drawn up on banks of the lagoon. The ubiquitous driftwood and coconut husks litter the beach. I wanted to see where the legendary 'Queen' Emma and the trader Thomas Farrell had begun their formidable careers. Two local boys had already paddled across to us from the settlement (no vehicles, a trade store, a school and a church). The canoe skimmed the surface, seeming to rely for floatation on surface tension, like a water insect. They perched on the delicate structure as if on a toy. The remains of Emma's house were on the south-west tip of the harbour, so we pushed off and applied the pliers to the throttle.

Late one night in 1873, the Princess Coe slipped off her light dress and dived naked into the dark Pacific from the black hull of the brigantine *Leonora*. Despite the prodigious amount of champagne she had drunk, the princess struck out strongly for the shore. The notorious 'blackbirder' and buccaneer Captain William Henry ('Bully') Hayes had just avoided a court martial for piracy and the two friends had been celebrating in Apia Harbour in Samoa. He had made a drunken pass at her which she did not appreciate. Thirty years older than the princess, he had been captivated by her charms and she in turn was attracted by this loveable rogue. Emma Coe would spend much of her adult life embracing or escaping redoubtable captains of one sort or another. The extraordinary story of 'Queen' Emma deserves an entire novel or movie, and I can only touch on the glittering highlights here.

She was born in Samoa in 1850 of a distinguished American father, Jonas Myndersse Coe and a Samoan mother, Le'utu, a princess of the royal Malietoa family. As a young boy Jonas had run away to sea from a puritanical home and as a man, enthusiastically embraced the *fa'a Samoa* or characteristic lifestyle and customs known as the 'Samoan way'. This led him into three marriages, resulting in at least eighteen acknowledged children. Emma was his favourite.

The first signs of her true nature revealed themselves at the Subiaco Convent for young Catholic ladies on the Paramatta river near Sydney. One night, bored rigid with dancing quadrilles, she decided to embrace her Samoan origins. She persuaded the other girls to massage her with coconut oil, donned a short *lavalava*[1] and strung some berries round her neck. In the dormitory before a statue of the Virgin Mary, complete with a burning candle, she performed a lascivious dance that left nothing to the imagination. She later asserted her breasts were at their most perfect at that age of thirteen. The nuns called her 'a little heathen' and she was expelled. Her spirited and independent nature, her panache, her sheer sexual style, would accompany her throughout life. She went on to her uncle in San Francisco, but by the age of nineteen had run away from him too, back to the pagan freedom of Samoa.

Under pressure from her father, she married the stolid Scots-

---

[1] Worn throughout Melanesia, a *lavalava* (Samoan) is similar to a *laplap* (Pidgin) – a calf-length piece of cloth wound around the body, falling from the waist.

man, James Forsayth. He owned a trading schooner, and was one of the few educated gentlemen in those parts. Marriage did not prevent her from enjoying the sweet delights of pigeon-trapping with the passing Earl of Pembroke, cooing bewitchingly with him in a leafy Samoan bower. According to the young Earl, who arrived in Apia on the yacht *Albatross* in 1870, 'La Coe' had a superb classical figure, her Polynesian beauty charmingly softened by her half-American ancestry.

As she matured, she attracted lovers of all types who wove fantasies around her volcanic yet seductive nature. One admirer (though not perhaps a lover) was a gay English aristocrat named Lyttelton. Some rogue traders attempted to sell him to a cannibal chief in a fit of resentment against his irritating superciliousness. A brilliant but dissipated graduate of Brasenose College, Oxford, he was fastidiously rejected by the cannibals as being too thin to be of interest.

Emma was developing a subtle and ruthless business brain. German traders had been coming to the South Seas in increasing numbers since 1856 when the old German trading company, Johann Cesar Godeffroy und Sohn, established an agency in Samoa. This event had far-reaching consequences for the commercial exploitation and politics of the entire South Pacific region. The firm began shipping dried coconut meat (copra) to Europe for pressing. Previously, the oil had been pressed on the islands and stored in sections of bamboo cane, but had often gone rancid en route to Europe. The new process revolutionised production and resulted in profound cultural changes. Orderly plantations were established and coconuts began to be grown on an industrial scale.

The race for Pacific colonies became urgent among the Great Powers. Emma's natural advantages made her vulnerable to the roaming collection of bankrupts, forgers, alcoholics and suave tricksters that wandered Samoa and the South Seas in the late nineteenth century. Her personal reputation often came into question and she increasingly sought assistance from a South

Seas entrepreneur named Thomas Farrell. He was a red-bearded, implacable, Irish-Australian trader 'whose stormy past and moral qualities do not bear a closer investigation,' according to Truppel, an early German chronicler of the Bismarck Archipelago. Farrell suggested a 'partnership' that would save both her finances and warm his bed. Emma was astonishingly modern in outlook and was never one to refuse a lover if it meant losing a business opportunity.

In 1878, on a trip to the pretty port of Mioko in the Duke of York Islands, she first conceived of her New Guinea empire. Mrs Forsayth was not yet thirty, bubbling with life, sexual energy and boundless ambition. Her husband, poor solid James Forsayth, was believed lost in a typhoon off the China coast. The island gossips inevitably began to spin their webs around this un-attached coquette and made her life a misery. She left Samoa with Farrell in 1878 determined to return a triumphant business success and cast it in their faces. As she wrote in a letter to a friend, 'some day I shall return to this place and spit in the faces of the bitches of both sexes who now dare to insult me.'

Trade always followed missionary activity in New Guinea, and the Duke of York Islands were no exception. The redoubt-able Reverend George Brown had founded his Wesleyan mission at Port Hunter in 1875. He was shortly followed by Godeffroy und Sohn, who established a small trading station. Emma was now living openly with Thomas Farrell, and they began work-ing at Mioko for the Deutsche Handels und Plantagengesell-schaft der Südsee Inseln zu Hamburg, successors to Godeffroy's. This intimidating company became known, not surprisingly, as the 'Long Handle Firm'. 'Mr and Mrs Farrell', as they biga-mously styled themselves, soon began trading on their own account.

Coconut trees do not start to bear until the seventh year, not achieving their maximum yield until the thirteenth. Tom was an impatient Irishman and never had faith in the fortune Emma believed she could make from plantations. She took a sensible long-term view of investments, but he welcomed the quick profits associated with labour recruiting. This was an infamous trade and reflected badly on their whole joint enterprise.

When a recruiting schooner sailed to an island, the captain would fire a signal cannon or set off dynamite to announce their arrival. Local men would come down to the shore with nothing but their betel-chewing equipment in a plaited bag and sign on voluntarily in a process known as 'beach pay'. It could be dangerous on these wretched cockroach-infested copra vessels, and for many recruiters the returns rarely justified the effort involved. Younger captains took it up simply for the adventurous life it offered. The labourers usually began with nothing but normally left employment, as was commonly said, 'healthy and wealthy'. The boys would invest their hard-earned money in brightly-coloured calicos and *laplaps*, pouches, knives, belts and gleaming key-chains which they distributed lavishly among their relatives when they returned home.

Before regulations were implemented after the German annexation in 1884, labour was often recruited in a violent manner by the infamous 'black-birders' who raided the coast searching for able-bodied men, press-ganging them from their canoes, throwing them headlong into the holds of schooners, sometimes firing on them murderously from above to silence resistance. The Irish physician, Dr James Murray, owned the notorious Melbourne brigantine *Karl*, and was responsible for the murder of many Solomon Islanders. The historian Edward Docker reconstructs the essential spirit of 'black-birding' in a fictional scene:

> After a bit, Dr Murray came aft. Lewis, the second mate, said, 'What would people say to my killing twelve niggers before breakfast?' Dr Murray replied, 'My word, that's the proper way to pop them off.'

Some crews masqueraded as missionaries or even bishops to gain the confidence of local people through duplicity and surprise.

The first missionary Bishop of Melanesia, John Coleridge Patteson, was impersonated by a 'black-birder' who tricked some trusting warriors onto his vessel and sailed away. The real Bishop Patteson was unfortunately murdered in the Solomon Islands in 1871, when some villagers erroneously believed the recruiter had returned to capture more men.

These virtual slaves were sold on to unscrupulous managers. Thousands were shipped off to crippling work on the sugar plantations in northern Queensland or Fiji and many died in the cane fields. Extensive areas of New Ireland suffered male depopulation which unbalanced the sensitive cultural balance within the villages. Apart from the shortage of labour, many women suffered extreme sexual frustration from the absence of males. The older men left behind complained about being brought close to death by their sexual demands.

In 1882 Thomas Farrell had become nefariously embroiled in the Marquis de Rays fiasco, ending up with the old *Marquis de Rays* steamer, thousands of bricks, the altar of the projected cathedral, hundreds of jewelled dog collars and most of the luxurious possessions of the benighted colonists. This booty was all hoarded at Mioko. But the scandal also drew the attention of colonial Europe to this remote region. Entrepreneurs rushed to lay hold of suitable territory in Melanesia, with the Germans and British in fierce competition. 'Queen' Emma's career developed in tandem with Germany's aspirations, although she had arrived some time before. The potential profits were immense but so were the dangers. The Bismarck Archipelago had a notably virulent strain of malaria and the most murderous cannibals in New Guinea. One alarming report described a chief who had captured a husband and wife. The woman was added to his 'harem' and forced to help eat her late husband at the wedding feast.

All that remains from this operatic period in Mioko is a heap of rubble, once the steps of Emma's comfortable house. A small plunge-pool full of stagnant water lies hidden under thick vegetation. A spooky *dukduk ples*[1] was pointed out in whispers by my self-appointed guides, and a place where the monstrous *masalai*[2] are believed to lurk. Myriads of crawling insects skipped and jumped about the jungle floor, skittering over the coral sand. Everything under my feet seemed to be alive, and I felt sure my entire skin was crawling with bugs in the oppressive heat. For a few magical seconds I watched the rare butterfly, *Ornithoptera priamus miokensis*, a huge iridescent turquoise-blue subspecies endemic to tiny Mioko, dip and glide in the dappled shade.

On the return journey, a landscape of Homeric grandeur and antiquity, more than worthy of the travels of Odysseus, opened out in the passage. Beneath a tumultuous sky, the noble caldera of Rabaul's volcanoes lay across the horizon framed between two palm-covered islands. The vista of Tavurvur with its arching plume of ash, the mauve cones of Mother and Daughters distant across Blanche Bay, would have been described by that notorious enthusiast of volcanoes, Sir William Hamilton, British Ambassador to the eighteenth-century Court of Naples, as a 'sublime view', appealing directly to the heart. It possessed the exalted beauty of a painting by Salvator Rosa or a 'picturesque' landscape by Alexander Cozens. Flying fish leaped from the mirror of the sea like spinning shards of glass, skimming along beside the boat at tremendous velocity before slicing back beneath the surface. A soft tropical breeze fanned my face, sweet perfume drifted over from an island shore.

I jumped down onto the sand at Kokopo after my trip and was surrounded by a group of dinghy pilots. Rastafarian dreadlocks, cheap gold sunglasses, betel-devastated teeth – terrifying to look at but friendly and innocently curious about me.

I decided to complete my day with 'Queen' Emma by visiting the remains of her house called Gunantambu ('forbidden to

---

[1] A sacred grove where the men's secret society of the *dukduk* meets.
[2] A pidgin term referring to dangerous spirits who haunt rocky places, trees, ponds or whirlpools.

people') and her private cemetery or *matmat*. As the afternoon volcanic dust from Tavurvur was rising on the wind, I walked to the Kokopo museum. A local man rushed out of the door of the deserted building to greet me.

'*Yu noken lukluk lon hap, bikman!*[1] Two wallabies killed by a dog are there!' I looked away instinctively in some alarm.

'Is this a museum?'

'Museum is here, sir, but the dogs kill the animals very quick.'

'Oh, I see. When did this happen?'

'*Long wik igo pinis.*[2] Last night the dog kill a cassowary we had in this cage. *Lukim blut blo em!*[3] That's where he get in! Look here! *Waya i bagarap pinis!*[4] *Yu kam na lukim bodi, bikman!* The body! *Lukim bodi blong enimal!*[5] *Bodi yah!*' He insisted on showing me the mutilated corpse which was laid under a tree and covered with banana leaves. Flies were settling and decomposition had already begun in the terrific heat. The smell was overpowering. The feathers has been savagely torn out, the flesh shredded. The poor dog was probably driven to it. He handed me a bloody feather as a macabre souvenir. Certainly this was the most unexpected introduction to any museum in my experience. A few bedraggled parrots and pigeons in cages cringed away from my glance.

Scattered about the museum garden were large numbers of rusting Japanese guns, armoured personnel carriers and fighter aircraft. Of great interest was the stone mill-wheel and cannons from the Marquis de Rays debacle. Photographs of the German colonial period revealed the upright administrators of Herbertshöhe (Kokopo) and a refined feeling for colonial architecture. Hotels and mansions, the most elegant in the Pacific, such as the German Residency in Kaewieng (Kavieng), the Hotel Furst

---

[1] 'You can't look there, sir!'
[2] 'Last week.'
[3] 'See the blood!'
[4] 'The wire is broken!'
[5] 'Look at the animal's body!'

Bismarck, Emma's own Hotel Prinz Bismarck and the New Guinea Compagnie's Hotel Deutsches Hof. They catered for such picturesque colonial gentlemen as Dr Wilhelm Wendland, often seen riding in a gig, dressed in white tropical suit and dazzling panama, carrying a whip.

The visiting Count Rodolph Festetics de Tolna described a New Year's Eve party at Herbertshöhe given by Max Thiel, local agent to the pioneering German trader Eduard Hernsheim. He wrote of the exquisite European food and French champagne that were served on a table furnished with fine crystal and decorative bowls of pink orchids, the perfumed gardens illuminated with garlands of tiny flower-shaped lamps and illuminated balloons. Emma's Samoan *protégées* whirled to a waltz played by an orchestra, while the German warship *Falke* launched fireworks. He found the proximity of the savage anthropophagi rather titillating. He likened the voluptuous atmosphere to Tahiti, but was waspish in his assessment of what he considered to be the pretensions of the German colonials and 'Queen' Emma, referring to her as 'Frau K . . .' and intimating her crass 'theft' of the treasured possessions of the unfortunate de Rays colonists. Eduard Hernsheim himself played the flute and could recite Shakespeare in English. His own extravagant parties were long remembered in Neu Pommern.[1] He adored Emma and gave her the title she carried into legend, *Die Königin* (the Queen).

Emma's lover, the ruthless Thomas Farrell, died of tuberculosis in 1887 and he left her substantial property. But she had tired of him long before and had taken up with the handsome and cultured Dalmatian, Agostino Stalio, soon to be captain of her largest ship. He became the love of her life.

As the years passed, armed by now with a revolver thrust into her belt, Emma traded Manchester prints and Sheffield knives as well as coconuts. Her business interests proceeded apace and she

---

[1] German name for the island of 'New Britain'.

purchased large tracts of the best fertile land on the Gazelle Peninsula. Thousands of hectares passed into her hands from local clans for a few rolls of scarlet cloth, fathoms of *tambu*[1] and sticks of trade tobacco. Sensing the German annexation, she cleverly registered her purchases from local chiefs formally with the US Consul in Sydney using her American citizenship. Her competitive and shrewd business mind outraged the Neu Guinea Compagnie.

The basis of her great fortune had been comprehensively laid down by 1892, but it was not without help. Richard Parkinson, one of the greatest authorities on the culture of the Bismarck Archipelago, was married to Emma's beautiful Samoan sister Phebe. He was another of that fascinating group of 'demi-aristocrats' that wandered the South-West Pacific at this time. Born in the Duchy of Schleswig in 1844, he was the illegitimate son of the Duke of Augustenborg, who had not abandoned the boy but taken particular care with his classical education at Augustenborg Castle. Emma invited Parkinson to Mioko from Samoa, where he had somehow fetched up, and he began to survey the Gazelle Peninsula on her behalf. His expertise facilitated her setting up the first commercial coconut plantation in New Guinea at Ralum on the southern shores of Blanche Bay. Known locally as 'the German Professor', he assembled a small private army of Buka warriors, his 'black devils', to protect himself and pacify the surrounding area. He was an excellent violinist and played Mendelssohn and Schubert around the fire to calm any outraged cannibals he encountered.

Parkinson rapidly developed Emma's plantation and laid out his own. By 1895 she had 400 hectares under coconut palms compared to 260 planted by the Neu Guinea Compagnie. Three years later her holdings had doubled. The industry was phenomenally profitable and the demand from European soap manufacturers remained high. She employed over a thousand 'boys' at

---

[1] Tolai word for shell money. The basic unit of length is the fathom.

Ralum. True to the aristocratic spirit of agricultural improvement, Parkinson introduced many new crops such as tobacco, cotton, rice and coffee. The local people marvelled at his imported horses, donkeys, Brahmin cattle, sheep, poultry and goats. But his relationship with E. E. Forsayth & Co. was not always placid. Emma argued with him about the time he spent on his ethnological collections and natural history specimens. This passion ultimately culminated in his *magnum opus*, the monumental *Dreissig Jahre in der Südsee*. Parkinson died from 'a lingering illness' just two years after its publication, in 1909.

As Emma's ambitions and empire expanded, she built a mansion at Ralum near the present Kokopo named Gunantambu. It was famous throughout the Pacific for its luxurious furnishing and appointments. Each morning, accompanied by a native footman and her secretary, she would sweep down a grand flight of steps through tropical gardens. Two muscular men from the island of Buka would swiftly haul her substantial personage by rickshaw through groves of luxuriant shrubs along a specially-constructed carriage road to her offices.

She entertained on a lavish scale. There was enough champagne to bathe in, feasts of twenty-eight courses for the German naval officers, theatricals, dance, music. She dressed in the finest Parisian clothes, served imported delicacies on gold plate and drank the finest French wine. She even invented a primitive form of air conditioning. With Machiavellian intrigue, Emma contracted wealthy European marriage partners for her numerous imported female Samoan relations. Douglas Rannie, a Queensland Government Agent in the 1880s, describes the extraordinary scene Emma contrived at a dinner given by Hernsheim.

> She was dressed in white satin, with a long train which was borne behind her by half a dozen little dusky maidens, natives of the Solomon Islands, all dressed in fantastic costumes. Above her jet-black hair she wore a tiara of diamonds and her whole carriage

was queenly ... After dinner her little maidens stood behind her chair. Some fanned her, while others rolled small cigarettes, which she smoked between her sips of coffee and cognac. Dancing until the early hours ... Samoan war dances which ended not in the shedding of blood but rather of clothes.

The great house with its wide verandas and striped blinds was built on rising ground overlooking Blanche Bay. A flight of mouldering concrete steps leading down an embankment is all that now remains. Gunantambu was totally destroyed during the Japanese occupation. The terraces of exotic plants have become the greens of a golf links. The vivacious girls in light cotton dresses have vanished, their pagan laughter lost on the wind. The fort mounted with cannon from infamous Port Breton, the inlaid 'drinks altar' consecrated by the Archbishop of Milan, the gold-fringed flag of the *Republique de Nouvelle France* have all disappeared.

Emma was a liberated businesswoman and free feminist spirit long before her time. She would have been celebrated today beyond her wildest dreams. With penetrating business acumen, she sensed the coming of war in 1914. Now in her fifties, she was tired of administering her empire and suffering continual personal indignities. On one memorable occasion she had been trussed to a pole like a pig and was being carried off into the jungle as a gourmet trophy when saved at the last moment by Parkinson and his Buka boys.

Romilly, the British Deputy Commissioner, referred in his reports to the 'sweepings of the colonies' who were arriving in increasing numbers to recruit labour, giving local people strong reasons for bitterness. Men such as the notorious George de Latour took up residence in the Solomon Islands. At the entrance to his house each gatepost was surmounted with a human skull and two crossed thigh bones nailed beneath. A printed notice on a nearby tree read:

NOTICE

Dogs and Niggers are Forbidden to enter inside the Portals
Of those Gates
Any Dogs or Niggers found therein will suffer
the Penalty of Death

GEORGE DE LATOUR
British resident

The German presence continued to expand and Emma felt the atmosphere of control becoming oppressive. By 1905, the authorities were forced to set up a system of local law enforcement teams with a local official in charge, an interpreter and a medical orderly. The 'boys' were given a portrait of Kaiser Wilhelm II and a peaked cap to wear on duty. Emma finally decided to liquidate all her assets in New Guinea well in advance of the imposition of further controls. The hedonistic German magnate 'Rudi' Wahlen finally bought Gunantambu. As Emma had predicted, the entire plantation was expropriated by the Australian Government at the conclusion of the war.

Her last husband, Captain Paul Kolbe, was a Prussian cavalry officer, the marriage one of convenience rather than romance. Emma was feeling her age, but they began an extravagant tour of Europe together on the proceeds of her empire. He was younger than Emma and had a roving eye, causing her characteristically to worry more about the possibility of an embezzled family fortune than a broken heart. 'Queen' Emma and Paul Kolbe died within two days of each other in unexplained circumstances at Monte Carlo in 1913. Captain Kolbe was an inveterate gambler, there were contradictory reports of a mysterious German woman who claimed to be his wife, and then the final twist of a bizarre motorcar accident. Emma Coe died as sensationally as she had lived, having created a legend in nineteenth-century German New Guinea.

* * *

The difficulty of finding her private cemetery or *matmat* seemed insurmountable. I fruitlessly wandered along roads and up tracks in the terrific heat. A young boy appeared out of the long grass with a long-bladed knife and looked threateningly at me.

'I'm trying to find Queen Emma's cemetery.'

'I know where is! Up them steps.' His glowering mood was magically transformed into friendliness, and he pointed with his bush knife to a bend in the road ahead, where a few overgrown steps led up to the top of a hill.

'I come with you and I show you, *bikman*!'

He wildly cut away the undergrowth to clear a path.

Emma certainly made sure her cemetery had a fine view for the souls of her departed friends. The vista over Simpson Harbour is as majestic as any nineteenth-century traveller could wish. All the volcanoes and the island of Matupit can be seen through groves of banana plants, flame trees and coconut palms. A concrete slab with a gaping hole is an austere reminder of where Emma and Paul Kolbe's ashes had lain until the end of the First World War, when they were removed to Sydney by her son. The most striking monument in the *matmat* was to her lover, Agostino Stalio, captain of her vessel the *Three Cheers*, killed in the Faed Islands in 1892 on a punitive expedition to avenge the murder of her brother, John Coe.

As I returned to The Hamamas, fine ash was blowing over the town. Power was fluctuating at the hotel as the volcanic dust penetrated the insulation and short-circuited everything. Fax machines spontaneously started up, bells rang, screens and lamps blinked as if operated by phantoms. After a shower I wandered down Sulphur Creek Road to the Kaivuna Hotel to watch the replay of the Melbourne Cup. The bar spectacularly overlooks the rumbling volcano. Expatriates, much the worse for wear, shouting and moving a set of mechanical horses while the Cup replayed on television, Tavurvur vomiting ash in the background, was a piece of truly absurdist theatre.

18. A Moveable Feast

It was time for me to meet one of the 'big men' of Matupit. Meli ToPaivu is the traditional and cultural leader of the Tolavit clan. He is a large man with a strong presence. Warmth and friendliness radiate from him. Meli is the son of Anton ToNgarama of Matupit who was highly respected in the community for his artistic ability and profound understanding of the Tolai culture. He was respected as an honest man, and his word was accepted without argument when settling disputes. Meli is Matupit's most successful indigenous entrepreneur or 'big man'. He has wide business interests, ranging from a betting shop and service station to a healthy interest in developing tourism, but his wealth is not obvious from his simple lifestyle.

I first encountered him and his new young wife on a rusty fire escape rising from the ashes of Rabaul's waterfront. He quickly ushered me from his modest flat into an equally modest office. He was anxious to show me his fax machine and indicate that I could use his e-mail.

'It may be difficult to find something for you,' he replied to my enquiry about ceremonies that might be staged while I was in Rabaul.

'We only have them at the weekends. Modern life, I'm afraid. And if someone important dies, of course.'

'So there's no chance of seeing the male cult figures of the *dukduk* or *tubuan*?' The subject seemed alien in this office, fax machine and computer monitor coated in a layer of volcanic dust, faded posters peeling from the wall. I knew the Tolai were less resistant to change than other indigenous groups, except for their most treasured and fundamental institutions, shell money or *tambu*, and the *tubuan* and *dukduk*.

'I must guard the secret of the *tubuan*.'

'Of course, but can't you tell me anything? It's very interesting for a European.'

'All right! But not too much! It's very strong. Young boys have many stages of instruction in the secrets, from age five. Shell money, *tambu*, must be paid before the initiation. Then we can say, "You dance on the water." You cannot pretend or dangerous punishment comes. The *dukduk* is the symbol of our identity, so they must become proud men.'

'What is the difference between *tubuan* and *dukduk*?' I looked up at a photograph of the strange costumes. The entire body except for the legs is hidden beneath a conical mask crowned with feathers, the torso down to the dancer's knees covered with a 'ball skirt' of pandanus leaves.

'The *tubuan*, which is female, has the small plume and gives birth to the *dukduk*. It is male and has the tall plume, you can see. But both are in the men's secret society. The design of the face, the feathers and the top is secret because others might steal it. Only initiated men make the masks. Women are not permitted anywhere near!'

'But when can you see them?'

My limited time frame did not fit the natural Melanesian rhythm at all.

'Usually at funerals or religious services. A *tubuan* was raised to bury my father ToNgarama. I have knowledge of this magic.'

'People still believe in it?'

'Of course. They respect the power of these cult figures. If there is a justice case, you can invite the *dukduk* to give a judgement by giving shell money. They *tell*, not *advise*, in such cases.' I was about to thank him and leave the office disappointed when he exclaimed unexpectedly, 'Look. There's something for you! A wedding feast of one of my relatives in Matupit village. How could I forget? Would you like to go? Not for visitors usually.'

Much of the island of New Britain is rugged terrain, thinly populated by various tribal groups. The mountainous interior of the Gazelle Peninsula is home to the mysterious Baining people. The surface of their social life, according to the anthropologist Jane Fajans, is 'mundane and repetitive in the extreme, and holds little to seize the interest'. As might be anticipated, deeper analysis reveals complex value structures displayed most clearly in their astounding fire dances, which they refer to as 'play'. Bizarre *kavat* masks that have a 'beak' like a duck, with huge eyes and ears, dance at night to the urgent music of a male orchestra. The Baining launch themselves through fiercely burning bonfires, expressing a violent confrontation between forests spirits and the civilised village. The bar of The Hamamas was decorated with many of these outrageous masks.

But it is the Tolai of the Gazelle Peninsula who are the most numerous in Melanesian terms, numbering about 120,000. They are also a surprisingly affluent, worldly group, and have been prominent in politics before and since independence. The Tolai regard the volcanic eruptions as a *kaia* or a breaking-out of destructive spirit forces. Tremors and tidal waves are regarded as dangerous but natural occurrences. The volcanic ash has rendered their lands extremely fertile and the extensive culti-vation of bananas, yams and taro is obvious everywhere. The coconut is the most prized and versatile crop. It provides food and a cooling drink, the palms are used to roof houses and the husks to fuel fires, the dried meat as copra provides profitable oils, moreover it is symbolic in many metaphysical contexts. Trade is still carried on in *tambu* around which almost their entire life revolves.

In the nineteenth century the Gazelle Peninsula became central to the economy of the German Empire in New Guinea. The present identity of the Tolai has been moulded by this historical influence, but also the period as an Australian colony and under

Japanese occupation during the Second World War. One of the German flags marking the official hegemony over New Guinea was raised on the island of Matupit on 3 November 1884. The harbour had already been a busy port before the German headquarters were moved to Rabaul, a reclaimed mangrove swamp, in 1910.

The German trader, Eduard Hernsheim, made Matupit the base of his extensive trading operations in the Bismarck Archipelago. A shaded path led from the wharf to his delightful bungalow. There was a billiard room in a garden pavilion and a music room, whose white walls were artistically decorated with New Britain weapons and ornaments. The Chinese came to settle as businessmen and traders, opening restaurants, laundries, bakeries and stores in Rabaul's Chinatown. One famous hotelier named Ah Tam offered mineral springs and health cures to visitors from the wilder parts of the country. The massive role of Chinese traders in the making of 'modern' Papua New Guinea is a largely untold story, wedged as they were between the colonial white *mastas* who featured so prominently in political histories, and the subsistence indigenous population who hardly featured at all.

Despite early outbreaks of violence and punitive expeditions, the German administration was respected overall by the Tolai. Some villagers began to wear white tropical suits and even ties to indicate their easy adoption of 'advanced' European ways. The introduction of coconut plantations by the Germans was enthusiastically received. Although the villagers hated labouring on the plantations, they took up the production of copra on their own account with a commercial eagerness typical of their more sophisticated outlook. The 'big men' of the village liked to be paid in shiny silver marks which were heaped up and warmly admired rather than spent. Almost half the arable land was lost to the Tolai through 'purchase' by Europeans. Complex claims to sacred clan titles led to unpleasant disputes extending to the present day.

When the Australians took over in 1914 the Tolai suffered

under a new economic policy that determined the colony must become self-sufficient. The expropriated German plantations were poorly run by inexperienced Australian planters who were mainly ex-servicemen. The price of copra plummeted in world markets. The new colonists believed that the native must be kept in his place and resentment grew. This was a difficult time, but the Tolai continued to work in a more 'Westernised' way, warmed to the cash economy, and built modern houses. A Matupi was the first New Guinean to buy a motorcar.

The period of the Japanese occupation after 1942 was heartless and capricious in its cruelty – houses set on fire, beatings, coils of the sacred shell money thrown into Blanche Bay, bibles used as toilet paper, and most offensively to the Tolai, the Japanese custom of beheading native 'collaborators' according to *bushido*, the code of honour and morals evolved by Samurai warriors. The postwar period marked a return to Australian administration. Maslyn Williams in the 1960s noted in his book *Stone Age Island* that

> The forty thousand Tolai people of New Britain are the most compact, literate and vocal group in the Territory . . . But they are not a tractable group.

The sixties also saw one of the first independence movements in Papua New Guinea flower on the Gazelle Peninsula. The Mataungan movement used 'modern' political techniques to enlist support for the nationalist cause. One of the first flags of independence was raised by the radical Tolai.

A large forty-seater bus marked 'Paivu Tours' pulled up to collect me. I was the only passenger. Meli indicated that his smaller vehicles were laid up, but I suspect he wished to make a commanding entrance at the wedding feast. While he slipped into the bank, I asked the driver, 'Where are we going?'

'To a funeral. Someone's died.' This was news to me.

'Oh, really? I thought it was a wedding.'

'It could be a wedding. A sort of wedding-funeral.'

We drove up a winding road into the hills to the village or *gunan* of Matupit. Perhaps the village had been moved from Matupit island in the harbour after the eruption. The half-finished fibro houses that made up the 'village' were relatively new. There are few bush material houses in the local villages around Rabaul, quite different to Alotau where they are scattered throughout the town itself. The arrival of the bus caused a ripple of excitement, and great interest was taken in the solitary traveller. A large number of local people, mostly women and children, were sitting around on the grass with huge bunches of green bananas and cardboard boxes of frozen New Zealand lamb. I was introduced to certain 'big men' of the village and members of Meli's clan or *vunatarai*. He is related to the bride. We stood surveying the scene.

'The most important thing is to understand *tambu*. It's more important than life itself, my life or your life!'

He showed me some instantly. The strings are made of a number of small nassa shells, the tops of which have been snapped off in the eye of a coconut, and threaded onto a rattan cane. Each shell is well spaced from the next, unlike other forms of shell money.

'We measure it in fathoms, the distance between your out-stretched arms. When we have forty or so fathoms it is made into a coil or *loloi*. When we have one hundred fathoms we close the coil and hide it away! There used to be ceremonies when a large bamboo structure was built called a *leo*, where the coils were displayed. Not so common now when people need cash money.'

'Can you spend shell money?'

'On small things, yes. But it is mainly to give to my relatives when I die. A young man asked me to change six hundred fathoms into *kina* the other day, but I wouldn't do it. During our life we should save it up; only after death is the coil cut and given out.'

'And if you die without *tambu*?'

'That is the greatest shame. You must sweat for it, work for it. Much *tambu* shows you have lived a full life according to the Tolai custom.'

Many attempts were made by the German administration and the Catholic missions to stamp out *tambu*, but it has a tenacious hold over the Tolai imagination both as a commercial tool and a treasured ritual article. The people around us wanted to talk to the 'big man' so I went over and sat on the grass. A solitary white chair indicated where he would sit, elevated above the rest of us. Political power now rests in Port Moresby, so the sponsorship of such ceremonies does not establish the 'big man' as it did in the past, but great prestige still accompanies the staging of a *balaguan* or major ceremonial event.

The married couple sat on a woven mat in the centre of the proceedings, the wife feeding a baby and the husband talking to a young boy. They looked to be in their mid-twenties. She was dressed in a colourful shift and he in a new blue denim shirt and lime-green shorts. Meli had wandered over to me and began to explain that there would soon be the giving of *tambu*. Only through the passing of *tambu* can the position of the wife be secured.

'Why do they already have children? Is it a second marriage?' I asked, puzzled at the presence of children on the mat. He laughed.

'No! They got married eight years ago. There were a few financial problems, the volcano erupting and other things. We're having the wedding feast now.'

People began to present strings of *tambu* and green leaves to the husband and shake his hand. The pile of *tambu* rose before them and was cleared away at intervals. They both averted their eyes and perfunctorily held out their hands as the shell money was offered. There seemed to be an element of hostility in the air, no smiles of acceptance. The couple showed not the slightest pleasure at receiving the gift. I believe that among the Tolai, the giver puts the receiver under an obligation which is resented. The giver does not seek or even desire any verbal acknowledgement. A 'thank you' would be regarded merely as verbiage. The strong emotions the receivers experience are obvious in their begrudging attitude during this competitive exchange ceremony, where the bestowing of gifts incurs reciprocal debt. More 'normal' wedding presents were then presented (plastic buckets, plastic bowls, knives, umbrellas and woven mats). The same apparent lack of appreciation was obvious to see.

The Master of Ceremonies became vocal as is customary, and berated everyone in sight, particularly the children.

'Sit down! Sit down! Sing up! Sing up!'

The first dancing group appeared at the edge of the crowd in the back of a Toyota utility and ambled to the dance arena. A group of young men and boys dressed in dark blue *laplaps*, woven headbands and sporting a single white feather began a shuffling type of dance to some PNG sub-reggae music on a huge ghetto-blaster placed on the Toyota's roof. Glowering youths in cheap gold sunglasses, check shirts and caps leaned against the cab, turned up the volume and watched the shuffling progress without interest. The dance curiously lacked energy or emotional commitment. It was as though these young people had lost faith in their own cultural traditions. Nineteenth-century travellers also remarked on the lack of fierceness and fire in some New Britain dancing, but they did not see this. No women were performing.

Many of the dancers seemed to be high on a mixture of *buai* and marijuana. They found it amusing to lunge at me with threatening gestures, wildly waving bunches of leaves and knives, looking at me with unfocused eyes. 'Clown' figures appeared with leaves draped over foreheads and white powdered faces. Perhaps simple, bawdy humour was intended as most of the villagers laughed and shouted encouragement. Their baseball caps and European clothes gave an unpleasant hybrid impression, but I was probably sentimentalising my 'ethnic expectations'. A man appeared in a white boiler suit and a type of rubber horror mask you might buy from a magic shop. He frightened the children who screamed with delight and ran away to peep from behind their mothers. I was becoming increasingly depressed. Talcum powder was squirted on the dancers from plastic Johnson & Johnson bottles rather than the traditional 'slap' of lime. The dancers ran with rivers of sweat and eventually wandered aimlessly away.

During the dancing, mounds of food, mainly bananas and raw pork, were being frantically divided. Plaited baskets were filled with small pieces of uncooked meat and exchanged with great intensity by the women. People came over to me and violently tore off a side of shell money, throwing it at my feet and ran away. Clearly I was a guest of the 'big man' and was to be treated accordingly. Meli was furiously eating peanuts while a child scrabbled around under his chair picking up the husks. Pigs were being cut up with huge machetes and the head of one was dumped unceremoniously at his feet. Its dreadful snout pointed towards me, the raw meat of the severed neck collecting grass and flies. Old women puffed on their pipes and betel juice shot about the place in crimson arcs. Clearly, the decline of cultural expression through dance and music was well advanced, more so among the Tolai than other groups I had encountered in my travels.

The Tolai have the reputation of being the most forward-thinking group in Papua New Guinea. They are the ones known for taking up middle-class professional positions in Port Moresby, taking an active part in the politics of the independent country, being 'first' in so many areas of modern life. Cultural 'performances' remain impressive, but in Rabaul they appear to have neglected their *kastom* to a marked extent at the grass-roots level, apart from the continuing respect for the *tubuan* and *dukduk* and the widespread use of *tambu*.

The missions and various forms of fundamentalist Christianity have assisted in the disintegration of much traditional culture. At Rakunai, a village inland of Blanche Bay, there is a modern

Catholic cathedral. Fervent Christian belief saturates the Gazelle Peninsula, partly due to the activities of the United Church, but also stemming from the recent beatification of the Tolai, Peter ToRot, catechist and martyr. According to the writer of his *Life*, Father Caspar ToVaninara, Peter was a kind and gentle man who 'always worked even when nobody was watching him'. He beat his wife but that was a local Tolai custom and he always apologised after doing so. He was thirty when the Japanese invaded New Britain in 1942 and began to repress the Christian faith. He opposed polygamy, enthusiastically taken up by the Japanese, and was executed by lethal injection. 'ToRot is a very bad man. He is preventing people, who want to do so, from having more than one wife,' explained a Japanese officer after the execution.

Meli seemed to sense my unease and suggested we leave.

'I was not so happy,' he said as we climbed into the huge empty bus. 'Young people today are too lazy to keep the proper *kastom* alive.'

As we drove away, another foolish group with banana leaves over their faces and white clown powder emerged from behind a truck. They circled around a dancer holding a European doll. They poked their fingers in its eyes with howls of laughter and lifted up its skirt, flicking the eyelids open and shut. I was pleased we were leaving.

Meli was rather silent as we returned past Simpson Harbour, the setting sun flushing the palms pink. The ancient outline of the volcanic caldera ranged across the horizon and Tavurvur gently puffed out regular billows of ash. Some things never change.

My final day in Rabaul was a Saturday and Bruce invited me to a picnic on an old plantation by the sea. I accepted as I was curious to meet some of the expatriates who had 'stayed on' after the eruption. There had been a violent thunderstorm the night before with torrential rain. A strong *guria* (earth tremor) had shaken the hotel and I watched the curtains and lamps swaying wildly in my room. The lights flickered on and off. When I lurched onto my balcony that evening, ash was raining down and a smell of wet sulphur filled the air. Harsh bolts of lightning flickered in the sky. The next morning Tavurvur was particularly active and sweeping was energetically in progress at the hotel. As we set off I could see bright sparks flashing within the ash clouds spewing from the cone.

'The bloody volcano is in a bad mood this morning!' Bruce said harshly.

Most of the inhabitants of Rabaul treat Tavurvur as a member of the family, talking quietly to it, shouting at it, encouraging it to go to sleep. The mood of the volcano is assessed each day. Rabaul after the eruption resembles a prize-fighter struggling on the ropes.

The drive to the coast took us past many grand plantations that date from before the Second World War. Row upon row of orderly coconut palms march into the distance on either side of the highway. Cocoa trees are planted between them giving an impression of productive and stylish abundance. The tracks inside the plantations are rutted and pitted with deep puddles and trenches. Five or six Toyota Hiluxes were assembled in a coconut grove together with three horses and a foal. Some of the expats would go from here to the picnic spot by boat, some by car and the rest would ride.

The plantation at Tovorua is excellent for a picnic lunch, as there is a long safe beach, a reef for snorkelling and welcome shade. A carpet of vegetation covered the bases of the slender coconut palms, planted in endless rows of military regularity. Reef fish were already cooking on the fire and the steaming horses were being unsaddled and swum in the tropical water. An exotic, alluring girl of mixed race from New Caledonia was riding bareback through the gentle surf dressed in the briefest of red bikinis, her honey-coloured skin shadowed against the pink palms, toe-rings glinting in the sun. She was an underwater model for scuba equipment who had been working with a photographer for a month in the clear waters around Rabaul.

The conversation over lunch was desultory but the grilled fish was excellent.

'Bill Skate brought the country to its knees. Now in the next election, the opposition . . .'

'Elections! A chaos of disorganisation, beatings, intimidation and murder. You name it. The EU thinks it's too dangerous even to send observers. The PM might get to vote if he's lucky!'

'I was thinking that I should raise goats instead of copra. What do you know about goats Bruce?'

I mentioned the Tolai and their behaviour during the eruption. Angry disillusionment was mixed with the tired bitterness of those forced to 'stay on'. Many businesses were simply unsaleable after the eruption. Either you abandoned them or made the best of it. This group, with their patched legs and high spirits, were making the best of it. I admired their tenacity.

'I'm always disappointed that the Tolai can't consider the expatriate history as part of their own. All the great characters have gone. Except Bruce, of course!' Bruce looked over.

He had recently quelled a riot on the local football field by riding a horse into the fray. Now every time a goal is scored the crowd chant, ''Orse! 'Orse!' and Bruce must perforce do a lap around the perimeter to celebrate the score. The animal is so popular it might be elected to the local council.

'I'm finding it harder and harder to get that adrenalin rush!' Laughter all round. His bravado was notorious in the district. He had recently shot a huge crocodile that had been terrorising the Tolai living around Simpson Harbour.

I described the wedding feast in rather critical terms and was surprised when they leapt to the Tolai's defence.

'You are only judging the surface. The traditions of *tambu* and other beliefs resist everything.'

The underwater model slid herself voluptuously into the pink lycra diving skin, threw back her head and drank from a water pitcher, tossing her mane of dark hair as she buckled on her scuba tanks. The drift dive was along the coral wall that drops off hundreds of metres a little way out from the shore. The kaleidoscope of tropical fish was stunning and so were the graceful movements of the New Caledonian, as ravishing and unattainable as a shark. This is the cinematic South Pacific and unfortunately only a very partial reality.

I had one more appointment before leaving Rabaul, and that was a tour of the Coconut Products Oil Mill. The ash storm and sulphur soup from Tavurvur were cooking well by the time I arrived at the gates. A boyishly enthusiastic Arthur Gillard, Systems Manager, was to show me around. The smell of fetid pools of copra by-products which lay about caught in the back of my throat and mixed chokingly with the fallout from the volcano. As we walked along the dockside where the copra is unloaded, he pointed out some decrepit cargo ships moored in Simpson Harbour. The volcano erupting behind the aged derricks transported me back to colonial days.

'The smoke-dried copra from the villages varies a bit in quality. It's tough work husking nuts.'

In the factory, the copra is transferred by conveyor belt to storage bins and thence to a fiendish device containing 'tomahawks' and 'knives' that whirl at tremendous velocity and slice it for the presses. Coconut oil has now been replaced by other oils in many industries and the copra market has collapsed. Some of this equipment has recently been modified to produce copra meal destined for animal feed. Ignoble mounds of it fill the sheds.

The main part of the plant contains machines that would have been at home in the Fritz Lang film *Metropolis*. Pipes covered in scorched foil lead up to impenetrable heights, open concrete troughs containing crude coconut oil glisten in the dim light, steam escapes from tubes, electric motors whirr, sprockets click, chains and toothed wheels spin, open transport screws turn and threaten to trap my boots. Local men shovel mounds of copra meal while the fire-through boilers are being noisily cleaned with steam.

'You're lucky to see the presses disassembled,' Arthur shouted.

The volcanic ash continued to rain down, the cloying smell smothered me and the noise from the huge Mirrlees Blackstone generators was deafening. Arthur stomped through the pools of muck and seemed oblivious to the cinders blowing across the harbour. I could hardly breathe. He gave me a cheery wave and the tour was over.

Clouds of volcanic dust covered Rabaul as I drove back towards the hotel. Children endangered their health by sitting on the front steps of shops, gasping in the poisonous haze. Many locals were wearing masks including two crazy individuals trying to survey the road near the wharf. The measuring tape simply disappeared into the grey clouds. I was forced to stop as I was driving blind, yet other utilities sped past crammed with more children. The plastic sheeting stapled over the windows of houses flapped uselessly in the breeze. Ash crept into every crevice, gritty between my teeth, in my ears, up my nose and through my hair. It settled immovably on my clothes and skin. I was beginning to wheeze. A wild afternoon storm sprang up with thunder and bolt lightning, the wind blanketing the town in terrible grime, the sulphurous smell overpowering. Truly this was hell on earth. Rabaul is located at the junction of numerous ley lines, which may explain the heightened sense of life that takes possession of you here. As I showered that night, I watched, fascinated, as dark rivulets streamed off me and spiralled down the drain. The attempt to continue living in Rabaul is almost Homeric in its heroism.

All my plans for visiting the Trobriand Islands had fallen through yet again, and my next destination had unexpectedly become the former German Solomon Islands, Buka and Bougainville, home to people with the blackest skins in the world. The islands are a hotbed of simmering political and military discontent. At Rabaul's Tokua airport, soldiers in sunglasses and battle fatigues were waiting to board the aircraft. Many were officers carrying long rolls of maps and camouflage helmets.

There was a major earthquake, 7.9 on the Richter scale, just hours after I left Rabaul. The Duke of York Islands are now partly submerged and villages on the mainland are being prepared for evacuees. The islands are slowly sinking together with the remains of Queen Emma's Mioko mansion. I was fortunate to have explored the region before the inevitable wrenching of the tectonic plates.

19. 'No Trespassing Except
By Request'

Sign on Sohano Island
Buka Passage

An old piston-engined military helicopter painted red staggers above the palms and clatters across from Buka to Bougainville. The wreck of a tastefully displayed Japanese fighter under flowering wattle carries the graffiti, 'Why need women when u have the devil'. Manicured lawns and carefully tended gardens of the former District Commissioner's house run down to the edge of the cliffs, overlooking the swift-flowing waters of the passage. Flying fish arch in perfectly synchronised formation. I was on Sohano Island in Buka Passage, an achingly beautiful channel that divides the islands of Buka and Bougainville and joins the South Pacific Ocean to the Solomon Sea. I was contemplating three periods of modern history – the Japanese war, the Australian colony, and the civil war on Bougainville – none of which was distinguished by their legacy.

The splendid English navigator Philip Carteret commanding the *Swallow*, first observed the island of Buka early on the morning of 25 August 1767. Another expedition in July 1768 by the French ships *La Boudeuse* and *L'Etoile*, commanded by the explorer Louis de Bougainville, sighted the eastern coasts of Buka and Bougainville. He writes of his first contact with the natives of the place:

They came alongside of the ships, shewing cocoa-nuts, and crying *bouca, bouca, onelle!* They repeated these words incessantly, and we afterwards pronounced them as they did, which seemed to give them some pleasure.

The inaccessible islands remained isolated from Europeans until German influence arrived in the late nineteenth century in the guise of explorers, traders and labour recruiters. One of the best known was Richard Parkinson (Queen Emma's brother-in-law), who made many trips to Buka and Bougainville, trading, collecting specimens and making notes for his monumental work *Dreissig Jahre in der Südsee.* He also recruited the labour for his famous 'Buka boys' or 'tigers' who performed as his fiercely-loyal bodyguard in New Britain. The two islands were initially administered by the Neu Guinea Compagnie and did not officially become part of the German Empire until 1899. Following the defeat of Germany in the Great War, Buka and Bougainville eventually became part of the Australian-controlled Mandated Territory of New Guinea in 1921. Buka Passage was transformed into a savage theatre of war with the arrival of the Japanese in 1942, many of whom died of starvation and disease fighting the Americans and Australians. The islands evolved into the controversial North Solomons Province at independence in 1975.

I had arrived in the dry or 'relatively less wet' season. Magnificent canoes did not fly over the waves with singing paddlers to greet me, merely a large crowd at the airport to meet a plane only half-full. I easily identified my bags among the chaos of luggage. They were the only ones covered in volcanic dust. I sat on some planks thrown across the back of a utility on the rough ride into town, hanging on for grim death. Buka township is a mixture of tawdry trade stores and beer kiosks, gleaming new banks and a traditional village market with a curved shingle roof. The extreme blackness of the skin of the local people, their fine white teeth and dark eyes set them apart from the rest of Papua New Guineans. Numerous banana boats with snappily-dressed pilots are lined up on the shore, ready to take passengers noisily across to Bougainville or other islands.

The current runs as fast as seven knots in Buka Passage. It only takes about three minutes to reach the landing stage on Sohano, my 'neutral' destination. Uninhabited until colonial times, this small island divides the tide race into two channels. The colonial heritage was immediately obvious. Carefully-trimmed lawns roll down to the cliff edge above a coral shelf which extends out into the clear waters of the passage. Vermillion hibiscus, frangipani, magenta bougainvillea, ferns, orchids, and coconut palms are laid out on undulating lawns in the most English manner imaginable. Spectacular pigeons and parrots worry their cages.

The veranda of my room overlooks the opalescent water of the reef, a vista opening out across the swift current to the volcanic peaks of Bougainville. Sohano was the administrative capital of the region from the end of the Second World War to the 1960s. The lawns and tropical plantings extend over the entire 'upper' island, which has largely preserved its colonial architecture. At the northern end is an abstract Japanese war memorial, facing the majestic entrance to the passage. At the southern end, adjacent to the hospital, an ancient rain tree with a circumference of perhaps twenty-five metres extends its massive branches like a vast parasol. It is covered in a riot of parasitic vegetation – lichens, ferns and creepers with festoons of aerial roots falling like braided hair. The southern end of the island also commands a fine view of the islands and the forbidding jungle-clad mountains of Bougainville.

I was tired. I needed to rest. My concept of time was beginning to change. After months of travelling, I had ceased to be hopeful of plans coming to fruition and had abandoned my frenetic attempts to coordinate my objectives with airline tickets and visa requirements. Melanesians sleep when they wish and are never seen to hurry. Boat schedules are an imaginary concept. History means little, modern times being when the spirits of creation ceased to wander the earth. There is something to be said for such an expanded time-scale, free of the regulation of minutes, hours, days and weeks but a great deal of money is needed to function in such freedom. I needed a rest from this vagabond life, time to think and collect myself. Sohano seemed admirably suited to this purpose.

The violence on Bougainville just across the passage had made me wary. Seated on the veranda listening to the faint laughter of children, I was descending into tropical torpor. A soft breeze rustled the fan palms and an atmosphere of peace and tranquillity descended.

I wandered down to the landing stage and spent an hour watching the tropical fish light up the crystal water. An obese expatriate arrived by banana boat. He wheezed up the path from the wharf in breathless desperation carrying two cartons of lager on his shoulders. I presumed he was in serious need of leave from a difficult posting. Earlier in the morning I had spoken to Misak, the warm and talkative local manager of the 'lodge'. He looked askance at the alcohol going up. His daughter wandered by strumming a ukulele with a bright red fretboard, a crown of vine leaves wound around her head, hibiscus floating in her hair. She hummed and played in the background as we talked. Time was unfurling like a sail.

'Alcohol is the biggest problem,' he said, nodding his head in the direction of the guest labouring up the steep steps.

'Yes, but I suppose the poor bloke needs to relax. He doesn't look too happy.'

It was as if I was making excuses. Silence.

'Where were you born, Misak?'

'On the other side of Buka Island. I married a girl from one of those small islands.'

He gestured along the passage dotted with clumps of white-fringed green. He had clearly been handsome in his youth and retained his distinguished appearance, apart from an eye which had clouded over with a cataract. He was an educated man.

'Buka culture was unique in the Solomons, wasn't it?'

'We're very different from other provinces in Papua New Guinea, not just because we're famous for pretty baskets. In the past we were the wild men, the most feared cannibals. The coastal people were always at war with the people from the mountains. Visitors weren't welcome!'

He seemed to look through me but laughed good-naturedly.

'Do you have a totem system here too?'

'Of course. We have various birds as our signs – the frigate bird, the cockatoo, the hornbill, the pigeon.'

'Sounds similar to Milne Bay.'

'Really? I wouldn't know that. You're lucky. You can travel.'

'You still have traditional ceremonies, I suppose?'

'Not many. The young people aren't interested any more. We used to have the most terrifying dances in the Solomon Islands with music of flutes and drums, people dancing around the fire until they fell down.'

This criticism of the young had become a litany as I travelled the islands. He soon climbed aboard a banana boat and sped off in the direction of Buka. I trudged back up the steep path in a thoughtful mood. The expat was sitting on the veranda with his mountain of beer, clearly at the beginning of a long road.

'G'day mate. Come and wet your whistle.'

'Thanks.'

A tropical pause.

'Why are *you* here?'

'Bit of a break from over there.' He jerked his head in the direction of Bougainville. 'I'm a building supervisor in Arawa reconstructing the new healthcare centre.'

'What's the situation like over there?' I was planning to hire a four-wheel drive and go down the coast towards the capital.

'Things are getting back to normal. A few carjackings. Theft's increasing. We get electricity for about three hours a day now, but lights are twinkling in the villages above Arawa. When the town was looted, villagers set up their own hydroelectric power generators in the hills. Inventive crew. Plenty of water but it's polluted.'

He suddenly stopped talking as though he had said too much. He drank from the bottle with the same nervous, jerky movements I have noticed in many expats, his eyes flicking about the garden and over the passage.

'Was the Arawa hospital destroyed in the early 1990s then?'

'The insurgents burnt the whole bloody thing down, nurses' accommodation, operating theatres, everything. To replace it now would cost about sixty million dollars. Mindless violence! Complete madness! It's more peaceful now. The women calmed the violence. They demanded peace to bring up their children. Probably withdrew their favours!'

He laughed and sprung open another bottle.

'Our homes are littered with terrified hearts and shattered dreams,' Penagi uttered softly, emphatically.

It was a remote-controlled war, caused by outsiders. Its perpetrators sat back to laugh at the stupidity of the people hacking each other to death. The fighters themselves were mere players, whose destiny had been falsely raised on fractured hopes and promises. The war was complex; no solution would be easy.

Regis Stella
Gutsini Posa (*Rough Seas* – 1999)

In his violent narrative, Regis Stella, the Bougainvillean novelist, fiercely expresses the convoluted emotional trauma of the Bougainville civil war. The narrative is set against a landscape of erupting volcanoes, 'rough seas' and brutal death. It gives the lie forever to the apolitical Pacific paradise that inhabits the popular imagination.

The story of the Panguna copper mine and the politics of the Bougainville secession is a story as intricate as any creation myth. The record began in the 1920s, when gold and traces of copper were found deep in the mountains of Bougainville behind Kieta. The ore discovered was low-grade, and at that time the technology did not exist to extract the metal commercially. The finds remained merely of academic interest for many years. However, the 1960s saw rapid developments in extraction methods and increased demand for copper worldwide. The mining of huge quantities of mineralised rock containing one per cent copper or less became a feasible and potentially lucrative enterprise.

In the 1970s few Australian mining engineers had professional experience of or had even seen large-scale open-pit operations, let alone the Australian Administration or the local people. Eight years of evaluation and feasibility studies eventually led to seventeen fabulously profitable years for the mine. Production began on April Fools' Day in 1972 as a joint venture of the majority shareholder, Conzinc Riotinto of Australia (CRA),[1] and the Australian Broken Hill Corporation. The transnational company formed from this cooperation was known as Bougainville Copper Limited (BCL).[2]

At that time, the mine was considered a model of responsible development in the Third World. Employment and training were offered to indigenous workers and significant royalties were paid both to local landowners and the national government. Families

---

[1] The Rio Tinto Company was originally formed in 1873 to develop the Roman workings of a broken-down mine at Rio Tinto in Spain. In 1962 Rio Tinto Zinc (RTZ) was formed by the merger of two British companies, the Rio Tinto Company and the London-based mining company, the Consolidated Zinc Corporation. At the same time, Conzinc Riotinto of Australia (CRA) was formed by the merger of the Australian interests of the Rio Tinto Company and the Consolidated Zinc Corporation. In June 1997, RTZ became Rio Tinto plc and CRA became Rio Tinto Limited, both having enormous global mining interests as well as being major players in mining exploration, research and technology.

[2] BCL is presently owned by Rio Tinto (53.6 per cent), the PNG government (19.1 per cent), and public shareholders (27.3 per cent). The mine will never reopen and in 2001 the company announced its intention to divest its Bougainville assets.

in the vicinity of the mine were relocated and accommodation constructed. The new mining town of Arawa shone brightly in the sun, but the calm was not to last. Perceived affronts to the traditional concept of land use, combined with insensitive negotiations, led to a violent and politicised confrontation with BCL. Land is contemplated with metaphysical awe in Papua New Guinea, quite apart from its significance as patrimony and its value for gardening.

The labyrinthine hierarchy of rights to land usage posed insoluble problems in terms of compensation to the dispossessed villagers. A range of matters such as the valuation of individual pigs and coconut palms, ulcerated fish that caused protein deficiencies, loss of river water for the ritual washing of the first-born child – such 'esoteric' concerns were presented in a nightmare of complicated compensation claims. The sheer scale of the gash in the earth's surface and the enormous quantity of waste land being pumped into the Jaba river as 'tailings' incensed and horrified the owners of that sacred earth. They wanted the 'treasure' left in the ground for their children. Traditional beliefs had collided headlong with the juggernaut of European technology.

Bougainvilleans had always been defined by their skin colour, an intense black verging on blue. Bukans are considered to have the blackest skin in the world. They felt this was a fundamental difference which distinguished them from the 'redskins' (actually various shades of brown) who populated the remainder of Papua New Guinea. Skin colour has unified them socially but detached them politically. Mine workers descended in droves from the Highlands and the Sepik river. The Bougainvilleans looked down on such people, who were known to breast-feed piglets and who might threaten their women. The increased mobility of culturally-distinct groups in Papua New Guinea, all seeking work, caused a multitude of social integration problems.

Secession movements from Australian rule had begun long before Papua New Guinea gained independence in 1975. The islanders had always considered themselves ethnically part of the Solomon Islands, not merely the North Solomons Province of a newly-emergent state. Unity of opposition was difficult to achieve, as Bougainvilleans were split into numerous cultural, religious and regional factions. The so-called Village Governments which emerged in the 1960s emphasised the importance of unity through the revival of *kastom*, traditional community values and control over development.

As the years passed, 'secession' gradually became the rallying cry among young, educated Bougainvilleans, but some districts such as Buka favoured autonomous government within Papua New Guinea. Yet another group wanted ancestral spirits to be intimately involved in politics while the more isolated islands were ill-informed and undecided. Some villagers even believed money and 'cargo' would arrive through supernatural agencies invoked by politicians. But the fundamental issues remained land alienation, BCL, expatriate exploitation and the 'redskins'. Those who led the separatist shift from the national government, the leaders of the Bougainville independence movement, were all fierce idealists. Some like the prominent Father John Momis had trained as Catholic priests, others as Marist brothers. He wrote emotionally in 1987 to Paul Quodling, managing director of BCL:

> You have invaded the soil and places of our ancestors, but above all, your mind has invaded our minds . . . The modern corporation does not obey the natural rise and fall of life and death, as does our Melanesian tradition, which distributes a leader's wealth when he dies. The modern corporation obeys the ideology of the cancer cell, to ever grow and grow, without ceasing.

As the years passed, the travails of the mine became increasingly politicised and the independence movement became symbolically linked to its future. By June 1988 it was universally believed that the North Solomons Province was not receiving its fair share of the revenue generated from the mine. The New Panguna Landowners' Association demanded compensation for damage caused to the environment, which was rejected by the company. The national government set up an independent scientific enquiry into the mining operation. The exhaustive report was ferociously rejected on Bougainville as a 'white-wash'. Simple militancy was precipitated into revolution.

On the night of 26 November 1988, guerrilla war arrived in Papua New Guinea. Pylons at the mine were blown up with dynamite stolen by the ragged members of what was to become the Bougainville Revolutionary Army (BRA). Francis Ona, who had once worked as a surveyor at the mine, styled himself the 'Father of the Bougainville Nation' and established a hideout in the jungle. He directed the blame for unrest onto the 'white mafia' who, he said, had destroyed everything precious to Bougainvilleans, enriching themselves in the process. The BCL was likened to a 'wild pig'. Unrest escalated with random burnings and lootings. The mine became so unsafe it was finally closed down in May 1989. The financial loss to the nation was dramatic. Panguna had contributed over a third of Papua New Guinea's foreign earnings. Damage was so severe that the mine is unlikely ever to reopen.

The poorly-paid Papua New Guinea Defence Force (PNGDF), a small force established at independence in 1975, was sent in by the Prime Minister to restore order and confront the BRA. They were badly equipped and lacked discipline. 'We are just risking our lives on credit,' commented one soldier after his pay was delayed. Deep resentment bred fearful violence. The assassinations and killings of a cruel civil war soon began in earnest, with human rights violations on all sides. Rapes, murders, torture, the enforced licking up of the enemies' blood, the entire repertoire of horror was enjoyed by soldiers, police and insurgents. Villages suspected of harbouring 'collaborators' and 'revolutionaries' were torched and some twenty-four thousand villagers were displaced. Political indecision inflamed the chaos.

By early 1990 an intransigent Francis Ona was insisting on a 'war situation' signing himself 'Supreme Commander' of the BRA. Increasing numbers of mangled bodies of Bougainvilleans were arriving at Arawa Hospital. Bougainville was blockaded by the PNG Defence Force in April 1990, and the displaced villagers were reduced to roasting lizards for supper. District Offices were set alight and government buildings destroyed. Compensation claims by local villagers for the ruination of clan land by the mine became monumental, bordering on extortion. Ona still refused to deal with the national government. At Arawa on 17 May 1990, the BRA made a unilateral declaration of independence (UDI) and celebrated the formation of the Republic of Bougainville.

Over the next few years many fruitless attempts were made to come to a peaceful settlement. In 1997 the national government made the fatal decision to engage foreign mercenaries to assist and train the PNGDF. The Private Military Company (PMC) Sandline International specialise in what they term 'conflict resolution'. This operation was directed by Lieutenant-Colonel Tim Spicer, OBE, who had built a distinguished professional career in the Falklands, Northern Ireland, the Gulf War and Bosnia. Sandline was a mercenary force of ex-SAS officers and South African-trained personnel with access to guns, sophisticated electronic weapons and helicopters. The bill for a state of the art surveillance and strike package for the Papua New Guinea government, known as 'Contravene', was quoted at $US36 million for the initial trial period of three months. The government would pay for it by scandalous cuts in education, health, defence and the police.

The 'furnace-eyed' commander of the PNGDF, General Jerry Singirok, was incensed by this offer of 'military assistance'. In an action perilously close to the definition of a military coup, the officers defied the government and devised Operation *Rausim Kwik* (Pidgin for 'kick them out fast') to rid the country of the mercenaries. They skilfully captured the hardened professional force and forcibly deported them from the country.

Papua New Guineans were now confirmed in their suspicions

that their politicians were truly corrupt 'termites of greed'. A mob supported by the military shook the gates of the parliament building in Port Moresby demanding action. The then Prime Minister, Sir Julius Chan, and two of his ministers eventually resigned to the jubilation of the crowd. It was a triumph of the democratic spirit in the adolescent state.

Some twenty thousand lives were sacrificed in this murderous civil war. A new Bougainville Peace Agreement was signed in August 2001 between the National Government and Bougainvillean leaders, the twentieth since 1989. The autonomous government of Bougainville will be phased in over a period of five years after weapons disposal has been overseen by United Nations observers. Bougainville will continue to work within the framework of the existing Papua New Guinean Constitution. At the Peace ceremony, the then Prime Minister of Papua New Guinea, Sir Makere Morauta, and the Governor of Bougainville, John Momis, were carried by dancers through the ruined Bougainville capital Arawa, past heaps of uncooked vegetables, betel nut and pigs to the historic signing. Local women wept and sang. They had exerted strong emotional pressure on their menfolk to settle the conflict and played a significant economic role in supporting their families during the fighting and the moves towards peace. At sunset, feasting and dancing on a huge scale took place on Bougainville. For the moment blood has ceased to flow. At the time of writing, BRA rebel leaders have warned that the peace process is in jeopardy as compensation claims are argued in US courts, the only jurisdiction where Bougainvilleans feel they will get a fair hearing. We await with interest the claims of the other island provinces for autonomy now that Bougainville has set a precedent. The great cultural diversity of Papua New Guinea makes future claims for separate regional and cultural identities almost inevitable.

'Messing is at 6 o'clock!' the Buka boy shouted at me as I sat on the veranda.

On the dot of six, according to standard colonial orders, I appeared for dinner in the dining room of the Buka Lumen Soho Lodge, once the old Australian District Commissioner's house. I fell into conversation with a young Melanesian girl on the veranda. She wanted to be a journalist and was full of youthful energy. Glowering looks and death glances were cast in my direction by the serving boys. They seemed to have a malevolent sexual obsession with protecting her from me, a rapacious 'whiteskin'. I suppose a detached conversation with a woman is considered an impossibility by this society. Resentment simmered away. Along the walls there were many war photographs of Australian and Japanese troops. One sepia print depicted Japanese naval troops loading officers' swords onto a motor launch in Buka Passage. A sign on the wall read:

CHRIST IS THE HEAD OF THIS HOUSE, THE UNSEEN GUEST AT EVERY MEAL, THE SILENT LISTENER TO EVERY CONVERSATION

A vase of knitted flowers decorated the serving table. A model of a Bukan *Mona* canoe and a mysterious bell were displayed near the serving hatch. Inscriptions in Latin and German around the base of the bell indicated it had been cast in 1744 for the honourable 'Abbot Anthony' in a German monastery. The attractive young Melanesian girl was seated at a laid table, so I sat opposite her. She smiled graciously. A Bukan boy descended on me in a fury.

'Please, you sit here, not there! This is your seat! Move, please! Move, please!'

I felt quite threatened by his aggressive 'protection'. I was conscious of the same offence a black man might feel if he sat with a white woman in the Deep South. This was the first time I had been made conscious of my skin colour and it was very uncomfortable. The huge Solomon Islander at my 'assigned' table was in Buka buying old outboard engines for scrap. We had little in common and the dinner proceeded mainly in silence. Television blared at the end of the dining room, informing us that a woman had been raped and murdered in the hills above Port Moresby, an MP held up and robbed by *raskols* outside his home.

I asked for a beer and found that all drinks and sweets were padlocked in a type of cage that formed an annex to the dining room. The owner of the lodge, who had sole access to the keys, was ill with diabetes. The food was cooked very well but it was entirely English – boiled potatoes every night, coleslaw, packet soups, chicken stew, knuckle of lamb and fruit salad and ice cream. I felt sure the District Commissioner had enjoyed a similar menu. The Bukan boys who served us were barefoot and dressed in shorts. They carried in the courses with a distinct air of resentment. A distorting mirror of colonial times had come to life in this reversal of roles. The health official was uncommunicative and sat sozzled on the porch. I returned through the darkness to my room, skipping over the enormous toads that sat immobile on the lawn.

My days fell into a comfortable pattern of tropical lethargy. Rising at dawn I would brush the bodies of scores of mosquitoes from the sheet, shower, and wander in the gardens. The waters of Buka Passage are like a mill pond at dawn and reflect the bright gold of sunrise. Each morning I would walk along the groomed paths around the island, read Joseph Conrad's *Victory* or go snorkelling on the reefs of the passage. The drop-off is dramatic, the sapphire current swirling up from dizzying depths carrying huge shoals of tropical fish. Victoria Crowned pigeons, blue-grey with maroon breasts and spectacular white-tipped crests moved about the lawns. Green lizards climbed the pink palms.

The lower settlement of the island is populated mainly by refugees from the Bougainville war. It has a shanty-town appearance with suspicious areas of undrained swamp. Boys were playing football with a plastic bottle on a concrete area next to a manicured football pitch. Above this is a neglected colonial cemetery. I was about to note down some of the names, when I was attacked by two dogs and had to move pretty quickly. The

The Polish anthropologist Bronislaw Malinowski (1884–1942) founded the discipline of 'British' social anthropology. From 1915 to 1918 he forged a powerful new methodology in the course of his extended fieldwork in New Guinea on the Trobriand 'Islands of Love'. Photographed by his friend, the Polish artist and writer, Stanislaw Ignacy Witkiewicz ('Witkacy'), 1885–1939.

The Polish ethnographer Bronislaw Malinowski examining the *soulava* of a Tukwaukwa girl with a degree of imperial *hauteur*. He intended this photograph to be the frontispiece to his beautifully-crafted study, *The Sexual Life of Savages in North-western Melanesia*, published in 1929, but may have been discouraged by a prudish publisher.

Malinowski's photograph of the skull cave or *bulagwau* of the Tabalu subclan on the Trobriand Islands. The lower jaw bones of the skulls are missing as they were always kept by the family as relics of bereavement. I found that the sites of such mouldy caves still arouse fear, but respect for their metaphysical significance is dwindling.

The brilliant dancer Tobigawepu Mokagai at the site of Malinowski's tent, Omarakana village, Kiriwina, Trobriand Islands. The Polish ethnographer spent years on the 'Islands of Love', observing and being observed, the tent achieving iconic postmodernist status. Note the treasured *mwali* Tobigawepu is proudly displaying, gained during his *kula* trading.

The great Russian explorer 'Baron' Nikolai Miklouho-Maclay (1846–88), one of the country's true spiritual fathers, set off for New Guinea in 1870 searching for *homo primigenius*. Known to villagers as the 'Moon Man' because of his pale skin, he treated them with immense compassion and pioneered extended field research in anthropology.

Errol Flynn (1909–59) dressed theatrically in white when he arrived in the Bismarck Archipelago in 1926 as a 'sanitation inspector' for the Australian administration. Impetuous sexual exploits and dubious business ventures forced him to leave New Guinea in 1931, allegedly with diamonds lifted from an ageing *inamorata* concealed in the hollow handle of his shaving brush. 'There is no thrill like making a dishonest buck,' he commented later.

Graves of German priests of the Catholic Mission of the Holy Spirit at Alexishafen, near Madang. This mission station, established in 1905 with appalling sacrifice in human life from malaria and murder, became one of the noblest achievements of the German hegemony in New Guinea.

dogs are in fine shape here, the only place in Papua New Guinea where this is the case.

'Hello, there! Come on over!'

A dark figure in a royal purple *laplap* gestured furiously to me from a high veranda. He turned out to be Dr Thomas Sawa, the first Bukan doctor to graduate from the University of Papua New Guinea. He had fled the fighting in Arawa in his fibreglass dinghy together with his family some time in 1990 and has not been back since. I climbed up to his veranda where he was decorating a carved canoe paddle. Various female members of his family were clustered under the house sweeping away huge cobwebs. I noticed a heavily-annotated copy of Richard Parkinson's *Thirty Years in the South Seas* and Robson's *Queen Emma* on a table. I sat on a white plastic chair with one arm.

'Well, my dear fellow, make yourself comfortable and I'll get you a beer.' His educated English manner caught me completely off-guard. He was bare-chested, had an interesting face, and cut a fine figure against the palms and magnolias in his garden.

'So you're visiting Buka. Where were you before this?'

'Rabaul. I stayed at The Hamamas. What a heroic group of people are struggling there!'

'Well, I was staying at The Hamamas when the volcano erupted! Susie's uncle is John Christie, the first white doctor to graduate from a black university in Papua New Guinea.[1] I'm the complete opposite. Black on the outside and white on the inside! Ha! Ha! The graduation ceremony brought tears to my eyes. Yes, yes, certainly they're heroic.'

'Do you work at the hospital on Sohano?'

I seem destined to talk about my least favourite subject, the state of medicine in Papua New Guinea.

'Not really. The main hospital has moved to Buka. It's just lepers and tuberculosis patients on Sohano now.'

---

[1]   Susie McGrade is the wife of Bruce Cameron Alexander, manager of The Hamamas hotel in Rabaul.

'Leprosy! I thought leprosy had been wiped out years ago.'

'Do you know the main causes of deaths in the old days?'

'Well, malaria must've been one.'

'Certainly, but there were others. Malaria, snake bite, falling coconuts and *fires in the grass skirts*!'

His voice softened to a confiding whisper. 'Did you know they performed trepanning operations in New Britain?'

'When was this?'

'In the old days.' I felt sure he was about to tell me the operation was still performed.

'How was it done?'

'Suppose the skull had been crushed with a club. Well, the wound was first bathed in the sterile liquid of a young coconut before the incision was made. The surgeon used an obsidian splinter. A couple of assistants drew back the scalp. The pieces of broken bone would be removed and the brain observed to see if it was pulsating. If so, the man would probably recover. The scalp was washed again, tied back with grass . . .'

'Stop! Stop!'

'Calm down! Magic plants did the rest. Actually, speaking of magic, I should've become a missionary, not a doctor! I just love Wesleyan hymns. I'm preaching tomorrow – why not come along?'

Pressure was being exerted on him by the community to change profession. The fierce afternoon sun slanted across the veranda, but towels had been strategically hung on a wire to create some shade. More beers were welcome.

'My grandmother used to weave baskets. Do you know those famous Buka baskets? Well, they're originally from Bougainville. As a boy, I remember huge ocean-going canoes, carvings, idols and pots. What with the war and colonialism, Buka culture has almost gone, gone forever.'

He looked pensive and began to leaf absent-mindedly through Parkinson's heavily annotated book, as if searching for the past.

The sun was setting in a spectacular fashion over Buka Passage. Silhouettes of coconut palms were etched into a sky of improbable blue. Colour washes of rose madder, orange, old gold and grey spilled across the island as if from a painter's brush, the waters of the passage swirling copper. Dusk was coming on and Dr Sawa began slapping the mosquitoes settling on his arms and legs. We headed back to the lodge for six o'clock messing. I dared not be late. I might be shot or beaten. I rolled down the sleeves of my safari shirt.

'Yes, do that. For God's sake don't get malaria – it's everywhere – cover up your skin and take precautions.'

'What precautions do you take?'

'I don't take any,' came the illogical reply. 'No point. There are four different strains of human malaria. It's coursing through my blood right now most probably. Lies dormant in the liver. One strain causes almost instant death with shivering, coughing and vomiting.'

Thomas slapped his arms and legs all the way back to the lodge, where I left him deep in conversation with the health-care engineer. In the stiff atmosphere of the national dining room, I faced my colonial stewed knuckle of lamb and boiled potatoes while the beer cage was unlocked and news of more political corruption blared from the television. A cheque for four million *kina* had disappeared on its way to a provincial government. Now I could make more sense of the other sign in the dining room:

**NO CREDIT ALLOWED TO ALL GOVERNMENT
DEPARTMENTS
PAY BY CHEQUE OR CASH IN HAND BEFORE BOOKING IN**

I missed the UN helicopter to Torokina on Bougainville by two minutes. Earlier that morning, a banana boat ferried me across the passage to Buka town where I headed for the police station. As I wandered towards the fibro building I saw a police cell in the courtyard containing three prisoners sweltering in the heat. The cell had bars as thick as a lion's cage and was being watched by a couple of glamorous policemen in dark glasses.

A pleasant face looked up from the desk in the small air-conditioned office. I enquired about chartering a helicopter and was told it would be around seven hundred pounds sterling an hour. A bit steep.

'But there's a UN chopper leaving for Torokina soon. I'll ring the airport. You'd have to pay your share. Oh. And the Panguna mine is still off-limits. We can't fly over it.' A long telephone conversation ensued that veered onto other topics. I glanced idly at a newspaper on the desk and waited.

A man of Chimbu and Saposa parentage was beaten all the way to the Buka police station, where he was dragged into the safety of the cells.

His face fell as he told me the chopper was about to take off and I was too late.

'Anyway, I don't know how you would've got back within a week. They're staying on there.'

A large man dressed in an immaculate police uniform came into the office.

'This is our Police Commander, Joseph Bemu.'

'Hello, there! I see you're doing your job. Those fellows safely in the cells.' I gave a slightly complicit smile.

'Ha! Yes, them. Forty-four arrests in the last two weeks. Pretty good going. In Buka just like you can buy a bottle of coke, you can buy a bottle of beer here and consume it right there. Alcohol had been banned for twelve years until recently. A problem.' He clicked his tongue against his teeth.

I felt they needed to talk in private so slipped out into the baking heat of the town. Youths lay asleep, spreadeagled across the trays of utilities. On an impulse and despite the dangers, I decided to take a dinghy across to Kokopau on Bougainville and see if I could hire a four-wheel drive to explore the coast road towards Kieta. I suppose I was pushing my luck travelling alone, but I wanted at least to set foot in the place. Bougainville was one of the most developed and beautiful of the island provinces before the tragic war.

The boy I hired to take me across was a snappy dresser – cream shirt, blue striped shorts, blue neckerchief and wrap-around blue sunglasses. He was smoking a 'cigar' made from loosely-rolled leaves of local tobacco. We fairly flew across the passage. With his help I managed to persuade his contact to 'lend' me a Hilux for half a day for a significant financial consideration.

The road was atrociously potholed, but I soon left behind the ragbag collection of long, low buildings at Kokopau and was swallowed up by the jungle. The steering on the four-wheel drive was vague, to say the least, and fording the deeper rivers was perilous. If I got stuck in the pebbles or took the wrong track there seemed no one about to help me. The old coconut plantations were picturesque and overgrown, some of the palms soaring to incredible heights. Every now and again the burnt frame of a torched house or ruined village could be seen distantly through the undergrowth.

It was when I was out of the cab checking the depth of a stream that I first noticed a slight movement within the suffocating vegetation. I was sure something or someone was there. With the warnings of carjackings on this road from my nervous fellow guests at the lodge, I quickly climbed back into the cab. By the time I had put the Toyota in reverse, two men carrying assault rifles and dressed in motley semi-military garb were sprinting down the river bank towards me, gesturing violently and pointing their guns aggressively in my direction. I slammed my foot on the accelerator, lurching into potholes and slithering in the mud as I spun the vehicle to face the opposite direction. There was a terrific crash of gears as I thrust the truck into first and shot off back towards Kokopau. The rear-view mirror shattered and I vaguely thought the road must have been pretty rough to break that. A terrific bang came from under the car and it began to move crabwise towards the side of the road. I kept driving and safely rounded a bend. I simply kept going, frustratingly forced to slow down at the river fords, water cascading in fountains from the wheel arches. A few innocent villagers on the roadside looked at me as if I was mad. I reached the landing stage at Buka Passage and leapt out, more than a little shaken and dripping with perspiration. Gideon, who had 'lent' me the vehicle, ran over.

'What's happening? *Lukim tarak blong mi!*'[1]

I did look. A neat bullet hole had punctured the rear of the cab, shattered the mirror to exit just above the windscreen. The truck leant at a fantastic angle, like a man with a broken leg. Gideon was scrambling underneath.

'*Spring blong tarak ibagarap yah! Yu bai baim mi planti lon dispela!*[2]

I quickly offered him a couple of hundred *kina* in cash which calmed him down. A great crowd had gathered talking rapidly in Tok Pisin and whistling through their teeth, inspecting the damage and glancing accusingly in my direction. Although I had not killed anyone on the road, for which you can be jubilantly lynched in Papua New Guinea, I decided to make a speedy exit and jumped into a banana boat to escape to Sohano. As I sat sipping my beer on the veranda and looking out over the passage, my feelings about tropical islands were certainly becoming less romantic. The insecurities of war are a tortuous reality for these people.

'Have you heard of Mr Shit?'

I was sitting on the veranda having a drink with the former Head of Protocol for the PNG national government and former Ambassador to China, Larry Hulo. Late in the evening, a superb rose doré sunset was again developing over Buka Passage.

'Come on, Larry, it can't be his real name surely.'

'In PNG anything is possible, my good friend. Surely you have realised that by now.'

---

[1] 'Look at my truck!'
[2] 'The spring's broken! You'll have to pay plenty for this!'

'Fine. Who is he then?'

'He's in the fertiliser business, as you might imagine. Real name – Abba Bina, but someone thought Mr Shit was an excellent marketing ploy, in his manure business, you understand. He was in the army for a while, perhaps the defence forces, and then went into politics.'

'Surely he didn't use that name in the campaign.'

'Certainly did. The Electoral Commission allowed him to use it on posters – "Vote One, Mr Shit". I have one at home. A real conversation piece.'

'Come on, Larry. You can't be serious!'

'That's not all. Everyone knows politics equals money in our society. His campaign slogan was written on his business card: 'Chicken shit, horse shit, cow shit – but no bullshit.'

The unusually urbane Larry Hulo was born on Sohano or, as he more correctly prefers to call it, Suhana, which means 'meeting place'. There were thirteen Australian families living here when it was the colonial headquarters for the province and he attended the colonial school. Larry went to the University of Papua New Guinea with Sir Kina Bona, the High Commissioner in London. Later he spent some twenty-three years in the Foreign Service, learnt Japanese to become Deputy Ambassador to Japan and then Ambassador to China in 1991. He was now deputy Mayor of Buka.

'Not much sign of *kastom* in Buka these days, Larry.'

'Tragic. Tragic! The traditional culture has withered in Buka-Bougainville because of the civil war. It might return in some form if autonomy is granted, I suppose. Actually, the Catholic Church is incorporating indigenous culture into the liturgy more and more. That preserves something at least. Have you heard of the Hahalis Welfare Society?'

'No, not at all.'

'Ah, well, that's something to learn about, I can tell you! It's having a resurgence.'

'So many successful men and women return to their villages, don't they? Why did you come back to Suhana, Larry?'

'Look, it wasn't easy. I had to relearn my own language. I had had too much contact with expatriates in embassies. I've returned to rediscover my roots, rather belatedly I must say.' An enormous black beetle began a slow progress across the floor like a small prehistoric creature.

'But you were a politician. What do you think are the greatest dangers to democracy in PNG? Sandline made everybody pretty nervous of a military coup.'

He leaned forward in his chair. 'Do you think I am rich enough to be a politician? The political system is based on the clan and responding in kind. Most villagers live in family groups and think the national government is a world away. It hardly affects them. The Papua New Guinean's first loyalty is to his parents and relatives, then comes the clan and the village, then the district and province and finally the good of his country. If a man is successful he is expected to distribute his good fortune to his relatives and members of his clan. Do you understand the concept of *wantokism?*'

'Yes, it means loyalty to those who speak your language and live in your district.'

'Exactly, but the edges are more blurred than that. It can lead to nepotism and corruption, but we don't always see it like that. The Westminster system sees it like that. And you former colonials always forget the Asian factor, the Asian way of doing business which is so prevalent here. It's not at all the Anglo-Saxon way, you know.' I prodded the black beetle with my foot and it began to hiss, like a sea-valve, in the most disconcerting manner. Its serrated jaws clamped shut on my boot.

'I suppose people see politics as the path to riches and there is little accountability.'

'True. Politicians are constantly recycled in this country. Round they come again as if trapped in a revolving door. Seriously though, this must be the worst time financially since independence. Living standards are continuing to fall in real terms. People feel they must make the best of their opportunities. At the Lord Mayor's meeting for PNG I attended recently, the atmosphere was so heavy with urban reports of violence, unemployment, lack of resources . . .' He paused for a moment. 'Shall we walk a little around his haven of neutrality? I'm getting depressed just sitting here!'

We set off down the path across the lawn towards the austere white Japanese war memorial at the edge of the cliff. Arum lilies were flowering in naturalised groups and creamy miniature orchids hung in festoons from the rain trees. We stood together overlooking majestic Buka Passage, the low island of Buka on one side and mountainous, threatening Bougainville on the other. The setting sun gave a monumental grandeur to the extravagant scene. Motorized dinghies were urgently crisscrossing the rapid current of molten bronze. He sighed heavily.

'My job is to talk up the peace process – to create hope.'

I remained silent. Day swooned into the arms of tropical night.

'Such a strange rock.' I was looking over the cliff at a piece of Sohano that had broken off and settled some distance from the shore. Plants had taken root on it and grew like wild hair. Wild boys spun their boats through the narrow channel in the gathering dark.

'Ah, yes. Tchibo Rock is magical and has a legend, although I'm not sure of the details. Do you know what lies beneath these waters you admire so passionately?'

'No.'

'Bones. The bones of thousands of my people. Alleged spies, informants, traitors and innocents tied to car parts or pieces of concrete and dumped from helicopters in the passage. Children still find terrible things washed up on the beach as old scores are settled.'

Horror lurked beneath the beautiful surface as is so often the way with nature. We parted company, not without puzzlement on my side that such an educated and cosmopolitan man could return to Sohano and happily settle down after touring the world at the highest political level. Many successful entrepreneurs and 'bigmen' had supplied themselves with escape mansions in Cairns. But other eminent Papua New Guineans still yearned for the simplicities of village life, far from their plush offices in great cities, away from international airports, close to the soothing lap of waves on an island shore.

It was not until some months later that I read about the extraordinary Hahalis Welfare Society in an obscure thesis in the Fischer Library at the University of Sydney. The society was set up in the 1950s by a couple of Buka youths. Originally it was rather like a gardening cooperative that grew into a private company with its own trucks, stores, private roads, mainly dealing in copra. Communal work and the profits went into one pot. It possessed elected officers, a proper administration and all the paraphernalia the colonial authorities liked to see in a 'native' enterprise. There was a darker side though. The society began to resemble a 'cargo cult'.

The Germans used to call 'cargo cult' *schwarmgeister* or 'ghost enthusiasm' in the old days, to describe what appeared to them to be periodic bouts of irrationality in the local population. Such 'cults' thrive on dreams and emotional desires that can never be satisfied – rather similar to the Western consumer's irrational passion for goods, entertainment and unattainable love. In Papua New Guinea, a seer would emerge without warning who foretold that the huge quantities of goods, 'cargo', would arrive after a millenarian experience, a preliminary catastrophe, and that the people should prepare. Sexual rules were broken and hysteria could overwhelm the community, as in the famous outbreak of the so-called 'Vailala Madness' in the Gulf of Papua in the 1920s. The arrival of the Americans in the Second World War brought villagers into contact with unimaginable amounts of 'cargo'. Only supernatural intervention could explain it. To 'bring on the cargo', a local leader emerged who claimed to be in communication with the spirit world. In imitation of the American soldiers, 'offices' were set up with 'desks' and 'documents' on hastily-built 'airfields', carefully constructed to await the arrival of planes. When the cargo failed to arrive, the prophets of prosperity persuaded the villagers that the whites had wickedly diverted the goods from their rightful owners.

In other 'catastrophes', the people of Karkar Island near Madang believed their island would overturn and all those not drowned would emerge from the sea with white skin. One modern prediction was that certain individuals would acquire credit cards and be magically tattooed with the number 666, a reference in the Book of Revelations to the evil deeds of the Devil. Such people could then obtain infinite amounts of money from holes in the wall. One cult in New Hanover revolved around the former American President Lyndon Johnson, associated in the local mind with the ubiquitous 'Johnson' outboard engines on their banana boats. The people attempted to 'buy' him as a leader and when he failed to accept, they offered him a seat in their local government. The essential feature of any cargo cult is that the world can be dramatically transformed at the bidding of a suitable prophet.

On Buka, the Hahalis Welfare Society, although not strictly speaking a 'cargo cult' at its inception, had Christian and neo-pagan elements. It became defiantly anti-missionary and anti-Catholic. The members believed that full Christianity was being denied them as they were only given short catechisms rather than the complete Bible. More outrageous to the Marist fathers[1] were the 'baby gardens' where young women were visited by male members of the society. The missionaries were scandalised by what were in effect free brothels. Christian morals were scorned. The young men justified themselves by explaining they simply belonged to 'matrimonial clubs' for trial marriages. The leaders were finally excommunicated, but the missions lost hundreds of the faithful to the Welfare Society.

---

[1] Members of the Society of Mary, one of the earliest Roman Catholic missionary and teaching orders to set up in Papua New Guinea.

By the 1960s the society refused to pay tax which alarmed the Australian administration. A rebellion was quelled by hundreds of police dispatched from Bougainville and Rabaul. It was only then that 'The Welfare' began to turn cargoistic. Matters got decidedly out of hand. One of the leaders was locked up in a mental asylum for beheading his sixteen-year-old nephew to bring on the cargo. His doodles indicated cargo dropping from helicopters and whites being bitten by double-tailed snakes.

Rapid change in the ideological fabric of Melanesian society under European influence often gave rise to such mutant yet understandable forms of behaviour. The oft-quoted leap 'from the Stone Age to the technological age' was real enough. The sheer stress of breaking away from the past was often followed by the radical disruption of all previously-accepted norms of conduct. Many 'cargo cults' were a passionate yet distorted expression of a newly-forged identity, the faltering beginnings of nationalism.

Another night would have to be spent in Moresby before embarking on the last leg of my tour of the islands, the flight to Madang and then boarding the catamaran for the Trobriand Islands. I wandered the lawns of Sohano absorbing the peace and tranquillity for the last time. Returning to my room, I put a Chopin piano concerto on my miniature tape player and lay on my bed gazing over the veranda through coconut palms, across Buka Passage to magnificent Bougainville. The distant barking of a dog, the splash of a leaping fish, the rhythm of a *garamut* slit drum were carried fitfully on the breeze.

I was reading Conrad's novel *Victory*, with palm fronds rustling outside my room. Heyst was listening to the grotesque Zangiacomo's Ladies' Orchestra at Schomberg's Hotel in Sourabaya. He met his lover, the pale violinist Lena, in the darkened garden after the Japanese lanterns had been extinguished.

It is not easy to shake off the spell of island life ... The islands are very quiet. One sees them lying about, clothed in their dark garment of leaves, in a great hush of silver and azure, where the sea without murmurs meets the sky in a ring of magic stillness. A sort of smiling somnolence broods over them: the very voices of their people are soft and subdued, as if afraid to break some protecting spell.

'The boat's ready!' someone shouted.

I collected my things and switched off the music. I was about to start travelling again, that delightful suspension of security where chance remains the master.

The airport was full of Defence Force officers and thugs dressed in camouflage battle fatigues and ammunition belts. One was delicately carrying a pet bird in a small bush-material cage decorated with shells. All I could see of the 'pigeon' was a maroon throat and terrified eye. Crowds of men wearing big boots (the boots most men wore in Buka and Bougainville seemed inordinately large and aggressive) were spitting, hacking, sniffing and chewing betel. Grey-haired missionaries withered in the shade. Youths, their green military berets precariously perched on Rastafarian locks, lay asleep in the tremendous heat, draped over the trays of utility trucks like wet towels. One 'tough boy' with the blackest skin imaginable, seemed to have stepped directly out of a combat computer game. With some panache he was wearing tight, black leather shorts, a steel-studded belt, big boots with scarlet laces, mauve-mirrored sunglasses, gold chains, battle camouflage jacket and had entirely shaved his head apart from a round clump on the crown. The whimsical 'Dress and Insignia' of the PNG Defence Forces was triumphantly on display.

As the jetliner climbed above Buka, the sea below appeared a fathomless blue. The aquamarine reef and verdant islands fringed with silver were incredibly beautiful. The scene belied all the murderous horrors that had occurred there. The mysterious symbiosis of beauty and cruelty preoccupied me for much of the flight to Port Moresby. Battle helmets and long military maps kept falling into the aisles all the way.

20. *Auf Wiedersehn, Kannibalen*

My travels to the islands of Eastern Papua New Guinea had begun with a voyage on the replica of the sailing ship *Endeavour*, captained by an Australian, and were concluding on a modern catamaran, the M/V MTS *Discoverer*, captained by a Trobriand islander. These vessels represented opposing Western assumptions about the Pacific islands. The 'noble savage' of the Enlightenment had evolved through the 'ignoble savage' of the nineteenth century to the present-day 'vanishing savage' on cultural display for the tourist camera. I was endeavouring to complete an historical picture for myself through these layers of time.

The geographical region known as 'the Pacific' is a foreign concept imposed like much else on native populations, their own world being restricted to the sailing range of their canoes. The first contact with Europeans did not start out as a 'fatal impact', but soon became one. The *Endeavour* was a symbol of Enlightenment imperialism. This floating laboratory introduced the Pacific islands to European academics as a rich subject for botanical investigation, anthropological study and philosophical speculation. In the eighteenth century the Pacific islands were thought of as the idyllic haunts of savage yet noble races.

The softly-glowing landscapes of Tahiti, bathed in tropical light and sweltering under immense skies, were painted by the influential artist William Hodges, the official painter on Cook's second voyage. He brought to the English sensibility a new feeling for exotic sun-drenched climes, and in painting a revolutionary vocabulary of colour and light. Polynesia has never been the spiritual home of Christian Englishmen, and these new images encountered some resistance. His beautiful portraits in red chalk presented a deeper character and psychology than was previously thought to exist within so-called 'primitive' men. Apart from the paintings destined for connoisseurs, popular engravings by Italian neo-classical artists depicted 'savages', particularly Polynesian chiefs, as noble and dignified figures from Greek mythology, carefully and aesthetically draped. Greek heroes, like the Polynesians, had also navigated by the stars. The eighteenth-century French naturalist Abbé Armand David described the Pacific islands as 'a region of perpetual ecstacy'. But this was a golden age of tolerance and wonder, and of only occasional horror at the cannibalism, infant murder and sinful sexual practices witnessed by visiting mariners.

This idea of the 'noble savage' mutated in the nineteenth century into fear of the region and abhorrence of the people. Pacific it was not. For Europeans, the Pacific islands developed into competitive spheres of strategic and scientific interest as well as becoming a spiritual and physical testing ground for various denominations of Christian missionaries. Cannibalism and disease were ever-present threats. Cook suffered a 'heroic death', murdered at Kealakekua Bay in Hawai'i with daggers made from iron spikes originally given as gifts. The French explorer Lapérouse was considered to have simply disappeared in the Solomon Islands until, in 1827, an Irish sailor, Peter Dillon, spied a glittering sword hilt instead of a jawbone dangling from a savage neck on the island of Vanikoro, east of New Guinea. The imposition of imperial Western values, the introduction of explosive weapons, European diseases and European commerce reaped a harvest of thorns as native peoples understandably answered the rape of their land with violence.

Writers such as Joseph Conrad began to emphasise the morally-destructive influence of the tropics on white men. At the turn of the century, the Australian writer Louis Becke also wrote stories of moral decay and perversity among South Sea traders. He spoke perceptively of 'that strange, fatal glamour that forever enraps the minds of those who wander in the islands of the sunlit sea . . .' The Pacific islands finally came to be characterised by waning cultures, degenerate islanders and dissolute romantics fleeing the fetters of civilisation. And then, like a Phoenix rising from the ashes, modern tourism has reassembled their Arcadian status. The escape to 'paradise' has become an irresistible alternative reality. Paradise *is* an island, but then islands have always held a privileged position in the psychology of escape.

The remarkable catamaran I was to sail on had been created as an icon of tourism, a sign of exuberant and imaginative optimism, a statement of faith in the beauty and cultural fascination of the island provinces and the Sepik river. In the newly-independent state of Papua New Guinea, it gradually became a symbol of sustainable tourism and was lavishly equipped in the manner of a small luxury hotel. I was taking one of the rare voyages that this purpose-built vessel makes to the Trobriand Islands. Tourism has sadly declined by fifty per cent in recent years. Apart from groups of divers, I met one other 'traveller' on my three-month journey. Papua New Guinea, always the exceptional case, will be the last country in the world to attract travellers *en masse*. It has unfortunately retained the residual fear that accompanied its original survey, bolstered by constant reports of contemporary violence. Legends and myths continue to create a surreal landscape in the mind of any prospective visitor. These foolish misjudgements gave me the rare opportunity of visiting some of the most isolated and enchanting island communities on earth, a subsistence way of life scarcely changed for thousands of years.

We left for the Trobriands just before nightfall. Electric lights stop abruptly along the Rai coast, a sudden curtain of black marking the division of a civilised town 'striving for beauty' and the true beauty of an undeveloped shore. I was elated at the prospect ahead as we sliced across the Bismarck Sea, leaving

Madang at copper-blue dusk, the squadrons of flying foxes seeking their night roosts among the feathery palms.

The passage across the Bismarck Sea and into Vityaz Strait can be monstrous if the prevailing wind and swells are against you. I spent a wretched night unable to leave my bunk or stand in my cabin. I managed to have dinner before it became too rough and retired to bed with the phantasmagoric *Sorcerers of Dobu* written in the 1930s by the anthropologist R. F. Fortune. The vile weather kept me on the point of seasickness. I woke early to a forbidding scene of low cloud and rain drifting across the venomous-looking vegetation around Finschhafen.

In the 1880s, the profligate explorer and first *Landeshauptmann* (Captain-General) of the Neu Guinea Compagnie, Vice-Admiral Freiherr G. E. G. von Schleinitz, had enthusiastically named the capital after Dr Otto Finsch, the original Compagnie envoy. From the deck of the *Ottilie* he could not have known that the climate was lethal. A particularly nasty strain of malaria and smallpox together with constant *guria* (earth tremors) ultimately drove the dissipated employees of the Compagnie to alcohol and the indigenous women to solace their existence. They tried moving their headquarters further up the coast to Stephansort (the present Bogadjim), searching for a healthier environment, but to little effect. No one knew then that malaria was borne by the *Anopheles* mosquito. Blackwater fever, beri-beri and dysentery also took their toll. In desperation at the continuing fatalities, the Compagnie moved even further along the coast to Friedrich Wilhelmshafen (present-day Madang).

It was raining but warm. Finschhafen consisted of a wharf and a dismal collection of tin-roofed sheds scattered along the shore, the entire landscape liquefying in the drizzle. Most of the original town had been destroyed during the Second World War. Undoubtedly the sun would transform the scene into an idyllic paradise as it always does in the tropics, but not on this particular morning. The coastline looked poisonous and I had no difficulty imagining the daunting task facing the isolated German traders.

The fatalistic memoirs of members of the Compagnie describe a malaria-infested fever-hole with marauding cannibals. The scattered islands dissolved in the tropical downpour like cubes of green sugar, staining the sea silver-grey. Despite the downpour I went ashore in a river boat. Rain releases a multitude of perfumes from tropical vegetation and my stroll along the jungle path was like a walk through an aromatic salad. Pretty huts and friendly faces were everywhere.

The clouds slowly began to clear and the catamaran was soon pounding towards the Tami Islands, some twelve kilometres south-east of Finschhafen. The sound of singing was carried over from the shore long before I could see the inhabitants. The diminutive figures of women and children stood on the beach of the low coral island, welcoming us with a high-pitched song and the slow waving of bunches of croton leaves. I felt transported back to the Pacific of Captain Cook. Giggling children and smiling adults surrounded the visitors, sprinkling us with water from finely-carved wooden bowls. One beautiful girl with soft brown eyes and sensual lips, a delicate tattoo on her right cheek and a red hibiscus in her hair, took my hand. She led me into the village. Other women placed bunches of dried palm fronds in the waist of their skirts at the rear, like a horse's tail, and sang and danced ahead of us. There had not been an outside visit for eighteen months.

The village consisted of a motley collection of bush-material huts and prefabricated houses. At the centre a covered market had a few carefully-selected objects for sale. The Tami Islands are noted for their delicately-carved wooden bowls. These have replaced the expensive earthenware pots of former times. Fearsome fish-hooks carved from bone with an arc of dark tortoise-shell lashed to the shaft lay beside strings of shell money fashioned into necklaces. I was after a canoe splashboard. The outrigger canoes from Tami are unique. A small, carved splashboard shaped like the front of a house with a decorated pitched roof is placed over the open ends of the well of the canoe to prevent the ingress of water. Occult symbols and 'eyes' are painted on it to ward off evil spirits. They are long and elegant craft, often handsomely decorated. Building a canoe is such an arcane art in Melanesia that they all have an oral history. The board I finally purchased was made for a canoe constructed in 1979 by Matthew of Kalil.

Spirits are believed to wander everywhere in the islands, and Tami is no exception. The rare Tago Spirit emerges every ten years and has to be appeased by a cycle of festivities that lasts a full twelve months. These spirits make an appearance every day, chasing the women and throwing stones at the men. The ghost mask is a complex affair of white-painted bark sewn onto a rattan frame. The wearer looks through slits beside the nose rather than through the fiendish eyes painted onto the mask above the twisted, grinning mouth. It is crowned with a hemisphere of brown 'hair' and a white feather plume not unlike a *dukduk*. Once seen, such a mask is not easily forgotten, and after much good-natured haggling I obtained one for my collection.

The islanders sang us away. Accompanied by a number of graceful canoes, we made a joyful departure from Tami. The former German Empire in New Guinea melted away in the wake of the ship and I left, probably forever, the coast where I had encountered such extraordinary friendliness and warmth.

The night-crossing of the Solomon Sea to the island of Sim Sim was very rough. The catamaran did not ride fluently with the adverse swells but fought them hard. I was feeling queasy, tired of holding the walls, when the thumping was suddenly calmed and we hove to in the lee of an island. We had arrived at the outermost north-west settlement of the Trobriand Islands, the Lusançay Islands and Reefs. It had taken some fifteen bruising hours from Tami. I had slept in my clothes for a second night, having given up wrestling the bucking craft.

Sim Sim is a low coral island covered with coconut palms, ancient rosewoods and dense vegetation. A line of woven huts borders the narrow beach on a long promontory. They look fearfully unprotected and vulnerable on this shallow spit jutting into the open sea. I leapt from the landing craft into the water, the sand and pulverised coral soft between my toes. A *waga*[1] with a splendid *lagim*[2] was drawn up on the shore. The island is seldom visited and the welcome was low key compared with Tami – a simple shaking of hands. A few excitable children clustered round me clutching threadbare scraps of pink and scarlet cotton to their nut-brown bodies. The feeling of remoteness was overwhelming. The lives of these villagers were so different to mine, carried on at subsistence level as opposed to my complex 'technological paradise'. But there would be no privacy here, the slightest movement noted and watched.

I wandered about a village that showed not the slightest sign of Western influence apart from the ragged clothing. Coconut husks and the detritus of storms clogged the beach, jagged lumps of coral littered the ceremonial areas between the huts, canoes lay under the frangipanis like crescent moons, the stumps of old palms thrust through the sand like fingers of ancient bone. The scene had an unsettling primeval immobility, as if time had ceased. The total absence of manufactured refuse – plastic bags and bottles, beer cans, nylon fishing line, labels – was almost shocking. The wind blew strongly off the sea, sweeping through the tangled hair of the palms, the whole island straining against the gale. I had no difficulty in believing that sorcerers possess a power over the winds. The north-west monsoon is believed to blow from the mouth of the local spirit, Yarata.

Too soon we were negotiating a gale and heavy swell to Kaibola Beach in the north of Kiriwina, the largest island in the Trobriands. The captain of the catamaran was from Omarakana, the chief's village, and knew the treacherous waters around the

---

[1] The generic term in the Trobriand Islands for all types of sailing craft, now including European craft.
[2] The two ornately-carved transverse splashboards that enclose the well of the Trobriand canoe at both ends.

islands intimately. He assured us that this anchorage was the calmest in all weathers and at all seasons.

An islander announced our arrival by blowing a conch shell. Kaibola village is a fishing settlement, obvious from the number of canoes hauled up on the beach and the small number of yam houses. The contrast with Sim Sim could not have been greater. There were many artefact-sellers on the beach, lined up in serried ranks. Their trading energy and inveterate pestering of visitors have been severely constrained by the local council. But I like the colourful energy of barter, and the Trobrianders are passionate traders. As I walked down the ranks of chastened islanders, their eyes cast up from the ground beseechingly, I felt they had been emasculated by insensitive 'modern decorum'. Whenever I deigned to bend down to inspect a finely-carved bowl inlaid with discs of mother-of-pearl or a similarly carved yam house, canoe, walking stick, crocodile, fish or grasshopper, I felt uncomfortably paternalistic.

Most of the huts in Kaibola village were arranged around a central area in a careless circle. Pigs roamed among old ladies smoking bush cigarettes, children squealed with pleasure, scattering among the adults like billiard balls. Enormous cockroaches scuttled away looking freakishly shiny compared to their surroundings. The usual collection of wretched dogs whimpered as they were driven from any haven of refuge. Women and elderly villagers sat talking on their *kaukwedas*[1] or squatted on the ground holding their heads and gazing into space. Then for no discernible reason they would amble across the bare ceremonial area to talk and perhaps laugh at the intruders. To one side a woman was making *green doba*[2] by laying fresh banana leaves across a carved board and smoothing the surface, embossing it

---

[1] A Trobriand roofed platform providing shade in the village for sitting and socialising.
[2] Leaf money used by women and made from dried banana leaves bundled together. It is also used in bridewealth negotiations and festivities following a death. It is green before it dries to a parchment colour.

with a delicate pattern. Another woman seated on the ground among the pigs, dogs and chickens was deftly forming taro cakes and laying them in careful layers in a carved bowl. Late in the afternoon we headed back to the beach, passing our Trobriand captain wearing a nifty battered panama. I relented to pressure from artefact-sellers and purchased an old pig tusk.

'Very old. Very old! From war nineteen forty-three!' the old man enthused as I handed over my desperately-needed *kina*.

21. Under the Mosquito Net in Malinowski's Tent

'I need some potassium cyanide!'[1]

The Polish writer and painter, Stanisław Witkiewicz ('Witkacy'), was beside himself with grief. Just before the outbreak of the Great War, he was planning to set out for New Guinea with his friend, Bronisław Malinowski. They were in Zakopane in the Polish Tatra mountains, a centre for Polish intellectuals and musicians during the first half of the century. Witkacy's fiancée, Jadwiga Janczewska, believing him to have abandoned her, had committed suicide in a snowy defile two months before. She had belonged to the radical 'Young Poland' group of intellectuals, and such a histrionic act was quite in keeping with this neo-romantic literary movement, steeped as it was in eroticism and emotional extremes. During their tempestuous relationship, the pathologically-jealous Witkacy had tortured her with accusations of a murky love entanglement involving a 'scoundrel', the composer Karol Szymanowski.

---

[1] My modest dramatisation of the Malinowski–Witkacy friendship was built up from letters exchanged between the two and diary entries and essays by both. Malinowski's anthropological and personal writing was clearly influenced by the early friendship between the two luminaries, and as such I include an account of it. It is difficult for an outsider to comprehend the complex Trobriand culture without Malinowski's detailed studies and descriptions, and his influence has been profound. The Trobrianders admit this themselves.

Some late snow was drifting over the mountains and carved wooden houses of Zakopane.

'Cyanide! Are you crazy, Staś?'

Malinowski was torn with compassion. He had thought a tropical voyage might lift Witkacy's deep depression. He had invited him to a semi-official meeting of the British Association for the Advancement of Science in Adelaide, acting as a draftsman and photographer. Malinowski the scientist respected and envied his friend's artistic genius but this suicidal despair was burdensome. Witkacy tried to reassure him.

'Broneczek, I'm grateful, truly I am. No rash decisions. But cyanide's better than being roasted alive by savages!'

'I feel the same uselessness about going to New Guinea as you do,' Malinowski replied despondently, 'but everything's arranged. We have to go.'

They would shop together in London for tropical gear and photographic equipment. Witkacy had photographed and painted portraits of his friend on numerous occasions. Always pessimistic about humanity, he felt life was like taking part in a ball on the Titanic, which had sunk two years previously. In the end he stole some cyanide from a gold factory in Kalgoorlie in Australia but never took it, although he let it be known that he had licked the crystals once or twice.

They boarded the SS *Orsova* of the British Orient Royal Line in Toulon bound for Ceylon in June 1914. Both became horribly seasick and disliked the ship and the other passengers. The painter wrote many suicide notes during the voyage, most of which remained undelivered. His diary reveals that the 'monstrous beauty' of the tropics and the fantastic colours of Kandy were like sulphuric acid bathing his open wounds. Rickshaws hauled them along the flowering avenues to the Temple of the Holy Tooth. The heat was terrific. Witkacy put his Browning pistol to his temple at Domhilli but failed to pull the trigger.

News of the outbreak of war in Europe caused a crisis of conscience and they clashed over the role of partitioned Poland.

As a Russian subject Witkacy decided to return and fight in the Tsarist army against the Austro-Hungarian Empire. He served with distinction as an officer in the glamorous Pavlovsky Regiment and witnessed the Russian Revolution at first hand before returning to Zakopane in 1918. Malinowski, an Austrian subject, elected to continue his anthropological work. As an 'enemy' he was monitored by the Australian government, but was given financial assistance to continue his fieldwork in New Guinea, first at Port Moresby, then on the island of Mailu and finally in the Trobriand Islands.

The friends continued to correspond but never again came as emotionally close as in their youth. 'Nietzsche breaking with Wagner,' Malinowski rather grandly observed. Witkacy complained that Bronio had become 'Anglicised to the core'. On one occasion they had even decided that Joseph Conrad was a traitor. Malinowski went on to become the 'father' of functionalism in anthropology,[1] a British national and Professor of Social Anthropology at the London School of Economics. Witkacy for his part, became a celebrated avant-garde painter, intellectual and philosopher. He finally fulfilled his irresistible urge to suicide in a small Ukrainian village the day after Russia attacked Poland in September 1939.

This youthful friendship influenced them in manifold ways. They were both products of the fraught Polish intellectual climate of the time, fragmented personalities looking for personal identity. Malinowski in his 'secret' *Diary* written in Polish,[2] describes the landscape of the Trobriand 'Islands of Love' with the erot-

---

[1] 'Functionalism' is a belief in the practical application of something. In the social science of anthropology, it is the theory that all aspects of society serve a function and that their interaction is vital for the survival of that society.

[2] *A Diary in the Strict Sense of the Term* was written mainly in Polish but with frequent use of English and isolated words and phrases in Spanish, French, Greek, German and Latin, as well as Tok Pisin and the indigenous languages of Motu from Port Moresby, and Mailu and Kiriwinian from the Trobriand Islands.

icised eye of a photographer and colourist. His attempts to capture a moment of time are a superb literary and optical record. Tropical nature enthralled him. The *Diary* is a complement to his disinterested academic work and reveals the profound effect on his personality of actually being in New Guinea. It gives a rare glimpse of the volatile private life beneath the measured academic surface. Witkacy encouraged this eye in his friend and greatly influenced his aesthetic. Malinowski's ability to paint with words indicates he possessed the acute visual sense of a German Expressionist artist.

9 January 1918

After lunch read newspapers until 3.30; Then translated Navavile's *megwa*.[1] Then with Ginger on the lagoon.

The red sunset was just dying away under a belt of black clouds, the long stretch of mangrove reflected in the water also standing in rigid darkness against sky and water. The heavy soiled red light seemed to be oozing through the luminous patch in the West. It floated and trembled with the slow motion of the waves, and encased in the black frame of the clouds and shore, it seemed as stifled and stifling as the tepid air, which rolled along in clammy, indolent puffs. I felt it clinging to my bare skin rather than beating against it.[2]

Then Swedish gymnastics ... Strong feeling of dejection.

The *Diary* is a personal testament of his intimate emotional life during the assembling of the objective and 'detached' material of his seminal and supremely influential book, *Argonauts of the Western Pacific*. The work is subtitled *An Account of Native Enterprise and Adventure in the Archipelagoes of Melanesian*

---

[1] Navavile was the *towosi* or garden magician for Oburaku village; *megwa* is the generic term in Kiriwinian for magic or a magic formula. The reference is to the garden magic of this highly-respected member of the Trobriand community.

[2] Italicised paragraph written in English, *Diary*, p. 179.

New Guinea which reveals a great deal about the author's generous attitude to the native peoples he studied and depicted with
such charm.

Much of the impetus to travel to the islands of Eastern Papua New
Guinea came from my own reading of Malinowski in the snows of
Zakopane just after the fall of communism. Now by some dream
magic I was anchored off the Trobriand shore as the fulfilment of
a long-cherished fantasy of visiting the 'Islands of Love'.

The moon had risen over Kaibola Beach, the sand a ghostly
strip of white leading down to the headland called Bomatu. Black
palms were etched into the silver sky. Laughter and singing
drifted over the glimmering bay from the village; tiny fires glit-
tered deep within the dark vegetation. Bomatu is one of the three
horizons of Demwana, the paradise of the ancestors. When a
departed spirit heads towards Tuma, the island spiritual paradise
where Trobriand spirits repine, they must first linger at Bomatu,
a type of gateway to this paradise. This is where *Togilupalupa*,
a huge snake with a human head lives. *Togilupalupa* may also
be a goddess with rather large ears who can transform herself
into an exquisite beauty and be intimate with the men who pass.

The day began by travelling to Omarakana, the village of the
Paramount Chief and the most important village in the Trobriands. He had provided the only motor vehicle on Kiriwina, a
dilapidated four-wheel drive. As we set off, a violent clunking
came from underneath and clouds of dust rose inside the cabin.
There was tremendous play in the universal joints, the tyres were
bald, one almost flat, and the steering had so much free movement
the car gave the impression of a yacht tacking into wind. We
advanced crabwise with blithe negligence. Occasionally, the
vehicle would stop altogether, requiring the driver to disassemble
the petrol filter, take a mouthful of petrol and blow it through,
then reassemble the unit.

The Trobriand Islands are flat and somewhat featureless. The
central plain of Kiriwina is intensely cultivated which means an

almost complete absence of large trees. The shattered transmission did not bother our juggling driver, who laughed and twirled the steering wheel despite the perfectly straight track. We headed inexorably towards deep ditches while he deftly manipulated his lime pot and spatula. That a vehicle could proceed at all in this condition was beyond belief. A broken spring pushed through the seat, painfully jabbing into my bottom.

Malinowski spent many months in Omarakana gathering material for his anthropological studies. The visit meant a great deal to me, having spent long hours poring over his books. A blessedly cool breeze was caressing the village as it frequently does. Trade was king and villagers pressed us to buy their artefacts.

Paramount Chief Pulayasi greeted us formally and I noticed that the convention that none of his 'subjects' takes positions physically above him was still carefully observed. He sat on a chair on a grassy knoll, and the villagers and visitors gathered below. Times have changed since the irascible nineteenth-century Lieutenant-Governor, Sir William MacGregor, pulled a Trobriand chief down by the hair and sat on the 'throne' himself, shouting, 'No man in New Guinea shall sit higher than I!' Trobriand islanders have lighter-coloured skin than Papuans and show distinct influences from Polynesia. Similarly, the hierarchical system of chiefs is highly evolved in this matrilineal society. They are well known as expert sailors, their trading skills are renowned, and their superb decorative arts are celebrated throughout the country. I took no great interest in the conventional tourist artefacts displayed on woven mats and wandered around the village.

Many of the most important huts are exceptionally decorated and arranged in a circular pattern around a central ceremonial area. From ground level the precise plan is not evident, but the position of each house on the circle is carefully determined. The chief's house was raised on stilts and heavily decorated with symbols associated with the sea. The spiritual *udawada* or kingfisher birds are symbolised by dolphins painted onto narrow boards that form the arch of the gable. Striped decoration and other marine motifs such as angel fish are painted in red, white and black. Festoons of white *Ovulum ovum* shells hang over the balustraded porch, the house covered with a roof of dried sago palm. With great pride I was shown a faded blue fibro dwelling, roofed with corrugated iron standing next to this traditional structure. I was told this was to be his new residence. Again I suffered a sentimental nostalgia for the 'picturesque', the preservation of which meant so much to me but so little to those wanting a few modern comforts.

The facades of the chief's yam houses were similarly decorated. Painted waves symbolising the ocean and the spirits of the ancestors flow across the base. In the centre, most curious of all, is *kaisikalu*,[1] represented by the squid. This is a symbol of personal development through giving, as the squid expels its cuttlebone in order to grow. The yam houses in more humble villages were not decorated in this ornate way. Cast concrete slabs decorated with Trobriand canoe prowboards and splashboards mark the graves of former chiefs. I have encountered these sterile concrete graves with disappointment everywhere in Papua New Guinea, but local people see them as permanent memorials to their illustrious dead.

Around the periphery of the village lay many ordinary houses with plaited walls and immaculate gardens carefully tended under incandescent bracts of mauve bougainvillea.[2] Young girls completed homework in the entrance to huts while their tiny brothers pulled their hair. I spoke to two teenage boys, Alex and Israel,

---

[1] 'Most important to remember' in Kiriwinian.

[2] The plant 'bougainvillea' is named after Louis Antoine de Bougainville (1729–1811), the great French explorer, statesman and mathematician. He set out on a voyage of exploration around the world aboard the frigate *La Boudeuse* ('the sulky one') in 1766, together with Philibert Commerson (1727–73), botanist to Louis XV. This compulsive collector of plants, described by his friends as 'a lunatic', suffered terrible seasickness on the voyage, writing in his journal, 'My dinner and supper are but loans, which I am exact in repaying one hour or half an hour afterwards.' He managed to collect this beautiful flowering climber on the coast of Brazil near Rio de Janeiro in June 1767 and loyally named it after his captain.

seated on the platform outside Israel's 'house'. It is the custom for young men to live alone in such small huts for a period as they journey towards manhood. They spoke excellent English as do many Trobriand islanders. They showed me the simple but eloquent grave of their brother. This was a long rectangle made of piled-up pieces of dead coral sprinkled with fresh white sand and decorated with rows of red hibiscus placed in old 'stubbies'[1] containing fresh water. The grave was bathed in cool filtered sunlight glowing through a blue plastic roof.

Tobigawepu Mokagai is one of the most brilliant Trobriand dancers, and has travelled with performances to London and Berlin. He told me how unhappy he was with the influence Christian teaching was having on the traditional art of dance. There had been no dancing in Omarakana for the last three weeks, as there was less competition in yams between villages on Kiriwina that year. He felt the enthusiasm was waning.

The *milamala* dances celebrate the period of the harvest when all the heavy work in the gardens is over. It is a period of joy and unrestricted sexual promiscuity. Songs are sung with much pelvic-thrusting and rhythmic clapping of hands on the buttocks.

'I don't feel like lying down, let's do it standing up!' they sing.

The fevered expression of this 'erotic bacchanalia' has earned the Trobriands the title of the 'Isles of Love'. Trobriand girls have perfect breasts that quite defy gravity. Their short grass skirts, layered in old gold, orange and magenta are tied with a knot at the side. As they swirl, the skirts slip slowly down their slender thighs. Faces delicately decorated, their skin is smoothed with charmed coconut oil and sprinkled with pollen that glitters like gold in the sunlight. Trobriand men are tall with elegant musculature and wear neat pubic sheaths of bleached pandanus leaves, fixed at the waist with a scarlet belt. Pandanus leaves are also tucked into finely woven armlets known as *bisila*. An exquisite corona of white cockatoo feathers decorates their hair. The

---

[1] A short or 'stubby' bottle used in Australia for beer.

singing is high-pitched and monodic, accompanied by drums and the low mournful wail of a conch shell. Some songs have the intensity of haiku poetry:

> *I yearn for you*
> *I wait for you to come*
> *I watch for your arrival*
> *I go searching for you*
> *My precious one*
>
> *Emerging, emerging*
> *Her radiant youth*
> *Like the full moon*
>
> *I yearn for her*
> *The young breasted girl*
> *Young brother, you stay back*
> *She has chosen me.*

The spiral of dancers winds and unwinds. Feathers, shells and leaves tremble in the eloquent black heron dance as feet are raised and lowered in an uncanny and poignant imitation of birds. Even the beautiful children aged five and younger perform the thrusting movements of sex and sing of making love. A child held an enormous leaf aloft while singing:

> I'll show my dance and I'll dance perfectly. It will look brilliant in the dancing area. Young-breasted girls standing in a line – watching me!

But traditional music has declined on the Trobriand Islands overall. No longer does the sweet Loloni flute summon a lover. It is rumoured that only one old man, far away, can still play the haunting Lilolila pan pipes. The dancers have been increasingly dissuaded from this exuberant and moving display by the restrictions of Christian modesty. Methodist missionaries strongly objected to dancing in the past which has resulted in significant cultural confusion in the present. The missionaries introduced cricket in an attempt to raise the moral tone – the conventional Victorian identification of Christianity with sport.

The Trobrianders modified cricket with a vengeance. A side can be made up of any number of players and each score is celebrated with wild erotic dancing of the most abandoned kind. Teams can be made up entirely of women which really sets the atmosphere aflame. The scorer slices fronds off a palm leaf with a machete. Magic spells are cast upon the opposing teams. The ball is a piece of ebony polished with a pig's tusk and serious injuries are not unknown. The game unleashes frantic passions in players and onlookers alike. A game played intermittently may last a year and is hugely entertaining.

Sex for the young is an amusing diversion on these islands but governed by carefully-observed rules and customs. According to anthropological sources, one of the most extraordinary beliefs supporting free love is that children do not result from sexual intercourse. The Trobrianders believe in reincarnation. The soul of the spirit-child waits at Tuma (the Trobriand Paradise) or floats on a piece of driftwood on the sea, awaiting the call to take a new human form. At an opportune moment the spirit enters the woman through her ear where a rush of blood takes it to her womb. It is difficult to determine whether such beliefs continue to be held, flying as they do in the face of modern education. I suspect the rational and irrational coexist in an imaginative and balanced hybrid form, a creative quality of Melanesian thought I have admired and envied above all others.

I was taken to the site of Malinowski's tent in Omarakana, an open area decorated with flowers, some distance from the centre of the village. The spot is marked with a stalactite taken from a cave. The tent itself has achieved iconic status in anthropology. Here he worked on his most notorious book, *The Sexual Life of Savages in Northwestern Melanesia*. This remarkable work of what might be termed 'literary anthropology' uses Trobriand sexual customs as a basis for detailed ethnological study. From the chapter entitled 'Erotic Approaches' an extract gives a hint of the mine of esoteric erotic customs in this volume:

> Another element in love-making, for which the average European would show even less understanding... is the *mitakuku*, the biting-off of eyelashes. A lover will tenderly or passionately bend over his mistress's eyes and bite off the tip of her eyelashes. This is done in orgasm as well as the less passionate preliminary stages.

Fearful of *tropenkoller* and recoiling from lascivious inclinations himself, Malinowski took physical exercise to the point of exhaustion. A deeply sensual and psychologically complex man, plagued by sexual guilt, he wrote in his notorious *Diary*:

Thursday 19 April 1918

> A pretty, finely built girl walked ahead of me. I watched the muscles of her back, her figure, her legs, and the beauty of the body so hidden from us, whites, fascinated me. Probably even with my own wife I'll never have the opportunity to observe the play of the back muscles for as long as with this little animal. At moments I was sorry I was not a savage and could not possess this pretty girl. In Kaulaka I looked around, noting things to photograph. Then walked to the beach, admiring the body of a very handsome boy who was walking ahead of me. *Taking into account* a certain residue of homosexuality in human nature, the cult of beauty of the human body corresponds to the definition given by Stendhal ... Resolve: absolutely never to touch any Kiriwina whore. But mental attitude to E.R.M.[1] bad. To be mentally incapable of possessing anyone except E.R.M.

*   *   *

---

[1] Elsie Rosaline Masson, his fiancée, a nurse and daughter of Sir David Orme Masson, Professor of Chemistry at the University of Melbourne.

I had begun to eat again after the thirty-six hours of rough weather from Madang to the Trobriands. The catamaran took only a short time to Keileuna, a large island west of Kiriwina. The village of Tawema reminded me of Sim Sim with its air of isolation and remoteness. The intense interest in us clearly indicated it has few visitors. Streamers of pandanus lay drying on the sand awaiting another of the many yam ceremonies. It was a bumper harvest this year as evidenced by the new yam houses that had been built. The largest yams were displayed in a type of measuring cradle, two monstrous specimens being over a metre long. The competitive spirit in yam gardening is very strong in the islands. We planned to walk through coral gardens and tropical rainforest to the village of Kaduwaga further south, a distance of a few kilometres.

The path through the rainforest was harsh under my soft European feet, pieces of coral bruising and unyielding, even through the soles of my boots. The wild exuberance of the vegetation on this flat coral island was astounding. Ferns cascaded across the path, strangler figs dangled like electric cables after a storm, old coconut palms rose to tremendous heights on thin trunks, stands of betel nut flourished vigorously. A special 'safe' strain of short coconut palms with orange nuts was scattered about. There was a profusion of butterflies, small blues, tiny browns with blue 'eyes' on the upper surface of the wings, fantastic emerald green and black monsters the size of small birds, *Ornithoptera priamus poseidon*, silently flapping into the mysterious interior. Pairs of brilliant red lorikeets bulleted through the treetops into the primeval greenery with coarse screeching and whirring wingbeats. Crimson-breasted fruit doves fed at the edge of garden clearings and dollar birds sat motionless on high branches in the sun. The air was cool and deliciously breezy from the south-east trades. The magic of nature gave way to the gardening magic of man whenever I encountered a coral garden.

Gardens and garden magic define the life of a Trobriand islander. He (and it is usually the men who garden) is basically a cultivator and harbours a burning passion for this activity. At least half his working life is spent in the garden, and in many ways the rhythm of his entire existence centres around associated activities. You must speak to yams gently and softly during the incantations that encourage growth. The yam is regarded as a human being metamorphosed into a plant. A man's entire prestige lies in his ability to manipulate such magic incantations, the garden magician being one of the most respected of men. As the practice of magic is considered quite normal in this society, his work is more of a public responsibility than an occult practice. Men begin gardening before sunrise when strong magic ranges abroad, before the light is bright gold. The Trobriand islanders have evolved an enviable spiritual relationship with their land. They pantheistically 'become' the bird, the plant, the stream, the leaf, as they garden.

These are not like European gardens in any sense. Magic rules here and not science. The position of the stars and planets determines the time of planting, the manner of planting, the visionary results of the planting. The cleared area of forest where the garden is laid out is surrounded by a carefully-constructed magic wall of sticks. The entrance to the garden is over a stile which also bequeaths magical powers. The appearance of the plot is of great disorder, discarded branches of dead trees lie about, huts and shelters covered with palm fronds are seemingly constructed at random. The humus of the forest makes up the earth of the growing beds.

Harvest time is variable in the Trobriands. At this season, normally early August, pride of place is given to the round pyramid of yams that sits under a shelter ceremonially marked off from the rest of the garden. It has the feeling of a sacred altar, a magical area. In one garden I visited, an entire proud family posed beside their magnificent conical edifice, the result of months of dedicated labour. The man stood almost to attention, bursting with pride. In this matrilineal society he should symbolically give all the produce to his sister, but in reality about half goes to nurture his own family.

Malinowski wrote a long and incredibly-detailed monograph on the Trobriand garden called *Coral Gardens and Their Magic* in which he explains the layout and significance of such gardens. His intricate diagram of a yam indicates that there is a Kiriwinian name for all the multiple tendrils, roots and even hairs of the plant. To the average European, a yam is a simple vegetable, but to the Trobriander, the humble tuber possesses a universe of referential meaning. A sentence in the introduction succinctly indicates the uncomfortable, almost bizarre, dilemma of a European intellectual attempting to rationalise the pagan hedonism embodied in the cult of the yam:

> ... the Trobriander would agree with Stendhal's definition of beauty as the promise of bliss, rather than with Kant's emasculated statement about disinterested contemplation as the essence of aesthetic enjoyment. To the Trobriander, all that is lovely to the eye and to the heart, or as he would put it more correctly, to the stomach, which to him is the seat of the emotions as well as of understanding, lies in things that promise to him safety, prosperity, abundance and sensual pleasure.

The children of Kaduwaga ran to greet me as I emerged from the forest and led me into the village. I wandered around lethargically, looking into huts and terrifying the little children with my panama. The pastor of the United Church showed me his modest house of prayer and painted concrete graves. A woman was plaiting a colourful mat while a wizened old man furiously pounded betel in the sun. There were many canoes placed on stilts above the shallows of the muddy lagoon and men repairing nets. Fishing is an important activity here. They fish for the sacred *Conus millepunctatus*, a cone-shaped shell which is used to make armshells called *mwali*, traded as part of the *kula* ring.

The season for fishing cone shell is known as *Vinanu*. The associated rites are complex and reflect the intense life natural objects possess for Trobrianders. Early in the morning, the fisherman's wife carefully packs his canoe with a coconut container of water and a yam. No one must enter the canoe at this time. The chief allots fishing grounds and the men invoke their favourite spirits with magic oil or leaves. Venturing out to sea unaccompanied is a test of a man's ability to be alone and a trial of his woman's loyalty.

At the fishing grounds, the men gaze intently into the still waters. If their magic is strong they will have found cone shell by mid-morning. Weak magic results in abandoning hope by mid-afternoon. Small birds will circle the house indicating a good catch. But there are dangers. The man will be attacked by a shark if the woman falls asleep while he is out fishing. If a man spies two shells lying point to point, it indicates his wife is flirting with another man. More seriously, if they are lying side by side, it means his woman is making love to another. Catastrophically, if he lifts a male shell and it squirts semen from its point, desertion is final and consummated.

Mwasila or magic surrounds all the activities associated with *kula*. Remarkable trading routes unite many far-flung islands with social bonds and reciprocal obligations over a vast area of ocean. A man owns for a short time, and in an alternating manner, two classes of objects that have no intrinsic value but accumulate great significance over time and through the personality of the temporary owner. One group of objects, the necklaces called *soulava*, are made from red spondylus shell, pearlshell, seeds and trade beads. They circulate in a clockwise direction around the islands that make up the *kula* ring. The other class of objects are the armshells or *mwali*, cut from a cross-section of a cone shell and decorated with *Ovum ovulum* shells, beads, pendants and dried leaves. They are traded counterclockwise.

Unlike the cone shell, when men dive for spondylus shell after summoning magic, they are testing the loyalty of the friendship their woman has with other females in the village. The men will benefit from this extended support. All spondylus shells appear similar until broken open. If a white lip is revealed, the friendship is not deep and the support weak. Red-lipped shells, the most valuable, indicate a deep and fulfilling friendship likely to support the man. No shells but lacerations and bleeding hands after fishing indicate the presence of evil in the female friendship.

The fishing of cone shell tests male solidarity and the fishing of spondylus shell the support of females for the man. Both *mwali* and *soulava* carry names, possess a history and the most famous carry an aura of hazardous romance. There are nine grades starting from the most humble and culminating in the most dangerous which can carry memories of sorcery, murder and death. Each possesses a 'living personality' and an identity.

It is said by the Trobrianders, 'Once in the *kula*, always in the *kula*', as men build up lifelong relationships of reciprocal obligation. The exchange is not based on need, but on a desire to build an emotional contract of indebtedness. The exchanged objects are the outward sign of a state of mind, the acquisition merely the culmination of a complex choreography of giving and receiving. The only purpose of these valuables or *vaygu'a*,[1] is to be possessed and exchanged, to circulate around the ring of islands. They are not retained by the purchaser as are commodities in Western societies. In Trobriand mythology the exchange also implies the recognition of physical beauty and is associated with powers of seduction. The valuables are proudly displayed in the villages or more rarely on the person. Malinowski writes in *Argonauts of the Western Pacific* that 'To possess *vaygu'a* is exhilarating, soothing, comforting in itself.'

Ceremonial ritual with rules and codes of conduct, often fiery, apply when two men are negotiating. They are exchanging matching values at white heat. Some anthropologists even see the *kula* as a substitute for war. The acquisition of a famous *mwali* or *soulava* leads to social distinction, pride and renown. The ceremonial takes place from fleets of highly-decorated, sea-

---

[1] A general term for 'native valuables', important in maintaining and demonstrating status.

going canoes called *masawas*,[1] which are able to cover great distances. When a fleet of canoes sets sail carrying *vaygu'a*, the moment of high drama is accompanied by the casting of magic spells. Mysterious chants increase the owner's beauty and his powers of seduction. Flying canoe magic speeds his path. A poetic song runs:

> I shall *kula* so as to make my canoe sink. My fame is like thunder,
> my treading is like the roar of flying witches.

The unique phenomenon of the *kula* ring and its attendant pagan dances are an engine of metaphysical power.

A hundred children skipped and sang through the shallows of Kaduwaga to say goodbye. Their joy was irresistible. A couple of the youngest were hitting their backsides and instinctively thrusting away in true *milamala* style in their excitement.

That night on the ship we examined some mysterious green stones, generally known as *beku*, which Tamara, a Russian art historian, had purchased. They were of great age, and were superbly polished and formed. Important men own at least one of these stone blades, streaked in different shades of deep green. When an owner dies they are passed on to the next generation in beautifully-carved ceremonial axe-handles. The important ones are named and are irreplaceable, as the greenstone is no longer mined on Woodlark Island.

I questioned the purchase of these objects by outsiders, as they are a birthright. Government legislation has made islanders so desperate for *kina* to pay for school fees and taxes, they willingly provide European collectors with opportunities to engage in all manner of illegal exports. Tamara had also obtained a fine *mwali*

---

[1] Each *masawa* is constructed by a group of men and the canoes are used communally according to established social rules. Each craft has its own name and a profusion of myth, magic and complex ceremonial accompany their building. They are decorated with beautiful *Ovulum ovum* shells (white, porcelain-like 'egg' shells), fringes and streamers.

and could scarcely contain her excitement. She coquettishly flicked her long blonde hair and questioned Matthew, our extraordinarily knowledgeable guide from the Madang Province. He possessed all the civilised charm of the professional educationalist.

'Matthew, darlink, do you think should I remove dis plastic things? Only dis armshell part is most valuable part of it.'

'I think you should read this book first.' His face assumed an expression of antique gravity. He showed her a volume devoted to the *kula*, gently laying the *mwali* on a side table. There was no further talk of removing anything.

Dawn in the Trobriands is a moment of transcendental beauty. It is easy to believe that the soul sets off from the headland of Bomatu at this miraculous time of day. A pewter sea unfolds to silver, shafts of white gold strike through the early morning clouds, deepening at length to the yellow gold of the fully-risen sun. Canoes slip out for fishing, remaining motionless like warming water skaters on a still pond.

We had an early breakfast off Kuiawa Island, about an hour south-west of Keileuna by catamaran. The village was carelessly scattered along the beach beside a cerulean sea of unimaginable clarity. Local people were more aggressive here, and pressed us to buy shells the size of basketballs, also an illegal trade. Despite being predominantly a fishing community, the yam harvest had been spectacularly successful and many new yam houses had been built. Deep psychological security comes to a village when a yam house is dark with fat yams and light cannot pass through the wall beams.

Many *mwali* were on display, hanging from the thick rope customarily looped through the armshell. An ancient couple with skin like withered fruit sat on a platform with a brightly-coloured, new grass skirt between them. I was offered a whale's tooth by a young boy with curly brown hair. The rich curls of Kuiawa are quite distinctive compared with the frizzy halos of Kiriwina. Too soon the catamaran was heading towards an island so small I scarcely believed it could support a population.

Manuwata was a tiny, windswept paradise devoted to fishing, the sense of remoteness extreme, even more so than Sim Sim. Tropical detritus littering the shore, canoes with fluttering red pennants, pastel-blue sky meeting emerald lagoon and straining palms, a scene of timeless exoticism. Children were everywhere, screeching outrageously, taking my hand, stroking my skin, feeling my pockets, fascinated by my very existence. This is a poor fishing community. The slim canoes with ornate scrolls and spirals carved on the prow are reminiscent of Polynesia. A solitary figure was carving a digging stick on the beach, the central desert of the village relieved only by the occasional wandering pig. Chickens panicked past with coloured identity streamers tied under their wings. A young boy held a live seagull by the tips of its wings.

As the wind freshened, I lay on the beach after swimming. Two canoes using palm fronds held as sails beached nearby. The child mariners sat on the sand and watched me with pure intensity. They possessed nothing except a few scraps of cloth, no electricity or radio, no watch, not even a proper sail to their canoe. In the islands, a shell, a leaf, a drop of magic oil, a wasp's nest, a fish, a useless armshell or necklace was given or received with eloquence. A spell was cast in a garden or on the sea, an entire life focused on fish, spirits and yams. This unique vision of the world had come about because they chose to make it so, chose to invest humble natural objects with metaphysical power. Westerners had chosen a more bloody and technological path. Theirs is a different set of miracles, a colder magic and somehow less human. Almost breathless, I watched the ancient harmony of man and nature play through the eyes of these children of the remote Solomon Sea, in wonder because I believed such a harmony no longer existed.

22. Farewell to That Strange and Fatal Glamour

The weather blew up and we were forced to anchor for the night in the lee of the island of Nubia. The passage was rough to the Amphlett Islands from the Trobriands across the Solomon Sea. My first view was of forbidding volcanic peaks swathed in low cloud. Huts clung desperately to the shallow beach or were glued to the bastions of jungle rearing above. Coconut palms clutched the shore, the trunks leaning over the sea at crazy angles. Ancient rosewoods with massive tangled roots gave it the feel of a blasted Elysium. The scenery was truly Wagnerian in intensity and grandeur. The contrast with the flat Trobriands could not have been greater.

After cruising a little we landed on Nabwageta among a number of *wagas* stranded on the beach. The dramatic weather alternately drenched us with rain and smothered us in sunshine. The people did not seem particularly happy to have visitors, and approached us with a sullen and unfriendly hauteur. Bougainvillea was flowering in almost painful magenta and pink, the wet leaves and freshly-washed perfume of the jungle marvellously invigorating. The usual starving dogs cringed at my glance. Most of the villagers were seated out of the rain under their houses. The dwellings are raised on stilts and the narrow beach forces the inhabitants to build in narrow 'streets' that impose a high degree of intimacy on the people. Malinowski was 'spellbound' by the picturesque villages of the Amphletts. Writing in his *Diary*, he came to the important conclusion that his ethnology was actually bringing the islanders of the Amphletts into existence: 'Feeling of ownership: It is I who will describe them or create them.'

The island is most famous for its large, beautifully-proportioned hemispherical pots made of fine clay. The clay comes from Fergusson Island. Many pots lay on planks under the houses. They are mainly used for cooking various types of taro dumplings. Burnt pewter-grey by the fire, they are simply decorated around the rim, some with an embossed circlet strangely suggestive of Scandinavian designs. I watched two women fashioning them from clay without a wheel, working from the rim upwards, skilfully closing the pot at the upturned base. The cooking pot and the three stones on which it stands are deeply symbolic in Massim society. The hearth is where men and women meet and around which family life revolves. Known as *kailagila*, the hearth is associated with a faithful partner. The sailor on *kula* must naturally have a safe place to return to after his voyages. The Amphlett pot is thus closely woven into the metaphysical imagery of the whole Massim region.

The Amphletts are also part of the *kula* ring. Tamara managed to persuade a woman to part with a named *soulava*, but the villager was so terrified of the reaction of the community to this breaking of the ring for *kina*, the article had to be 'secretly' delivered to the catamaran by canoe. Island villagers have severely-limited opportunities for making money, depending on visits from occasional boats. Provincial governments now insist that burdensome school fees and taxes be paid in cash, so desperation prevails, leading to the breaking of traditions.

Malignant clouds whirled across the sea at low altitude as the weather worsened. The Trobriand captain guided us through the treacherous waters, south-west towards Fergusson Island. The Amphletts mark an extraordinary geographical transition from the flat coral islands of the Trobriands to the wild, volcanic rocks of the D'Entrecasteaux Group. Perhaps it is this fearsome landscape that gives these islands their reputation for witchcraft, sorcery, and the ruthless cannibalism practised in the past. The noble profile of Fergusson pressed through the mist as the catamaran cruised along the impenetrable north-west coast before entering Moresby Strait. Villages were reduced to elongated lines of huts straggling along a shallow shore. Coconut palms and delicate eucalypts precariously attached themselves to sheer volcanic slopes. Above the treeline, sacred gardens were ranged above the villages on savage inclines rising hundreds of metres to the summit of craggy, infolded escarpments. This is fertile land for garden magic but physically punishing for any human who tills that soil. The fecund valleys steamed with cloud swirling around precipitous slopes, as if the island was in primeval upheaval. The wind increased and we bucked on a sea of white horses. The air turned yellow, the screens of vegetation became even more threatening: nature on a sublime scale, voluptuous, intimidating, yet wholly magnificent.

We anchored at Wagifa, an inhabited volcanic rock just off Goodenough Island. Goodenough, named after a nineteenth-century English commodore, is a monumental and towering piece of melodramatic scenery taken straight from a Salvator Rosa painting. Three peaks form the 'dome' or crown, the highest being over 2500 metres, this on an island only forty kilometres long and twenty-five across. It is the steepest-sided island in the world. We landed on tufa and volcanic rock, wet and slippery-black with rain. Tiny crawling beetles seemed to cover everything underfoot. The houses here were far larger and more substantial than anything I had seen on other islands. They stood on stilts having split bamboo walls with complex internal bracing supporting the customary sago palm roofs. These were pitched with a gable, many houses having porches and verandas.

Despite Goodenough not being in the *kula*, large *wagas* were moored in the forbidding bay. I climbed a steep, rocky path through torrential rain, skittering through the mud and dank foliage to the village of Awanane. There I engaged in a little *kula* myself as an experiment. A rather pathetic, cracked armshell with minimal decoration, definitely not a true *mwali*, lay for sale on a mat. I had heard that it was customary to deride the object you desired so as to mislead the 'giver' to your true intentions.

'Do you *really* expect me to buy this terrible thing for twenty *kina*? Look at the cracks! It's absolute rubbish! Pathetic! A scandal!'

I stalked away and tripped over a volcanic rock. Shocked faces emerged all round the village. I continued in this absurd vein until they slowly realised it was an act, smiles spread across their faces like the sea breaking over a rocky shore, as I handed over my *kina* and pocketed my 'treasure'. The next village along the stony and shell-littered path was called Debani. Many people wore traditional dress, particularly male children who had miniature pubic sheaths and old women who were bare-breasted and wore long grass skirts. Fires had been lit under the houses, the villagers attempting to dispel the miasma and mist.

As I stood under a tree out of the pelting rain, I recalled the story of 'The Snake Woman'. This curious myth from Goodenough explains the island's exclusion from the *kula* ring. A woman named Ninialawata fell pregnant and when passing urine in the jungle gave birth to a snake. Terrified, she hid it away. She fell pregnant again, and this time gave birth to a human daughter. The snake lived secretly in the bush, stealthily fed by its 'mother'. The people round about lived in grave fear of it. The daughter, with all the intense curiosity of a child, pestered her mother to take her to feed her 'brother'. At last Ninialawata relented, on the condition the child did not scream when she saw him. Of course, as soon as the daughter spied the wretched reptile with its fierce tusks, she screamed so loudly she could be heard in the high mountains.[1] The snake felt greatly insulted and, in a

---

[1] The conundrum of a snake identified with a pig in this myth indicates the possible presence of a hybrid animal, hence the reference to 'tusks'. The word *mwatabawe* literally means 'snakepig' which further deepens the knotty mystery (Michael W. Young).

fit of pique, went on hunger strike. This was unfortunate for all concerned, as he was also the guardian of the local valuables and eventually decided to leave this inhospitable country. He threaded himself through all the *mwali* and *soulava* and took off for Muyuwa or Woodlark Island. As he left Goodenough that night, there was a terrible storm and deluge. The earth began to shake. So it was in this fashion that the Goodenough islanders lost their place in the *kula* ring to the people of Woodlark. For his part, the snake continued to wander the islands accompanied by judicious magic, at one point carving a river with his flailing body while escaping two crocodiles. His final, yet troubled resting place, is at the extremity of East Papua, Sudest Island, where his power has been restrained by an alliance of courageous men.

The south coast of Fergusson was gradually absorbed by leaden clouds; the swell became heavier until furniture and cups began to fly about the deck. It was the tail end of the wet season and the frigate bird was flying low – not a good sign. Waves broke over the bow of the light catamaran like a windswept waterfall. Then, without warning, the cloud miraculously began to break up and layer into disparate levels, dark grey billows moving low at tremendous velocity, white fleece lethargic at the higher levels until streaks of blue allowed the sun to reveal the landscape.

An imposing vista ranged across the horizon on a majestic scale, strangely reminiscent of the Scottish highlands. The rapidly-changing elements would have fascinated Constable, a painter so obsessed with rushing skies. The vaporous vortices swallowing the rugged peaks would have enthralled Turner. The grandeur of the D'Entrecasteaux islands is breathtaking. We anchored for the night opposite the Catholic mission station of Budoya. In this sheltered bay the sea almost instantly became a millpond and the catamaran ceased to move at all.

The hot springs at Dei Dei are an excursion into the age of the dinosaurs. Shortly after dawn the landing craft was slicing across the steely lagoon to a hamlet on the beach. Outside a miserable hut, a beautiful island girl in a magenta smock was feeding an Eastern black-capped Lory, a spectacular parrot with bright red breast and green wings. The cordylines used in Trobriand garden magic, collected on Fergusson and used ceremonially all over Oceania, grew profusely around the hamlet. The narrow path to the springs meandered through waist-high kunai grass, pandanus, eucalypts and coconut palms. The air was absolutely still and the sky bright but overcast. The heat and humidity were stifling. Birdsong filled the air and lorikeets shot across the path into the jungle canopy like screeching projectiles. Myriad blue and brown butterflies danced about. My three 'guides' were aged about eight and ran in front of me with great excitement. They carried large pitcher plants into which they had put sprays of fuchsia-coloured miniature orchids and star ferns. They began to sing, spin around and urgently beckon me forward.

I was being led into prehistory. The first sign of thermal activity was a steaming stream flowing beside the path reeking of sulphur. The banks looked like an artist's palette of crushed, burnt umber and yellow ochre. Some villagers were cooking food in a pool of boiling water. Sections of the stream were still and sinister, reflecting the lush growth, the aerial roots of palms rotting in the chemical brew. Spiders as big as my panama waited for prey. Fumaroles appeared beside the path and boiling pools of mud signalled dangers close beneath.

The springs themselves presented many and varied faces. Dusky lakes of sulphurous water contained marooned, rotting pandanus palms that reared up in the heat like surprised mutants at some unspeakable meal. A thin, black crust covered seething cauldrons, like bizarre pastry. It crunched worryingly underfoot, threatening to give way at any moment. There were no barriers here, no warnings, no litter, just a primordial landscape in an untouched state. My eyes stung with salty sweat. I balanced precariously on a bridge of poles across a sulphur stream to reach a huge geyser at the edge of a steaming lake. The geyser gushed intermittently from a deep, green pool through pillows of rock encrusted with a hellish cocktail of elements. The children began an incantation to 'Susulina', the wood spirit of the place who guards this infernal spot.

'Come out! Come out and show yourself, Susulina! Show yourself!'

Each time the chanted incantation ended, the vent exhaled gouts of boiling water in fountains. The scene at Dei Dei with the young boys holding orchids and raising the spirit of the hot springs with ancient magic was the closest I had ever come to the authentically primitive, either in nature or in human culture.

As we ate some eggs that had been cooked in a nearby thermal pool, I looked about me and waited for *Tyrannosaurus Rex* to crash through the undergrowth. The children sang all the way back to the hamlet, laughing and dancing. Tamara, the Russian collector, was examining another ancient *beku*, a superfine greenstone adze, columnar in shape with a perfect edge. It had been found by local divers at the bottom of the bay, probably originally used for canoe-building – expensive and of museum quality.

I decided to make the long trek to Budoya mission along the coast. The overcast weather made walking laborious in the motionless air, oppressive heat and humidity but the exotic land-scape compensated for the effort involved. My companion was one of the crew from the catamaran.

'See that type of pandanus palm there?' he said, grinning. The plant looked exactly like the rest of the species to me. 'If you take the pith out of the trunk here,' he pointed to the weeping base of the plant.

'Yes?'

'Grind it up.'

'Yes?'

'And rub it on your penis.'

I was getting alarmed by now.

'You wake up in the morning and it is *twice* the size! Fantastic!'

He erupted into gales of laughter and spat great goblets of betel juice into the jungle.

I was suddenly surrounded by five young girls from the mission school all wearing Coca-Cola T-shirts. I steered tactically away from the plant and we fell into animated conversation about England and London. I took their photograph and their names and promised to write. The mission station was immaculately laid out with a school, a maternity hospital and a Catholic church. The delivery rooms and wards were clean and well-kept, four proud mothers tending their sweet Melanesian bundles. As I passed the packed church on the way to the wharf to rejoin the ship, an Irish priest was deeply involved in the transubstantiation of the body and blood of Christ, the consecration of the Catholic Mass. His rich brogue hypnotised the congregation with the words of the old Latin rite:

> Take, and eat ye all of this . . . For this is my body . . . Take, and drink ye all of it . . . For this is the chalice of my blood.

And the bells rang thrice. During the consecration of a Mass I had attended in New Ireland, two practitioners of the occult danced in full-feathered costume on either side of the altar as the chasubled priest raised the host. Times are certainly changing.

The 'perilous' island of Dobu had been nestling in my imagination for years. Malinowski wrote of it as 'the land fraught with dread beauty and mythological associations'. The power of the occult is respected all over Melanesia, but was especially strong in the D'Entrecasteaux Archipelago and more particularly on the small volcanic island of Dobu. Their neighbours, the Trobrianders, feared, loathed and admired the Dobuans at once. The southern Massim, known as *koya*, was considered a place of ruthless cannibals, head-hunting, grim legends, flying witches and sorcery. Certainly, on that particular morning, as the catamaran gingerly negotiated the shoals, the ragged weather and jagged landscape magnified any sense of malevolence that might have accumulated around the misty peaks. The poet William Blake would have placed Dobu prominently between the black and the white spiders, the eternal battleground between good and evil.

The southern section of *kula* includes Dobu in the circle of islands, but the context of *kula* has broadened since Malinowski's first reports. The *mwali* armshells and *soulava* necklaces are all traded by the Dobuans. Accompanying this, there is a more conventional trade in pots from the Amphletts, greenstone from Woodlark, carved bowls from the Trobriands and various domestic items. Even water, transported in sealed bamboo tubes, is traded for cooking. The inland people cook their food in salt water from the coast and the coastal people cook in the spring water from the mountains. The utilitarian and non-utilitarian aspects of *kula* have thus become more balanced. The spirit of *kula* replaces antagonism with a spirit of cooperation and personal enmity is replaced with friendship.

The sides of the volcanic crater of Dobu are deeply fissured and valleys radiate down to the sea from Mount Gulebubu. The usual groups of vivacious children came to greet and overwhelm us on the beach. The first sounds I heard were not the shrieks of cannibals or the swish of flying witches but childish laughter and hymns coming from the large concrete church, the first of this material to be built in Papua New Guinea. I wandered along the coast under huge rain trees and pink frangipani to a memorial by the shore. It read:

The First Methodist Service held by a
Pioneer Party
Under
Rev. W. E. Bromilow D.D.
June 19th 1891

The houses were nothing like the Trobriands, arranged around a central dancing area. The Dobuans live in what are almost detached suburban houses with individual flower gardens, washing on the line, a pig snuffling in its poke. The houses stood along 'streets' and gave an impression of intense family privacy. I had read that in the past families were hermetic and did not communicate with one another. The villagers were exceptionally friendly and crowds accompanied our 'progress'. Some young boys were anxious to show me a crumpled photograph sent from Europe of a female anthropologist who had worked on Dobu. She was wearing only a grass skirt, her naked breasts exciting particular comment. Luxuriant palms and broad-leaved tropical plants were in abundance, butterflies glided towards nectar-bearing flowers, a tiny yellow bird with a blue throat, a jewel of life, sang its heart out. The climb up Mount Gulebubu was arduous, but the spectacular view over the Dobu Straits and Normanby and Fergusson Islands was worth the tortuous tripping over concealed roots.

The *longlong*[1] man of Dobu, his Rastafarian locks flying, came out to greet me from a tumbledown shack. He limped out, as his knee was swollen to the size of a football. He pressed upon me some wild lemons and a number of pages covered in 'secret' formulaic writing, whilst explaining in mime that fluid had rushed from his body to his knee and exploded out over the sea.

A tremendous squealing suddenly erupted and the naval commander travelling with us disappeared beneath a sea of gesticulating children. They rushed to the water's edge in excitement and then back again. He had a digital camera and was showing them their image on the screen, which resulted in unbelievable tumult. And so the magic of modern technology had briefly vanquished the ancient magic of sorcery and witchcraft. Offshore, kite fishermen were quietly luring the longtoms to entangle their needle teeth in the sticky balls of spider web suspended from kites floating on updraughts behind their canoes.

Dobu seemed peaceful, beautiful and fervently Christian. My

---

[1] Pidgin for 'mad' or 'insane'.

mind attempted to reconcile this Eden with the world of savage unseen forces described by anthropologists writing in the past. In 1930 the anthropologist R. F. Fortune wrote a sensational account of Dobuan sorcery and witchcraft in his book *Sorcerers of Dobu*. One story he recounts is of a gigantic testicle appearing like a ball of fire in the heavens above Dobu followed by a flying witch as she carves her way across the night sky. Volcanic crystals and flames shoot from her genitals as she hurtles across the island. A cannibal song rises from the sinister jungle celebrating the tender folds of the vulva, reportedly a favourite cannibal morsel.

> *Your vulvas we eat*
> *With a crowd of my mother's brothers*
> *Our assembly feast*
> *A cooking pot full around.*

Christianity has made great inroads into pagan belief and the government has now made sorcery an illegal practice. Alleged sorcerers have been murdered in some parts of Papua New Guinea. Recently, in Chimbu province in the Highlands, two men were set alight with kerosene by a village mob who suspected them of killing an old man with incantations. Others have had their hands and legs tied and have been thrown over cliffs and from the tops of mountains. Women suspected of witchcraft have been chopped up with axes and bludgeoned to death with tree branches. The islands however remain peaceful oases where such violent responses are extremely rare.

As I boarded the catamaran for the last time, the weather began to change again for the worse. I stood at the stern as we sailed between Fergusson and Normanby Islands, Dobu perfectly framed between them. A magnificent primordial landscape under the arc of a rainbow unfurled itself like an immense painted canvas. I had no difficulty in imagining a more pagan past. As we coasted Normanby sailing south-east, snow-white cockatoos accompanied us, skimming and screeching through the coconut palms. A salty mist swallowed us, villages clustered in darkening tunnels by the shore. The Trobriand captain skilfully negotiated a constricted passage between an enormous vegetated rock and an offshore island. Children ran about like ants, frantically waving. The swell of the Solomon Sea launched us into blind space and the rain descended in torrents. Alotau was many hours' sailing and we would arrive in the harbour late at night. Normanby faded in grey sheets of drizzle as we headed towards East Cape. I never saw the primeval coastline of Papua New Guinea again.

Epilogue

The catalogue screen in the Humanities 2 reading room at the British Library flickered as it searched for musical references. The search had thrown up something quite unexpected in the area of the Trobriand Islands. It was six months since I had returned to London from Papua New Guinea. I had fallen into a regular pattern of research and was accumulating a mountain of extraordinary material. In a way I felt I had never left the enchanted island provinces. The catalogue entry read:

LIST RECORDING:	**C46/1397**
Collection Title:	Malinowski Cylinders
Recordist:	Malinowski, Bronisław Kasper
Other Copy No:	C46/1
Collection inventory:	Wax cylinders: 5
Purpose of Original:	Anthropological Study
Contents note:	Dobu
Country:	Papua New Guinea
Format:	Cylinder, black wax
Dubbing ref. No.:	0313W

'Malinowski cylinders'? My eye scanned the screen. What were *cylinders* exactly? And made of *black wax*? The reference to Dobu seemed as mysterious as the island itself. I continued to stare at the screen but it provided no further clues. I submitted a request for the material and waited a week for a phone call to say the cylinders could no longer be located.

'Was a copy made of them?'

'There were seven cylinders actually, sir. They were moved from the Museum of Mankind when it closed. There's no trace of them. We believe a dubbing was made some years ago but we can't find that either.'

'Could you keep looking for me?'

'Of course, sir. We'll let you know.'

In the interim I had discovered that these mysterious objects were probably 'Edison' cylinders, a form of acoustic recording from the turn of the century. A shiny, spinning cylinder instead of a flat disc. Could Malinowski possibly have taken a recording machine to the Trobriands? I had read nothing of this in all the voluminous material I had consulted. I imagined the ancestors of those islanders I had so recently left, singing or speaking into a huge acoustic recording horn, perhaps outside the famous tent or before the chief's yam house in the centre of Omarakana village. He would have almost certainly played their voices back to them. Did they imagine their souls had been snared in a strange box with a huge metal flower? It must be alive! Look at that black part moving so quickly! The Trobrianders must have regarded the spinning black cylinder as magic or sorcery. Their metaphysics could explain such an outrageous phenomenon in no other way.

I scoured the *Diary* once again for references. In some excitement I read the entry for 3 June 1918:

> I must overhaul the phonograph and Viteku must sing for me . . .
> Then I wound up the phonograph, although I could get nothing
> out of it – Viteku won't sing for me.

Edison had invented a 'clockwork phonograph' with a two-spring motor by 1897. This was soon redesigned to use a single spring and a better speed regulator giving the cylinders a four-minute duration. This was just before the invention of disc recording. Yes, it was certainly possible. My heart missed a beat.

The phone rang one evening, some three weeks later. Someone from the Department of Ethnography at the British Museum. At this hour?

'Sorry to ring you so late, sir, but I've been working in the archive and thought you might like to know we've found your Malinowski dubbings. The tape had been lost since 1984.'

'Marvellous news! What are they exactly?'

'We can't tell you that, I'm afraid. There are no dubbing sheets describing the material.'

'What happened to the original cylinders?'

'Well, the cylinders aren't in the boxes that seem to refer to them. Some might be in the Royal Anthropological Institute Collection.'

Despite her obvious conscientiousness, my archivist was becoming tetchy. She was tired and the cat had not been fed in her Lewisham flat.

'Would you like to make an appointment to hear the tape, sir?'

'Of course! Next week would be fine.'

And so it was that I excitedly climbed aboard the 73 bus in Oxford Street, the route I had abandoned in boredom and disgust so many months before. Even in London, matters associated with Papua New Guinea were eluding my grasp and slipping through my fingers like fine sand. The tendrils of the suffocating jungle extended across the great oceans to England. The unpredictable and the ill-defined still swirled around me like a clammy mist. That familiar anticipation of the unexpected which followed me throughout my journey in the islands was again pumping away.

I was shown into a panelled listening booth at the British Library by a polite young man who demonstrated the controls of a complex tape player. I could record the tape and then examine it at will on playback.

'Oh, you might find these useful, sir. The dubbing sheets.'

'Really? I thought they were lost!'

'I don't know anything about that, sir. I was simply given them to pass on to you.'

Diffused lighting illuminated the leather-covered desk in an atmosphere of scholarly luxury. A red diode came on. The surface noise of the first Edison roll spinning at 150 rpm came through the headphones. One of the most plaintive, wailing chants I had ever heard carried me unresisting into the distant past. The dubbing sheets told me it was the story of Gumabu, a Trobriand islander sailing on *kula* whose canoe had been driven by a storm onto the island of Dobu and wrecked. He had been cooked and eaten by the Dobuan cannibals in a ritual sacrifice. This was his lament composed some six generations ago in what is now an archaic dialect.

> *Gumagabwe kinane,*
> *Siga odabana wakoya,*
> *Gumagabwe*
>
> *Oluvabuse Tokinana*
> *Oinapali guyau*

The translation runs:

> Gumabu
> You have been cooked by the Dobuans
> On their stone altar atop of the mountain
> Gumabu
>
> You arrived at Tokinana
> During the time of giving the chief *kula* gifts

The other cylinders were equally eloquent, the cracked discs and surface noise contributing to the other-worldly atmosphere of voices straining through the static of the past. A single voice sang a monody, reminiscent of Japanese temple music, with perhaps the beat of a drum and a stringed instrument? What was this? A contemporary version of the song?

Guyawe omuwaga
Silai walai
Guyawe omuwaga

Chief your canoe
Got stuck on the reef
Chief your canoe

I sat in the booth for a long time, captivated by the magical lamentation and aura of sorcery leaping across space and time. Malinowski had captured an essence here, trapped a spirit in his magic machine. And now the miracle of modern technology had restored the spirit to life. The genie was out of the bottle, hovering around my head.

'Would you like me to burn a CD of this for you, sir?'

Someone was speaking to me. I looked up in a trance. His face was as if seen through shallow water.

'A CD, sir. Shall I burn one? Shall I burn one for you?'

'Burn? Burn? Oh, yes. I can put it in my archive.'

A recorded lament, written for a man eaten by cannibals in a faraway place at a faraway time. The recording was like opening a window onto the sound world of another planet. I tapped the square CD case in my pocket almost with affection.

Afterword

Risks abound when using as source material the dramatic accounts of sorcery and witchcraft throughout Papua New Guinea described by anthropologists between the wars. Many of these studies unfairly demonise the inhabitants of the notorious Black Islands. This fearsome image persists unjustifiably to the present day, and times have changed considerably under the influence of extensive European contact and the spread of Christianity. During my travels throughout the island provinces I attempted to accommodate the dramatic literary accounts of the past with what I could discover from an all-too-brief visit. In many cases, the old anthropological descriptions could have been of life on another planet compared to the joyful peacefulness I experienced and observed. Certainly the drug-induced erotic scenes, states of undress and imaginative titillation I witnessed on Bondi Beach after my return to Sydney were significantly more 'pagan' than anything I saw in Papua New Guinea.

Extraordinary mythologies were once accepted as true, and 'feverish' customs were certainly practised. Cannibalism of a dreadfully entertaining type did take place. But oral memories are fragile affairs, inherently prone to distortion. Little oral history has been written down except by Western scholars. Native peoples are now ironically turning to anthropologists as cultural authorities to educate them in their own faltering memory of *kastom*. The Papua New Guinean philosopher and Speaker of the Parliament, Bernard Narakobi, comments perceptively:

'Melanesians are walking in the shadows of their Western thinkers and analysts . . .'

The peoples of the island provinces are no longer 'savage and primitive tribes' but more like innocents attempting to come to grips with the savagery of our own times, vastly more horrifying and destructive than the most imaginative sorcerer or Top Gun Witch could conceive of living on an isolated island in the Solomon Sea. As contemporary world atrocities mount, there has been talk of sending missionaries from Melanesia to Europe and America in a timely reversal of historical roles. The idyllic social harmony I encountered is a world away from cluster bombs, dividends and globalisation. Suicide bombers and genocidal maniacs pose far more inhuman dangers than the heroes of Melanesian myth. In the Highlands and Port Moresby, cannibalism has been supplanted by the graver threats of alcoholism, unemployment, rising taxes, a resurgence of tribal warfare, murder, rape and political corruption. The island provinces on the other hand, remain remote Elysiums crammed with joyful children.

The Melanesian sees the human as an interdependent part of the entire spiritual, plant and animal world. Unlike most Europeans, the Melanesian inhabits a parallel universe of magic. The old indigenous cultural beliefs and *kastom* of the Black Islands are profoundly complex and often unsettling in their strange detail, but it would be a mistake to draw too many parallels with present-day beliefs and practice. However, unless the past is absorbed and understood, the exponential changes now thrust upon Papua New Guineans will only give rise to further confusion of cultural identity and an increasing lack of self-confidence in modern nationhood. The claims of the irrational are still a deeply serious part of Papua New Guinean life, but a beautiful, sometimes desperate, creative attempt is being made to balance this with the practical exigencies of the contemporary world.

Brief Chronology of
Significant Historical Events
in Papua New Guinea

From evidence in recently-excavated caves in New Ireland, habitation probably began about 50,000 years ago. Possibly the world's first agriculturalists originated in Papua New Guinea around 4000 BC.

1526 Portuguese explorer Jorge de Meneses named the west coast of the main island *Ilhas dos Papuas* or 'Island of the Frizzly Heads'.

1545 Spanish explorer Inigo Oritz de Retes named the north coast *Nueva Guinea*.

1700 Buccaneer William Dampier sighted New Britain and New Ireland.

1768 Louis de Bougainville sights Buka and Bougainville.

1792 Antoine Joseph Bruny D'Entrecasteaux explores the Bismarck Archipelago.

1871 Russian 'Baron' Nikolai Miklouho-Maclay lands and establishes a 'laboratory' at Garagassi on Astrolabe Bay.

1873 Captain John Moresby names Port Moresby after his father Admiral Sir Fairfax Moresby.

1874 London Missionary Society established at Port Moresby.

1876 Luigi D'Albertis explores the Fly River in the spirit of grand opera.

1878 German trader Eduard Hernsheim sets up the first copra trading settlement at Nusa, New Ireland.

1878 'Queen' Emma Forsayth and Thomas Farrell arrive in Mioko in the Duke of York Islands between New Britain and New Ireland. They manage the trading station of the German company Godeffroy und Sohn.

1880 First colonists of the ill-fated Marquis de Rays expedition land at the fever hole of Port Breton in New Ireland.

	GERMAN NEW GUINEA	BRITISH NEW GUINEA	
1884	**May** The banker Adolph von Hansemann founds the Neu Guinea Compagnie in Berlin. **3 November,** *Reichsflagge* raised at Matupit in Neu Pommern (New Britain). **4 November,** flag raised to annex Bismarck Archipelago on Mioko in Neu Lauenburg (Duke of York Islands). **12 November,** flag raised at Finschhafen in Neu Guinea.	**6 November,** British Imperial Government proclaim their Protectorate at Port Moresby. Sir Peter Scratchley appointed Special Commissioner – dead from malaria within three months.	
1885	Finschhafen made capital of Kaiser Wilhelmsland and the Bismarck Archipelago.		
1892	Capital of Deutsch Neu Guinea transferred to Stephansort (Bogadjim).	Labour Ordinance protecting Papuans from slavery on sugar plantations in Australia.	
1899	Administration transferred from the Neu Guinea Compagnie to the German Government. Albert Hahl appointed Governor.		
1906			
1908			
1910	Headquarters transferred from Herbertshöhe (Kokopo) to Rabaul in New Britain.		
1914			

PAPUA	MANDATED TERRITORY OF NEW GUINEA
Australia accepts responsibility for the British Protectorate and renames it Papua.	
Hubert Murray appointed Lieutenant-Governor.	
Bronisław Malinowski begins fieldwork in New Guinea until 1918.	Australian forces occupy Rabaul and take over the German colony.

	GERMAN NEW GUINEA	BRITISH NEW GUINEA	
1920 1921			
1942			
1949			

PAPUA	MANDATED TERRITORY OF NEW GUINEA
Papuan men forbidden to wear clothing on the upper part of the body.	League of Nations confers on Australia Mandate for the Territory of New Guinea.
In August Japanese land at Milne Bay.	January bombing of Rabaul begins. In July SS *Montevideo Maru* torpedoed in error with the loss of 1000 lives.
Papua New Guinea Act combining Papua and the Mandated Territory into a single entity of the Territory of Papua and New Guinea.	

THE TERRITORY OF PAPUA AND NEW GUINEA

1958   First Highlands Local Government Council established at Goroka.
1964   First General Election to the House of Assembly.
1966   University of Papua New Guinea established at Port Moresby.
1971   Name of the Territory of Papua and New Guinea changed to Papua New Guinea with adoption of national flag and emblem.

PAPUA NEW GUINEA

1972   National Coalition Government formed with Michael Somare as Chief Minister.
15 August 1975   Constitution adopted.
15 September 1975 5.15 p.m.   Australian flag lowered.
16 September 1975 10.25 a.m.   Bird of paradise flag of the Independent State of Papua New Guinea raised on Independence Hill, Waigani, Port Moresby.

Bibliography of Principal Sources

GENERAL

Albertis, Luigi Maria d', *New Guinea: What I Did and What I Saw* (2 vols) (London, 1880)

Allen, Benedict, *Into the Crocodile Nest* (London, 1989)

Bassett, Marnie, *Letters from New Guinea* (1921). With a Postscript and some Notes added in 1969 (Melbourne, 1969)

Becke, Louis, *By Reef and Palm* (London, 1894)

Becke, Louis, *Wild Life in the Southern Seas* (London, 1897)

Beehler, Bruce M., Pratt, Thane K. & Zimmerman, Dale A., *Birds of New Guinea* (Princeton, 1986)

Bodrogi, Tibor & Boglar, Lajos, *Opuscula ethnologica memoriae Ludovici Biró sacra (Memoriae Ludovici Biró)* (Budapest, 1959)

Bret, David, *Errol Flynn: Satan's Angel* (London, 2000)

Brown, George, Rev., *The Pacific, East and West* (Sydney, 1902)

Brown, George, Rev., *Melanesians and Polynesians* (London, 1910)

Burridge, R. G. L., *Mambu (Cargo Cults in New Guinea)* (London, 1960)

Cayley-Webster, Capt. H., *Through New Guinea and other Cannibal Countries* (London, 1898)

Chalmers, James, *Work and Adventure in New Guinea* (London, 1885)

Chalmers, James, *Pioneering in New Guinea* (London, 1897)

Chalmers, James, *James Chalmers: His Autobiography and Letters* (London, 1902)

Champion, Ian, *Across New Guinea from the Fly to the Sepik* (London, 1932)

Coates, Brian, *The Birds of Papua New Guinea: including the Bismarck Archipelago and Bougainville* (2 vols) (Alderley, 1985–90)

Conrad, Earl, *Errol Flynn: a Memoir* (London, 1979)

Conrad, Joseph, *Victory: an Island Tale* (London, 1915)

Conrad, Joseph, *Heart of Darkness* (Oxford, 1985)

Corbin, Iris Annie, *Tamate the Fearless* (London, 1932)

Craig, B., Kernot, B. & Anderson, C. (eds), *Art and Performance in Oceania* (Bathurst, 1999)

Dodwell, Christina, *In Papua New Guinea* (Yeovil, 1983)

BIBLIOGRAPHY OF PRINCIPAL SOURCES

Dorney, Sean, *Papua New Guinea: People, Politics and History since 1975* (Sydney, 2000)

Edmundson, A. & Boylan, C., *Adorned: Traditional Jewellery and Body Decoration from Australia and the Pacific* (Macleay Museum, Sydney, 1999)

Eri, Vincent, *The Crocodile* (Brisbane, 1970)

Flannery, Tim, *Throwim Way Leg: Tree-Kangaroos, Possums, and Penis Gourds. On the Track of Unknown Mammals in Wildest New Guinea* (London, 1999)

Flynn, Errol, *My Wicked, Wicked Ways* (London, 1960)

Forrest, Captain T., *A Voyage to New Guinea and the Moluccas from Balambangan, Including an Account of Magindano, Sooloo and Other Islands 1779* (London, 1780)

Garrett, John, *Footsteps in the Sea. Christianity in Oceania to World War II* (Geneva, 1992)

Geertz, Clifford, *Works and Lives* (Stanford, 1988)

Gellner, Ernest, *Language and Solitude. Wittgenstein, Malinowski and the Hapsburg Dilemma* (Cambridge, 1998)

Gibbes, Wing Commander Bobby, *Sepik Pilot* (Melbourne, 1971)

Goodale, Jane C., *To Sing With Pigs is Human* (Seattle/London, 1995)

Grimshaw, Beatrice, *The New New Guinea* (London, 1910)

Haddon, A. C., *Canoes of Oceania: the Canoes of Melanesian Queensland and New Guinea* (Honolulu, 1937)

Hawskesworth, John, *An account of the voyages undertaken by order of His present Majesty . . . performed by Commodore Byron, Captain Wallis, Captain Carteret, and Captain Cook . . .* (3 vols) (London, 1773)

Herdt, G. & Stephen, M. (eds), *The Religious Imagination in New Guinea* (New Brunswick/London, 1989)

Herdt, G. & Stephen, M. (eds), *Ritualised Homosexuality in Melanesia* (Los Angeles/Oxford, 1984)

Hides, Jack, *Papuan Wonderland* (London, 1936)

Hiery, Hermann J. (ed.), *European Impact and Pacific Influence* (London, 1997)

Higham, Charles, *Errol Flynn: the Untold Story* (New York, 1980)

Hogan, Nigel, *Causes and Effects of Political and Administrative Corruption in Papua New Guinea* (unpublished dissertation, November, 1999)

Howe, K. R., *Nature, Culture and History: the Knowing of Oceania* (Hawaii, 2000)

Hurley, Frank, *Pearls and Savages* (New York, 1924)

Israel, Paul, *Edison: a Life of Invention* (New York, 1998)

Lawson, Captain J. A., *Wanderings in the Interior of New Guinea* (London, 1875)

Lindstrom, Lamont, *Cargo Cult: Strange Stories of Desire from Melanesia and Beyond* (Honolulu, 1993)

Leahy, Mick, *The Central Highlands of New Guinea* (*Journal of the Royal Geographical Society*, vol. 87, London, 1936)

Leahy, Mick, *The Land that Time Forgot* (London, 1937)

Leahy, Mick, *First Contact* (Ronin Films, Canberra)

Levi-Strauss, Claude, *Tristes Tropiques*, trans., John Weightman and Doreen Weightman (London, 1973)

MacFarlane, Reverend Samuel, *Among the Cannibals of New Guinea; being the story of the New Guinea Mission of the London Missionary Society* (London, 1888)

MacGillivray, John, *Narrative of the Voyage of HMS Rattlesnake, Commanded by the Late Captain Owen Stanley, RN, FRS etc. during the years 1846–1850 including discoveries in New Guinea and the Louisiade Archipelago* (London, 1852)

Mackay, Colonel Kenneth, *Across Papua: being an account of a voyage around, and a march across the Territory of Papua* (London, 1909)

Marriot, Edward, *The Lost Tribe* (London, 1996)

May, Erskine, *Treatise on the law, privileges, proceedings and usages of Parliament – 22nd ed.* (London, 1997)

McCosker, Anne, *Masked Edem. A History of the Australians in New Guinea* (Ringwood, 1998)

Monckton, C. A. W., *Some Experiences of a New Guinea Resident Magistrate* (London, 1921)

Moran, Michael, *Point Venus* (Sydney, 1998)

Moresby, John, *Discoveries & Surveys in New Guinea and the D'Entrecasteaux Islands: a cruise in Polynesia and visits to the pearl-shelling stations in Torres Straits of HMS Basilisk* (London, 1876)

Moresby, John, *Two Admirals: Sir Fairfax Moresby, John Moresby* (London, 1913)

Murray, J. H. P., *Papua or British New Guinea* (London, 1912)

Narakobi, Bernard, *The Melanesian Way* (Institute of Papua New Guinea Studies, Port Moresby, 1980)

Nelson, Hank, *Taim Bilong Masta – the Australian Involvement with Papua New Guinea* (Sydney, 1982)

Nelson, Hank, *Black, White and Gold: Gold Mining in Papua New Guinea 1878–1930* (Canberra, 1976)

O'Neill, Tim, *And We The People. Ten Years with the Primitive Tribes of New Guinea* (London, 1961)

Oram, Nigel, *Encyclopaedia of Papua New Guinea* (Melbourne, 1972)

Parkinson, Richard, *Thirty Years in the South Seas* (1907); English translation (Bathurst, 1999)

Parsons, Michael, *The Butterflies of Papua New Guinea: Systematics and Biology* (Sydney, 1999)

Rainer, Chris, *Where Masks Still Dance: New Guinea* (Boston/London, 1996)

Roberts, Jan, *Voices from a Lost World* (Sydney, 1996)

Sampson, Jane, *Imperial Benevolence: Making British Authority in the Pacific Islands* (Honolulu, 1998)

Sillitoe, Paul, *An Introduction to the Anthropology of Melanesia: Culture and Tradition* (Cambridge, 1998)

Sillitoe, Paul, *Social Change in Melanesia: Development and History* (Cambridge, 2000)

Smith, Bernard, *Imagining the Pacific in the Wake of the Cook Voyages* (Melbourne, 1992)

Smith, Michael French, *Hard Times on Kairiru Island* (Honolulu, 1994)

Souter, Gavin, *New Guinea: the Last Unknown* (London, 1964)

Stewart, Gloria, *Sepik Art* (Sydney, 1972)

Stow, Randolph, *The Visitants* (New York, 1981)

Swadling, Pamela, *Plumes from Paradise* (Port Moresby, 1996)

Waiko, John Dademo, *A Short History of Papua New Guinea* (Melbourne, 1998)

Wallace, Alfred Russel, *The Malay Archipelago: the Land of the Orang-utan, and the Bird of Paradise* (New York, 1869)

Williams, Francis Edgar, *The Vailala Madness & Other Essays*, ed., Erik Schwimmer (London, 1976)

Wilson, Robert, *Voyages of discoveries round the world . . .* (London, 1806)

HMS *ENDEAVOUR*

Endeavour Millennium Tour, Australian Maritime Museum. Published for HM Bark Endeavour Foundation (Sydney, 1997)

Beaglehole, J. C. (ed.), *The Endeavour Journal of Joseph Banks* (2 vols) (Sydney, 1962)

Beaglehole, J. C. (ed.), *The Journals of Captain Cook: the Voyage of the Endeavour 1768–1771* (Cambridge, 1952)

O'Brian, Patrick, *Joseph Banks* (London, 1987)

Carey, Caroline, *Tales of Trial Bay* (Sydney, 1993)

MILNE BAY PROVINCE – Samarai, Kwato, Trobriand Islands and D'Entrecasteaux Group

Abel, Russell W., *Charles W. Abel of Kwato: Forty Years in Dark Papua* (New York, 1934)

Abel, Charles William, *Savage Life in New Guinea: the Papuan in Many Moods* (London, 1901)

Baker, C. & Knight, G., *Milne Bay 1942: the Story of 'Milne Force' and Japan's First Military Defeat on Land* (Loftus, NSW, 1992)

Beran, Harry, *Betel-Chewing Equipment of East New Guinea* (Aylesbury, 1988)

Beran, Harry, *Mutuaga – A Nineteenth Century New Guinea Master Carver* (Wollongong, 1996)

Billig, Otto & Burton-Bradley, B. G., *The Painted Message* (New York, 1978)

Brune, Peter, *The Spell Broken – Exploding the Myth of Japanese Invincibility. Milne Bay to Buna-Sananda 1942–43* (Sydney, 1997)

Clowes, Cyril A., *The Clowes Report on the Battle of Milne Bay, 1942* (Loftus, NSW, 1995)

Ellen, Roy (ed.), *Malinowski Between Two Worlds: the Polish Roots of an Anthropological Tradition* (Cambridge, 1988)

Fortune, R. F., *Sorcerers of Dobu*, intro. by B. Malinowski (London, 1932)

Grubb, Norman Percy, *C. T. Studd, Cricketer and Pioneer* (London, 1933)

BIBLIOGRAPHY OF PRINCIPAL SOURCES

Horner, Frank, *Looking for La Pérouse: D'Entrecasteaux in Australia and the South Pacific 1792–1793* (Melbourne, 1995)

'Konteksty', *Exhibition Catalogue Malinowski – Witkacy. Photography: Between Science and Art* (Warsaw, 2000)

Kwato Extension Association, *Title Letters from Kwato Mission, Samarai, Papua*

Leach, Jerry W. (director), *Trobriand Cricket: an Ingenious Response to Colonialism* (Video recording) (Ronin Films, Canberra, 1976)

MacLaren, Les (director), *Kama Wosi: Music in the Trobriand Islands* (16mm film) (Australian Film Commission, Institute of PNG Studies, Boroko, 1979)

Mahony, Elizabeth, *The Queen of Sudest: the Life and Times of Elizabeth Mahony* (Bowraville, NSW, 1993)

Malinowski, Bronisław, *Argonauts of the Western Pacific* (London, 1922)

Malinowski, Bronisław, *Sex and Repression in Savage Society* (London, 1927)

Malinowski, Bronisław, *The Sexual Life of Savages in North-Western Melanesia* (London, 1929)

Malinowski, Bronisław, *Coral Gardens and Their Magic* (London, 1935)

Malinowski, Bronisław, *Malinowski among the Magi: the Natives of Mailu*, ed. with an introduction by Michael Young (London/New York, 1988)

Malinowski, Bronisław, *A Diary in the Strict Sense of the Term* (New York, 1967)

Malinowski, Bronisław, *The Story of a Marriage: the letters of Bronisław Malinowski and Elsie Masson*, ed. Helen Wayne (London/New York, 1995)

Malnic, Jutta, *Kula: Myth and Magic in the Trobriand Islands* (Sydney, 1998)

Micinska, Anna, *Witkacy: Life and Work* (Warsaw, 1990)

Pfund, Kurt, *Islands of Love: Portrait of the Trobriand Islands* (Adelaide/Port Moresby, 1972)

Pollock, John Charles, *The Cambridge Seven. A Call to Christian Service* (London, 1955)

Taaffe, Stephen R., *MacArthur's Jungle War: the 1944 New Guinea Campaign* (Lawrence, Kansas, 1998)

Wetherall, David, *Charles Abel and the Kwato Mission of Papua New Guinea 1891–1975* (Melbourne, 1996)

Witkiewicz, Stanisław, *Listy do Bronisława Malinowskiego* (Warsaw, 1981)

Witkiewicz, Stanisław, *The 622 Downfalls of Bungo or The Demonic Woman 1910–11*, ed. Anna Micinska (Warsaw, 1972)

Young, Michael W., *The Ethnography of Malinowski: The Trobriand Islands 1915–1918* (London, 1979)

Young, Michael W., *Magicians of Manumanua: Living Myth in Kalauna* (Los Angeles/Oxford, 1983)

Young, Michael W., *Malinowski's Kiriwina: Fieldwork Photography 1915–1918* (Chicago, 1998)

EAST NEW BRITAIN PROVINCE – Rabaul, Duke of York Islands

Boyd, Bill & Lentfer, Carol, *Maunten Paia: Volcanoes, People and Environment* (Lismore, 2001)

Brown, George, Rev., *Some New Britain Customs* (Melbourne, 1901)

Brown, George, Rev., *Notes on the Duke of York Group, New Britain and New Ireland* (Journal of the Royal Geographical Society, Vol. 47, London, 1877)

Cilento, P., *My Life* (Sydney, 1987)

Danks, Benjamin, *In Wild New Britain: the Story of Benjamin Danks from His Diary*, ed., Wallace Deane (Sydney, 1933)

Dutton, Geoffrey, *Queen Emma of the South Seas* (Melbourne, 1976)

Epstein, A. L., *Gunantuna – Aspects of the Person, the Self and the Individual among the Tolai* (Bathurst, 1999)

Epstein, A. L., *In the Midst of Life – Affect and Ideation in the World of the Tolai* (Los Angeles/Oxford, 1992)

Epstein, A. L., *Matupit. Land, Politics, and Change among the Tolai of New Britain* (Canberra, 1969)

Fajans, Jane, *They Make Themselves: Work and Play among the Baining of Papua New Guinea* (Chicago, 1997)

Hiery Hermann, J., *The Neglected War: The German South Pacific and the Influence of World War I* (Honolulu, 1995)

Lauer, Sue, *Pumice and Ash: an Account of the 1994 Rabaul Volcanic Eruptions* (Lismore, 1995)

Pembroke, George, *South Sea Bubbles* (Melbourne, 1872)

Powell, Wilfrid, *Wanderings in a Wild Country, Or, Three Years amongst the Cannibals of New Britain* (London, 1883)

Powell, Wilfrid, *Observations on New Britain and Neighbouring Islands, during Six Years' Exploration* (Proceedings of the Royal Geographical Society, London, 1881)

Pullen-Burry, B., *In a German Colony* (London, 1909)

Rannie, Douglas, *My Adventures Among the South Sea Cannibals* (London, 1912)

Robson, R. W., *Queen Emma* (Sydney, 1965)

Sekiguchi, Noriko (writer & director), *Senso Daughters* (Video recording) (Tenchhjin Productions Australia, Siglo Co. Ltd. Japan and Institute of PNG Studies, Boroko, 1990)

Stone, Peter, *Hostages to Freedom* (Yarram, 1994)

NEW IRELAND PROVINCE – Kavieng, Libba, Tsoi Island (Mansava)

Baudouin, Alexandre, *L'aventure de Port Breton et la colonie libre, dite Nouvelle France: souvenirs personnels et documents* (Paris, 1883)

Brown, George Rev., *A Journey along the Coasts of New Ireland and Neighbouring Islands* (Proceedings of the Royal Geographical Society, London, 1881)

Gunn, Michael, *Taxonomic Structure and Typology in the Malagan Ritual Art Tradition of Tabar, New Ireland*. Paper delivered at the Fifth International Symposium of the Pacific Arts (Adelaide, 1993)

Gunn, Michael, *Ritual Arts of Oceania: New Ireland in the Collections of the Barbier-Mueller Museum* (Milan, 1997)

BIBLIOGRAPHY OF PRINCIPAL SOURCES

Halliman, Pet, *Revelation of the Malagans: the Ritual Art of New Ireland* (Florida Gardens, Queensland, 1990)

Kohnke, Glenys, *The Shark Callers: an Ancient Fishing Tradition of New Ireland, Papua New Guinea* (Yumi Press, Boroko, 1974)

Lincoln, Louise, *Assemblage of Spirits: Idea and Image in New Ireland* (New York, 1987)

Michener, James A., *Rascals in Paradise* (London, 1957)

Mouton, J. B. O., *The New Guinea Memoirs of Jean Baptiste Octave Mouton*, ed. with an Introduction by Peter Biskup (Canberra, 1974)

Niau, J. H., *The Phantom Paradise. The Story of the Expedition of the Marquis de Rays* (Sydney, 1936)

O'Rourke, Dennis (director), *The Shark Callers of Kontu* (O'Rourke & Associates Filmmaker, Lindfield, NSW, 1982)

Owen, Chris (director), *Malagan Labadama: A Tribute to Buk Buk* (Film) (Institute of PNG Studies, Boroko)

MADANG PROVINCE

Finsch, Otto, *Madang (Friedrich-Wilhelmshafen) in 1884*, trans. Christiane Harding, ed. Mary R. Mennis (Madang, 1996)

Miklouho-Maclay, Nikolai, *Travels in New Guinea: Diaries, Letters, Documents* (Moscow, 1982)

Miklouho-Maclay, Nikolai, *New Guinea Diaries 1871–1883*, trans. C. L. Sentinella (Madang, 1975)

Webster, E. M., *The Moon Man: A Biography of Nikolai Miklouho-Maclay* (London, 1984)

NORTH SOLOMONS PROVINCE – Buka and Bougainville

Blackwood, Beatrice, *Both Sides of Buka Passage: an Ethnographic Study of Social, Sexual, and Economic Questions in the North-Western Solomon Islands* (Oxford, 1935)

O'Callaghan, Mary-Louise, *Enemies Within – Papua New Guinea, Australia, and the Sandline Crisis: the Inside Story* (Sydney, 1999)

Oliver, Douglas, *Black Islanders: a Personal Perspective of Bougainville 1937–1991* (Victoria, 1991)

Quodling, Paul, *Bougainville – the Mine and the People* (St Leonards, NSW, 1991)

Rimoldi, Max, *Hahalis and the Labour of Love: a Social Movement on Buka Island* (Oxford, 1992)

Stella, Regis, *Gutsini Posa (Rough Seas)* (Port Moresby, 1999)

GERMAN NEW GUINEA

Bache, Hans, *Das Deutsche Kolonialbuch* (Berlin, 1926)

Kotze, Stephan von, *Aus Papuas Kulturmorgen* (Wiesbaden, 1899)

Neuhauss, R. Prof. Dr Med., *Deutsch Neu-Guinea in Drei Bänden* (Berlin, 1911)

Sack, Peter & Clark, Dymphyna, *German New Guinea – The Annual Reports* (Canberra, 1979)

BIBLIOGRAPHY OF PRINCIPAL SOURCES

Schultz-Ewerth, Dr E., *Deutschlands Weg zur Kolonialmacht* (Berlin, 1934)

Nachrichten für Kaiser Wilhelm-Land und den Bismarck-Archipel 1891–1898 published by the Neu Guinea Compagnie

Deutsches Kolonialblatt published by the German Imperial Government from 1890

Amtsblatt für das Schutzgebiet (Deutsche) Neu Guinea published by the Imperial Government from 1909

Deutsche Kolonialzeitung published by the German Colonial Society

Denkschrift über die Verhandlungen des Reichstages uber den mit dem Reich wegen Ubernahme der Landeshoheit geschlossenen Vertrag (Berlin, 1896)

CANNIBALISM

Arens, William, *The Man-Eating Myth: Anthropology and Anthropophagy* (Oxford, 1979)

Barker, Francis; Hulme Peter; Iverson, Margaret (eds), *Cannibalism and the Colonial World* (Cambridge, 1998)

Bjerre, Jens, *The Last Cannibals* (London, 1956)

Bjerre, Jens, *Savage New Guinea* (London, 1964)

Darwin, Charles, *Journal of Researches into the Geology and Natural History of the Various Countries visited during a Voyage of HMS Beagle* (2 vols) (London, 1846)

Haining, Peter, *The Flesh Eaters: True Stories of Cannibals and Blood Drinkers* (London, 1994)

Holthouse, Hector, *Cannibal Cargoes* (Adelaide, 1969)

Langsdorff, George von, *Voyages and Travels in Various Parts of the World during the Years 1803–1807* (London, 1813)

Lestringant, Frank, *Cannibals: the Discovery and Representation of the Cannibal from Columbus to Jules Verne* (Oxford, 1997)

Macleod, Helen, *Cannibals are Human. A District Officer's Wife in New Guinea* (London/Sydney, 1962)

Miller, 'Cannibal' Charles, *Cannibal Caravan* (London, 1950)

Powell, Wilfrid, *Wanderings in a Wild Country, Or, Three Years amongst the Cannibals of New Britain* (London, 1883)

Rannie, Douglas, *My Adventures Among the South Sea Cannibals* (London, 1912)

Sagan, Eli, *Cannibalism: Human Aggression and Cultural Form* (New York, 1974)

Sanborn, Geoffrey, *The Sign of the Cannibal* (Durham and London, 1998)

Sanday, Peggy, *Divine Hunger: Cannibalism as a Cultural System* (Cambridge, 1986)

Sutton, Leslie, *Crocodiles, Cannibals and Cream Teas* (Hoddesdon, 1998)

Viaud, Pierre, *The Shipwreck and Adventures of M. Pierre Viaud*, trans. Mrs Griffith (London, 1771)

Index